Series in Biomedical Engineering

Founding Editor

Joachim H. Nagel

This series describes the applications of physical science, engineering and mathematics in medicine and biology. The books are written for graduate students and researchers in many disciplines including medical physics, biomedical engineering, radiology, radiotherapy and clinical research. The Series in Biomedical Engineering is the official book series of the International Federation for Medical and Biological Engineering.

The IFMBE is an association of constituent societies and organizations which was established in 1959 to encourage and promote international collaboration in research and practice of the profession as well as in the management of technology and the use of science and engineering in medicine and biology for improving health and quality of life. Its activities include participation in the formulation of public policy and the dissemination of information through publications and forums. The IFMBE as the only international organization and WHO/UN accredited NGO covering the full range of biomedical/clinical engineering, healthcare, and healthcare technology management, represents through its 50 national and international member societies more than 120.000 professionals involved in the issues of improved health care delivery.

The IFMBE is associated with the International Union for Physical and Engineering Sciences in Medicine (IUPESM). Through the IUPESM, the IFMBE is a member of the International Council for Science (ICSU). http://www.ifmbe.org

More information about this series at https://link.springer.com/bookseries/7752

Vladimir Medved
Editor

Measurement and Analysis of Human Locomotion

Springer

Editor
Vladimir Medved
Faculty of Kinesiology
University of Zagreb
Zagreb, Croatia

ISSN 1864-5763 ISSN 1864-5771 (electronic)
Series in Biomedical Engineering
ISBN 978-3-030-79687-7 ISBN 978-3-030-79685-3 (eBook)
https://doi.org/10.1007/978-3-030-79685-3

This Springer imprint is published by the registered company Springer Nature Switzerland AG
The registered company address is: Gewerbestrasse 11, 6330 Cham, Switzerland

Preface

Human locomotion may be a subject of interest in many areas such as sports, medicine, performing arts, and also, assuming a slightly broader perspective, ergonomics, traffic sciences, didactics, etc. Modern approaches to studying gross human movement and locomotion are characterized by the possibility of application of various engineering technology solutions enabling, in effect, a physics- and physiology-based quantitative grasp of the phenomenon. It is here where this book focuses, its aim being to introduce, succinctly describe, explain and illustrate uses of this sort. Among application areas mentioned, its primary orientation is towards those in medicine and in sports. The book has been motivated by the need for lecture notes of a course carrying the same name within Medical Studies in English at the University of Zagreb, running for more than a decade.

Starting from a historical perspective, the idea of understanding human locomotion by applying technical measurement devices and incorporating measurement data into physical representation of gross body movement is presented and explained, an approach known as inverse dynamics. Within this approach as a kind of an umbrella concept, components of measurement systems including relevant signal and data processing methods are described. Modern instruments to capture body movement by measuring its kinematics, kinetics and surface electromyography (sEMG) are thus described; all systems being used dominantly—if not exclusively—in a movement analysis laboratory setting.

Focusing mainly on human posture and gait, but including also selected examples of movement patterns exhibited during performance of kinesiological and sports activities, the book attempts to present essentials of biomechanics and biomedical engineering approach to this subject matter. It illustrates how data collected and elaborated by modern engineering technology attainments can complement traditional expert knowledge of a medical doctor or a kinesiologist. The examples of this book's practical application might be, for instance, in evaluation of efficiency of human gait, in evaluation of local skeletal muscle fatigue in physical exercise, in biomechanical assessment of traumatological conditions requiring orthopedic treatment, and the like. The book is applicable in the fields of sports, physical activities, as well as in medical diagnostics and rehabilitation.

Measurement and Analysis of Human Locomotion is written by a multidisciplinary author team from the University of Zagreb, a Zagreb-based clinical hospital center, Kinematika—Polyclinic for Orthopedy, Physical Medicine and Rehabilitation in Zagreb, University of Split, and University of Salerno, including engineers, medical doctors, a kinesiologist and a sport scientist. The book might be used in planning and executing research endeavors, and in particular in a clinical context as a reference for various diagnostics procedures.

Zagreb, Croatia Vladimir Medved

Acknowledgement

Springer Nature is to be thanked for accepting this book project and conducting it towards realization. Major actions and coordination were provided and have contributed to successful production of the book. Among the staff at Springer, in particular Christoph Baumann, Karthik Raj Selveraj, Parimelazhagan Thirumani, and Coral Zhou, each in his/her respective capacity, have given special help and support; it was a delight to work with them.

Vladimir Medved
Zagreb, Croatia

Contents

Contributors

Daniele Albano Department of Human, Philosophical and Educational Sciences, University of Salerno, Fisciano (SA), Italy

Boris Božić Department of Neurosurgery, Sestre milosrdnice University Hospital Center, Zagreb, Croatia

Mario Cifrek Faculty of Electrical Engineering and Computing, University of Zagreb, Zagreb, Croatia

Nikica Daraboš School of Medicine, University of Split, Split, Croatia

Igor Gruić Faculty of Kinesiology, University of Zagreb, Zagreb, Croatia

Vladimir Medved Faculty of Kinesiology, University of Zagreb, Zagreb, Croatia

Maja Mirković Kinematika-Polyclinic for Orthopedy, Physical Medicine and Rehabilitation, Zagreb, Croatia

Marko Pećina School of Medicine, University of Zagreb, Zagreb, Croatia

Tomislav Pribanić Faculty of Electrical Engineering and Computing, University of Zagreb, Zagreb, Croatia

Marija Rakovac Faculty of Kinesiology, University of Zagreb, Zagreb, Croatia

Rodolfo Vastola Department of Human, Philosophical and Educational Sciences, University of Salerno, Fisciano (SA), Italy

Chapter 1
Introduction

Vladimir Medved

Abstract The field of biomedical engineering is introduced and shortly described. Roles played by the fields of biomedical engineering and of biomechanics in the development of the study of human locomotion are discussed. Current status of education of biomedical engineers, worldwide and in Croatia, is commented. Overview of the book follows, pointing shortly to the subject of each chapter. The topics covered are: history of locomotion study and current research methodology; evolution of human gait; elements of biomechanics of the lumbar spine; measuring kinematics and kinetics of human movement; elaboration of 3D photogrammetry; pedobarography; kinesiological electromyography; gait analysis; critical issues in orthopedic traumatology; on sportive movement patterns study. Measurement and analysis of human locomotion is a multidisciplinary and interdisciplinary topic that requires cooperation among biomedical engineers, medical doctors of several specialties, and kinesiologists.

1.1 Biomedical Engineering and Biomechanics in Studies of Locomotion

Human locomotion, a subject of interest to the fields like medicine, sports, dance, etc., always concerns certain body movements, meaning moving of the body as a whole and/or relative movements between body segments. Natural approach to the study of a phenomenon of this kind is therefore by visual observation. The observer is supposed to be an expert for certain class of movements and/or locomotions (a physician specialized in physical and rehabilitation medicine; a trainer (coach) for certain sport discipline, for instance) and his visual analysis of movement performed is a kind of qualitative evaluation. Human movements usually occur at a relatively high speed compared to the "sampling speed" of man's visual apparatus, amounting to cca 25/s. It is therefore understandable that potential means of measurement are desirable that

V. Medved (✉)
Faculty of Kinesiology, University of Zagreb, Horvaćanski zavoj 15, 10000 Zagreb, Croatia
e-mail: vladimir.medved@kif.hr

V. Medved (ed.), *Measurement and Analysis of Human Locomotion*,
Series in Biomedical Engineering, https://doi.org/10.1007/978-3-030-79685-3_1

could objectively register and capture a movement pattern in a way, enabling later analysis and processing of the acquired kinematic record in an exact manner. This stresses the importance of technical instrumentation, devices and methods able to realize this task.

In general, measuring physical quantities like displacements, velocities, accelerations, forces, etc. is a task for engineers to perform. Development of the field of engineering; mechanical, with rather ancient roots; electrical and electronic, which on the other hand are truly modern, and other, has enabled successful realization of this task. In this realm it is the field of biomedical engineering that encompasses, among other, also the design and the construction of apparatus needed.

Arising around the mid of the twentieth century, the field of biomedical engineering has undergone and is still undergoing rapid development [1]. Relying in the beginning primarily on physics, biophysics and electronics, many engineering disciplines were involved later, and today it is an area of dynamic development and very diverse, truly multidisciplinary. Biomedical engineering means applying methodology of exact and technical sciences to solve problems of biological and biomedical systems. On one hand, technical (engineering) approach is used for modeling and simulation, hence a better understanding of the functioning of respective biological system(s) and, on the other, it is used for realization of technical devices (instruments) and procedures for medical diagnostics, therapy and prevention. Biological system may be accessed at several levels: from global to sub-cellular.

The Whitaker Foundation (founded in 1975 and lasted until 2006; https://en.wikipedia.org/wiki/Whitaker_Foundation) has been among major United States foundations instrumental in supporting biomedical engineering since its modern beginnings (we emphasize modernity, as traces of essentially engineering approaches to biological systems and mechanisms can be followed a long way back in time). It has outlined the following dominant sub-areas of biomedical engineering: biomedical instrumentation, biomechanics, biomaterials, systems physiology, clinical engineering and rehabilitation engineering. The Biomedical engineering Handbook [2] basically respects the above, rather conservative division, breaking the field further down into more than 20 areas within which are covered, for example: biomedical sensors, biomedical signal analysis, visualization (of internal organs and tissues), biotechnology, tissue engineering, artificial organs, medical informatics, ergonomics, artificial intelligence, etc. (It is worth to mention that the first graduate program in biological—as differing from biomedical—engineering has been launched in 2005 at MIT, signifying a shift towards a more basic level of study, and reflecting scientific progress in molecular biology, human genomics, systems biology, and the like.).

Profiting from this rich palette of biomedical engineering technologies, studies of gross human movement and locomotion have moved beyond the, previously mentioned, kinematic level of observation and analysis and have progressed towards the neuromuscular [3, first edition by Verne, Thompson and Inman, 1981] and the control levels of the locomotor system function. The science of biomechanics assumed an important role. Special attention has been directed to biomechanical properties of bones, joints and skeletal musculature, and, further, to the control properties of the nervous system. The book: "Muscles, reflexes, and locomotion"

[4], authored by McMahon, reflecting systems analysis methodology in approaching human neuromusculoskeletal system and explaining mathematical models of muscle contraction that appeared around the middle of the twentieth century, was recognized as an instant classic in the field and still remains vital [4; Fig. 1.1].

Measurement technique of electromyography (EMG) has enabled a sort of window into the neuromuscular system operation [5, 6, 7]. Bio and neuro-cybernetics approach by Vodovnik [8] has contributed to the field of neuromuscular control, marking initiation of the Ljubljana school of cybernetics in medicine. Achievements of this kind make the components of a comprehensive theoretical and experimental framework for the study. A notion of kinesiology, the premier science of human movement, has assumed a bioscientific profile [9], in addition to the one belonging to social sciences, existing traditionally. Apart from and in addition to the biomechanical approach, the functioning of the human body has commonly been studied also from the physiological and medical aspect [10, 11], therefore the physiology of activity (exercise physiology) has contributed substantially to the study of movement and locomotion.

Fig. 1.1 "Muscles, reflexes, and locomotion" (1984) by Thomas A. McMahon [4] front cover (by permission of Princeton University Press)

In parallel to the progress at a theoretical level in fields like biophysics, biochemistry, and biomechanics, forming of a special kind of experimental laboratories has taken place, and appropriate facilities have been established. This is, first, a laboratory for human movement analysis, sometimes named biomechanics laboratory, and next, a human physiology laboratory, meaning laboratory for exercise physiology testing (pulmonary and cardiovascular systems). Both types of facilities are aimed at enabling proper conditions to measure particular aspects of gross human movement and locomotion in a controlled environment and with a satisfactory degree of non-invasiveness. Instrumentation employed in facilities of this kind, although typically a subject of constant technological improvement (hardware and/or software wise), remains fundamentally the same and has been standardized to provide a working environment for reliable accumulation of experimental data of high accuracy, to be used in research and/or in (clinical) diagnostics. There are centers of excellence in this realm in the world that lead in implementation of standard and introduction of new measurement protocols. So, naturally, the professional areas oriented to the study of human movement, like kinesiology, physical medicine and rehabilitation, sports medicine, orthopedics, neurology, pediatrics, etc. are gradually adopting improved or new technical solutions into their "working repertoire" to their benefit.

To operate working environments of this kind requires a proper personal staff, available at a full time and/or a part time basis, and of a multidisciplinary character. For clinical gait analysis laboratory, for example, such a staff should include: director, manager, gait analyst, biomechanist, technician, clinician and clerical officer [12]. Biomedical engineers can play a significant role in categories: technician, biomechanist, gait analyst and director.

Historically, as far as education of biomedical engineers is concerned, it is worth to mention that in Germany, already between the two World Wars, and in the United States especially after the World War II, independent university programs were established designed to educate biomedical engineers. So, in the United States, in 2000ies there were more than 90 departments offering biomedical engineering programs at universities [1] which, as a rule, provided education at the graduate and post graduate level. Today, in the United States, according to some sources, there are over 300 accredited biomedical engineering schools and university departments offering biomedical engineering programs providing education at several levels up to a Ph.D. Rapid proliferation of the field is evident.

In addition to solid engineering knowledge and knowledge in the relevant biological and medical fields, biomedical engineers are expected to possess the above-average ability of fostering inter-disciplinary communication to be able to function successfully in the development, implementation of standard and procurement of new—often technically highly sophisticated—diagnostic and therapeutic devices and procedures. Biomedical engineering (bioengineering, clinical engineering, including medical physics) today is an established profession with a growing labor market.

In Croatia, although there are no formal educational programs for biomedical engineering yet, an initiative at the University of Zagreb has been launched in 2012: the Coordination Committee for Development of Biomedical engineering was formed. In fact, the Croatian Biomedical engineering and Medical Physics Society

(CroBEMPS) has been active for decades and has been organizing numerous scientific symposia and workshops. It assembles professionals, scattered around various university departments, clinics and institutes, that foster an interest for biomedical engineering, each in his/her respective sub-field [13]. The Society is a member of the International Federation for Medical and Biological Engineering (IFMBE). Professor Ratko Magjarević of the University of Zagreb is former President of the IFMBE Administrative Council (2012–2015) and also a President Elect to step on again in 2022. CroBEMPS encompasses, among other, also Division Biomechanics. There is a tradition of the field of biomechanics in Croatia being developed mainly in realms of the Faculty of Mechanical Engineering and Naval Architecture and the School of Medicine, including also faculties like the Faculty of Mining, Geology and Petroleum Engineering, and other institutions. It included medical doctors and engineers and it typically addressed problems like bone biomechanics, artificial hip design [14], while recent efforts include, for instance, solutions such as customized implants production (pelvic surgery). Only in 2019 was an adequate program designed with the goal to develop the first university program in biomedical engineering, satisfying the standards for profession training and qualification. Through educational workshops, teaching staff from various departments of the University of Zagreb will participate in efforts to improve their didactic competencies and skills [15].

Currently, teaching activity in the area of biomedical engineering and physics in Croatia is accomplished mainly through a number of graduate and post graduate mandatory and elective courses, at several university departments. At the University of Zagreb, at the graduate level, human locomotion studies are being taught at the Faculty of Kinesiology, and also—via elective courses—at the Faculty of Electrical Engineering and Computing and at the School of Medicine [16]. A human locomotion measurement textbook has been written [17], as were also an orthopedic traumatology textbook [18] and a more general biomechanics textbook, co-authored by a multidisciplinary author team of the University of Zagreb and the Josip Juraj Strossmayer University of Osijek [19; Fig. 1.2].

A lecture held at the School of Medicine in Zagreb may be noted, given by the author of this Chapter, under the auspices of the Croatian Biophysical Society (invitation by Selma Supek) entitled: „Biomedical engineering and biomechanics in the study of locomotion" (in Croatian), on November 21, 2001. It marked an initiation of a number of similar lectures in the time to follow, aimed at presenting to various audiences the work in the field of locomotion biomechanics being conducted at the Faculty of Kinesiology, University of Zagreb. Properly equipping the Biomechanics Laboratory at this institution which took place afterwards has provided a necessary boost to the field.

The Laboratory was equipped as a concerted effort. A part of instrumentation has been given at disposal from the Faculty of Mechanical Engineering and Naval Architecture by courtesy of Professor emeritus Osman Muftić (1934–2010) (Chap. 4, Fig. 4.2). A series of projects conducted at the Faculty of Kinesiology, sponsored by the Croatian Ministry of Science, Education and Sports, contributed financially to buying and renewing the majority of equipment. Professor Branka Matković, acting as Vice-Dean for Science, was instrumental here in placing priority to modern

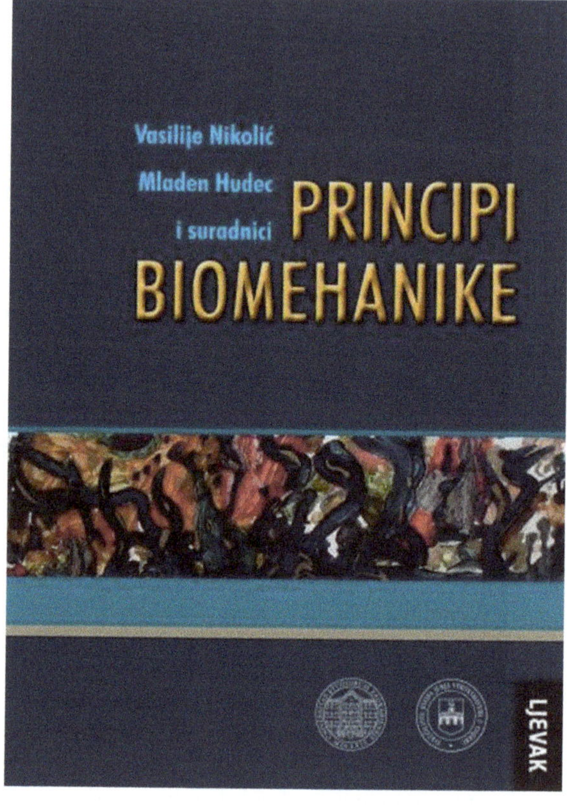

Fig. 1.2 "Principi biomehanike" (2011) (Biomechanics principles) by Vasilije Nikolić and Mladen Hudec (eds) [19] front cover. Painting: Vasilije Nikolić: *Danse macabre*, oil, 1993. (by permission of Naklada Ljevak)

measurement equipment for the Laboratory. The firm Polimedika, of Zagreb, further, has been involved in importing part of new instrumentation and securing legal prerequisites for import. A research program, again sponsored by the Ministry of Science, Education and Sports, entitled: „Minimally invasive measurements and technologies in biomedicine " (in Croatian), conducted at Faculty of Electrical Engineering and Computing, led by Professor Stanko Tonković, contributed financially. The University of Zagreb contributed financially through the research support funds. The Faculty of Kinesiology contributed financially as well.

1.2 Overview of the Book

In continuation, in Chap. 2, a glance on history of the study of human locomotion is given. Major landmarks and most significant contributors are mentioned, that have pursued the development of the approaches intended to get to know the phenomenon of gross human body movement and locomotion. Time period since Antiquity until our days is encompassed. Contribution of each new arising technology (in modern times these were, for instance, high-speed photography, digital computers, new types of sensing, etc.) is outlined. Chapter ends with pinpointing elements of currently attained methodology and paradigms for studying the phenomenon, combining biomechanical approach behind modern measurement, modeling and diagnostics systems, with exercise physiology approach comprising measurements and testing of pulmonary and cardiovascular systems' performance.

In Chap. 3, human gait, a plantigrade mode of locomotion, unique in animal kingdom, is discussed from the evolutionary perspective. Human bipedal locomotion is a very complex behavior. The evolution of upright posture and obligate bipedal gait—a central event in the evolution of humankind—took several millions of years and its incentives are still (and surely will be in the future) a subject of scientific debate. Some of the evolutionary milestones are still recognizable in the development of mature gait in toddlers. To provide an overview on evolution and development of human gait, this chapter starts off with a short description of modern human gait modalities—walking, running, and, (arguably) skipping. The broad topic of evolution of human locomotion is first approached by a comparison of the musculoskeletal anatomy of the humans and extant apes. Positive and negative aspects of bipedalism are then tackled, followed by an overview of existing theories on human evolution and theories on evolution of human bipedalism. The chapter finishes off by describing the development of mature gait in toddlers, an amazing period in human life in which we could argue that 'ontogeny recapitulates phylogeny'—the development of the species is reflected in the development of a single person.

Chapter 4 focuses to a specific part of human body; a lumbar spine segment with surrounding tissues, biomechanics of which is crucial for maintaining the upward erect body posture and moving the body in the intended direction by means of walking, running and other locomotor modalities. The pelvic segment possesses a pivotal role in the walking process, with biomechanics of the lumbar spine being crucial for realizing dynamic spatial (3D) movements. Uprising of humans on lower legs during evolution has challenged this body part by subjecting it under big mechanical strain, as best witnessed, unfortunately, by neurosurgical praxis treating spinal injuries resulting from mechanical overloads, a consequence of sports training or occupational tasks. Zagreb school of biomechanics, appearing in second half of twentieth century, is introduced shortly. Approaches of anatomy, orthopedics and mechanical engineering were integrated. Initially viewed as essentialy a static construction, the spine was modeled subsequently as a dynamic mechanical structure. An archival example of in vitro type measurements of mechanical properties of lumbosacral joint speciments is presented in some detail, in the context of evaluating specific surgical interventions. Interlaminectomy has been found a procedure of choice in the treatment of disc hernias of the lumbar spine. Non-invasive in vivo measurements of spine kinematics during normal gait were performed later.

In Chap. 5, selected technical solutions aimed at measuring kinematics and kinetics of human locomotion are succinctly pointed to and illustrated. Among kinematic measurement systems the marker-based ones—as the most common—are put forward, making mention of related signal and data processing. Wearable type sensor measurement systems are commented next. Further, kinetic measurement instruments; a force measuring platform (force plate) and a pedobarograph are addressed shortly. Few examples of measurement results of sportive and normal locomotions are commented to illustrate the methods. These measurement instruments, and related measurement protocols, are to be regarded as an integral part of an inventory at disposal to modern student of human movement.

Chapter 6 is intended for a technically inclined reader in the first place, such as an advanced student of engineering, or a practicing instrumentation designer. It further elaborates on crucial points of close range photogrammetry technique (a component of kinematics measurement systems described in the previous chapter), which has been successfully adapted for biomechanical purposes during last decades. The focus is on describing measurements of kinematic parameters, e.g. spatial position, velocity and acceleration. To this end a photogrammetric approach is introduced where spatial three-dimensional (3D) measurements are derived from 2D camera images. The principle how a camera models 3D world into 2D image is described with the camera model parameters, and computation of parameters through the camera calibration procedure is presented. Since 3D kinematic systems operate on stereo principle, it requires imaging of the scene with minimum of two cameras, and their spatial relationship is given in the form of epipolar geometry. The focus is further turned to the usage of 3D kinematic systems which is frequently defined by certain measurement protocol, tailored to analyzing a specific movement. Several popular possibilities where 3D kinematic system is jointly used with other types of devices are pointed out. Additionally, an inverse dynamics approach is briefly described and explained, allowing a derivation of internal kinetic data (joint forces and torques) combining an output of a 3D kinematic system with inertial body segment parameters. At the very end, several topics related to the research challenges and the anticipated development of 3D reconstruction systems are presented.

Figure 1.3 shows trends in twentieth century electronics and automation impacting the field of human motion measurement and analysis.

Up to this point, a physics-based methodology of the study of gross body movement, based on multi segment rigid body system kinematics followed by inverse dynamics calculation, has been presented. The treatise further advances towards

Fig. 1.3 Trends in twentieth century electronics and automation impacting the field of human motion measurement and analysis

Electronics and automation

- Semiconductor physics
- Developement of sensor technologies
- Digital electronics (Negroponte N.: Being digital. Vintage Books, New York, 1995.)
- Automatic control
- Telemetry
- Video and consumer electronics
- Automation of measurement and data acquisition processes
- Modeling, simulation and animation of human movement

additional measurement methods (Chaps. 7, 8 and 9) and addresses specific areas of application (Chaps. 10, 11 and 12).

Chapters 7 and 8 concern a specific, quite handy measurement method of pedobarography. In spite of its existence in modern form for a few decades already, relevant actions towards its standardization were undertaken rather recently. In Chap. 7 biomechanical modeling approach of the foot incorporating realistic geometric and material properties of both skeletal and soft tissue components is pointed to. This may serve to generate simulated contact stress distributions, to be compared with measured pressure data obtained using pedobarography. The method may assume several technical design possibilities employing various types of pressure sensors, the common feature of these systems being that the process of measurement occurs via a contact of a firm (sensor) and a deformable (foot sole) surface. Critical issues regarding standardization of measurement equipment, measurement protocol, and signal processing and interpretation are commented in some detail. An attempted standardization consensus for device's performance suited to biomedical instrumentation standards is documented. A short end view on the instrument device in Zagreb laboratory is put forward.

In Chap. 8 results are shown of clinical applications of the method of pedobarography, sometimes combined with computerized shoe insole design and manufacture. Pedobarography can be used in clinical practice in many areas, as a diagnostic tool, in monitoring of the effect of therapy, both surgical and nonsurgical such as physical therapy and rehabilitation of injuries, painful conditions, degenerative and rheumatic diseases, habilitation of the children with developmental disorders. It can also be used as a combination of diagnostics of gait disorders and computer-aided design (CAD) with computer-aided manufacturing (CAM), resulting in producing orthopedic insoles. There is a broad spectrum of possibilities using orthopedic insoles. A number of clinical findings in orthopedics and sports medicine (sports traumatology) are shown and interpreted. The use of orthopedic insoles in the most common diagnoses in orthopedics as there are Flat transverse arch and metatarsalgia, Flat feet and High-arched foot (pes cavus) are presented. The application of orthopedic insoles in different entities in sports medicine like there are Plantar fasciitis, Achilles tendinitis/tendinosis, Flexor hallucis longus muscle dysfunction, Posterior tibial muscle dysfunction, Peroneal muscles dysfunction, Anterior impingement syndrome of the ankle, Sesamoiditis, Stress fractures, Sever disease, Patellar tendinitis/tendinosis (jumper's knee), Osgood-Schlatter disease and Groin pain are described.

Chapter 9 brings the reader's attention to the neuromuscular level of treatise, meaning that the motor control aspect comes into focus. In the chain of information transmission—neural impulses serving as information carriers signal conduction takes place via neural pathways, followed by transmission of excitation to muscle fibers. Skeletal muscle function is presented as seen through its bioelectrical manifestation; electromyographic (or myoelectric) signals (EMG), the signal being modeled in a form of an interference pattern. The method of electromyography is succinctly explained next including technical aspects of signal detection, amplification and registering. In kinesiology, primarily surface electromyography (sEMG) is used. A classical repertoire of signal processing methods in time and in frequency domain is presented next, with interpretations commonly used in biomechanics and kinesiology. It is known as kinesiological electromyography, with principal applications in the evaluation of movement skill and of local muscle fatigue.

Gait analysis can be considered a backbone of modern clinical locomotion biomechanics, and it makes the subject of Chap. 10. Basic methodology, integrating kinematic, kinetic and myoelectric measurements in course of a subject's walking and subsequent interpretation of results is summarized and illustrated mainly with findings from our Zagreb and Salerno situated laboratories. In addition, pedobarography and portable oxygen consumption measurement system make desirable components of an equipment inventory for gait analysis. Physical examination and observational analysis precede measurement of several gait trials in a laboratory. Interpretation of measurement findings is the next step. Modern tendencies, pursued worldwide, incorporate individualized (subject-specific) neuro-musculo-skeletal modeling into a clinical procedure. Non-linear analysis of gait data emerges as yet another methodology of gait analysis. Alternative modern tendencies are finally pointed to whereby classical measurement instrumentation setup confined to laboratory environment is substituted by portable, so-called pervasive multisensor measurement solutions offering freedom of movement and aiming to replace classical approach, albeit with no success yet.

Modern orthopedics and traumatology are nowadays impacted ever more by a field of bioengineering. The next Chap. 11 gives reflections of an orthopedic traumatologist on the area of human locomotion. An orthopedic traumatologist has a goal to treat the injuries or diseases of patient's locomotor system with the highest respect to its biomechanical features and relationships. It is very important for him/her to understand the individual anatomical characteristics and biomechanical performances of each patient's joints. Therefore, empirically investigated values are described of locomotion capacity and mobility of individual regions of the locomotor apparatus of the lower extremities of an average patient, known so far. Practical examples of treatment of patients with locomotor system injury are presented through surgical treatment techniques commonly used by the orthopedic traumatologist, including the use of computer-aided surgery (CAS) technology when appropriate. In this way, the surgeon tries to bring the end result of treatment as close as possible to the imagined performance of the patient's locomotor system up to the moment of injury. Demonstrated are some practical applications of such treatment options through individual cases of treated patients. The chapter covers critical diagnostics and treatment aspects of hip, knee and ankle pathology through presented examples of individual cases.

Chapter 12 closes the book with the topic of sporting movement patterns study. Cross section of biomechanical, predominantly kinematics-oriented, research in the 'Zagreb kinesiology circle' is often categorized within division related to monostructural, polystructural, complex, and aesthetically-conventional kinesiological activities. Movement patterns in sporting activities are represented through reductionist-constructivist dualism. Reductionist sports, colloquially called 'gram-meter-second (GMS) sports', are essentially reduced to the goal of achieving results through international standards of measurement and measures, e.g. the SI system. Movement analysis is mostly performed in controlled space (either closed, indoors, such as in a motion analysis laboratory, or outdoors) and with technologies for 'extended' sensory-motor circuits, and motion control in general. Examples of javelin throw and sprinting discipline are presented. Constructivist sports, in addition to the physical manifestation of the individual movement, include additional aesthetic component, cooperation, opposition, or some achievement of an externally ruled goal. Movement analysis should imply both GMS dimension of movement and psychosocial and systemic elements, such as tasks, tactics, strategy, momentum, etc. Examples of tennis serve and team handball jump shot are presented.

As a practical addendum to the book, at the end several current standards for reporting of kinematic, kinetic, and EMG variables are reprinted (Appendices A-E).

References

1. Medved V (2006) Biomedicinsko inženjerstvo (Biomedical engineering) In: Barišić P (ed) Prvi kongres hrvatskih znanstvenika iz domovine i inozemstva, Zagreb-Vukovar 15–19 studenoga 2004. Ministarstvo znanosti, obrazovanja i športa Republike Hrvatske, Zagreb, pp 282–286
2. Bronzino JD (ed) (2000) The biomedical engineering handbook, 2nd edn. CRC Press, IEEE Press
3. Rose J, Gamble JG (eds) (2006) Human walking. 3rd edn. Lippincott Williams & Wilkins, Philadelphia, Pa
4. McMahon TA (1984) Muscles, reflexes, and locomotion. Princeton University Press, Princeton, NJ
5. Basmajian JV, De Luca CJ (1985). Muscles alive: their functions revealed by electromyography. Fifth edition. Williams & Wilkins, Baltimore, Md
6. Merletti R, Parker PA (eds) (2004) Electromyography, physiology, engineering, and noninvasive applications. IEEE Press, Wiley & Sons, Inc. Hoboken, New Jersey
7. Merletti R, Farina D (eds) (2016) Surface electromyography: physiology, engineering, and applications, 1st edn. IEEE, John Wiley & Sons, Inc.
8. Vodovnik L (1985) Nevrokibernetika (Neurocybernetics). Faculty of Electrical Engineering Ljubljana
9. Enoka R (2008) Neuromechanics of human movement. In: Human kinetics, 4th edn. Chicago, Il
10. Tipton CM (ed) (2003) Exercise physiology: people and ideas. Oxford University Press
11. Hollmann W, Strüder HK (2009) Sportmedizin: Grundlagen für körperliche Aktivität, Training und Präventivmedizin, 5. Auflage. Schattauer Verlag
12. Baker R (2013) Measuring walking: a handbook of clinical gait analysis. Mac Keith Press, London, UK
13. Magjarević R, Lacković I (2011) Biomedical engineering-past, present, future. Automatika 52(1):5–11
14. Ruszkowski I, Orlić D, Muftić O (1985) Endoproteza zgloba kuka, Medicinski fakultet Sveučilišta u Zagrebu (Hip joint endoprosthesis).
15. Lončarić S. Primjena Hrvatskog kvalifikacijskog okvira u području biomedicinskog inženjerstva (HKO-BI) (Application of Croatian Qualification Framework) —Projekt (2019)
16. Medved V (2007) From research to teaching human locomotion biomechanics: a Zagreb experience. In: Zanchi V, Revetria R, Cecchi A, Mladenov V, Zemliak A (eds) Challenges in remote sensing. Proceedings of 3rd WSEAS international conference on remote sensing (REMOTE' 07), Venice, Italy, 21–23 Nov 2007, WSEAS Press, pp 43–46. ISBN: 978-960-6766-17-6. http://www.wseas.us/e-library/conferences/2007venice/papers/600-138.pdf

17. Medved V (2001) Measurement of human locomotion. CRC Press Inc., Boca Raton, Fl
18. Pećina MM, Bojanić D (2003) Overuse injuries of the musculoskeletal system, 2nd edn. CRC Press Inc., Boca Raton, Fl
19. Nikolić V, Hudec M (eds) (2011) Principi biomehanike (Biomechanics principles). Naklada Ljevak, Zagreb

Chapter 2
History of the Study of Human Locomotion and Elements of Current Research Methodology

Vladimir Medved

Abstract A glance on history of the study of human locomotion is given. Major landmarks and most significant contributors are mentioned, that have pursued the development of the approaches intended to get to know the phenomenon of gross human body movement and locomotion. Time period since Antiquity until our days is encompassed. Contribution of each new arising technology (in modern times these were, for instance, high-speed photography, digital computers, new types of sensing, etc.) is outlined. Chapter ends with pinpointing elements of currently attained methodology and paradigms for studying the phenomenon, combining biomechanical approach behind modern measurement, modeling and diagnostics systems, with exercise physiology approach comprising measurement and testing of pulmonary and cardiovascular systems' performance.

2.1 Studying Human Locomotion: A Short Historical Review

The following, rather concise historical survey—a kind of a „snapshot" of the history of the study of human locomotion—is based on a number of literature sources [1–39], and only the most significant events and personalities are highlighted. Achievements are shortly documented that have contributed to biomechanical and medico-physiological approaches to human movement and locomotion.

As emphasized by Aurelio Cappozzo, giving a Wartenweiler Memorial Lecture at the XXV ISB meeting in 2015 in Glasgow [1] (ISB goes for International Society of Biomechanics), an interplay between arts and sciences in approaching human body

V. Medved (✉)
Faculty of Kinesiology, University of zagreb, Horvaćanski zavoj 15, 10000 Zagreb, Croatia
e-mail: vladimir.medved@kif.hr

V. Medved (ed.), *Measurement and Analysis of Human Locomotion*,
Series in Biomedical Engineering, https://doi.org/10.1007/978-3-030-79685-3_2

in motion has been constantly present during a jurney that began long time ago. Interest for human locomotor and movement activities may indeed be traced to the times of ancient civilisations, the most remarkable in this context being ancient Greek civilisation. It is visible in their arts, as kinematic representations of athletes appeared in artistic media such as paintings and sculptures, demonstrating an interest for human body and its movement, including sports activities. Greek artists developed faithfull methods for depicting motion, and their sculptures manifested good knowledge of body surface anatomy and illustrated well the dynamics of movement.

Hippocrates (460–370 B.C.), a Greek physician, a father of Western scientific medicine (Rothschuh 1973, citation after 2) beleved in the principle of causality, and has been commited to forming medicine a rational science. His emphasis on observation and experience of the senses pioneered the use of rational scientific method in the practice of medicine. He even had some notion of what we call metabolic processes in relation to exercise [2]. Plato (427–347 B.C.), a Greek philosopher, believed for the world of senses to be a faint representation of reality while essential information is contained in ideas. His school of thought influenced strongly all aspects of human knowledge. Aristotle (384–322 B.C.; book: „De motu Animalium", 344 B.C.), a Greek philosopher, observed and analyzed movement of animals qualitatively; he considered interaction betweeen an animal and its surrounding to be of importance. (This, in fact, is a rather modern attitude, properly reflected in today's approach to human movement biomechanics). Seeing animal's body as mechanical system [3], he was the first to describe the activity of muscles and movement in the joints during locomotion.

Claudius Galenus-or Galen (ca. 129–200, according to 2), a physician and surgeon to gladiators in ancient Roman Empire, a follower to Hippocrates, was aware of muscular action and traumatology. His primary interest was in human anatomy and he considered the study of anatomy to be the foundation of medical knowledge. As performing dissections on human corpses was forbidden, he experimented on lower animals such as apes and sheeps. He travelled widely, exposing himself to a variety of medical theories and discoveries, and also studied at the great Alexandria Medical School. Galen wrote extensively ("On the natural faculties", "De motu musculorum"). As a philosopher, his writings were influenced by earlier Greek and Roman thinkers, including Plato, Aristotle, and the Stoics. His contribution to the Hippocratic understanding of pathology was immense. His works on the circulatory system were also much ahead of his time. Galen demonstrated that contraction is the specific activity of muscles [4] and is considered an initiator of myology—a science on muscles. Long time after him his medical doctrine prevailed, and he was considered an utmost authority on medical theory and practice in Europe up to about 1400 years to follow.

During Middle Ages not much has happened with regard to the subject of the study of human locomotion. The dominance of Church didn't stimulate science, but on the other hand it enabled storage of information on ancient knowledge in books, as well as forming of first schools and universities. The Medical School of Salerno, whose first coming into being is not precisely known, however it was already mentioned of being ancient in ninth century and reached the top of its influence in twelfth century, was the first medical school in the World in modern Western civilisation, the forerunner of modern university medical schools.

In Persia, meanwhile, Greek books on mathematics and other areas have been translated and so the respective knowledge has been saved for future generations. Avicenna (980–1037), an islamic philosopher and neoplatonist, mentioned by Gruner (citation according to [5]) („A Treatise on the canon of medicine of Avicenna", London, 1930) wrote „The book of healing".

In Renaissance, a more thorough insight into the natural phenomena begun to be obtained thanks to the revival of ancient Greek and Roman artistic, scientific and cultural traditions. Both analytical insight as well as experimental approach were introduced into scientific pursuit. This provided the foundation for development and diversification of natural sciences, a process going in continuation until our days. Leonardo da Vinci (1452–1519), an Italian painter, sculptor, builder, and scientist, has done among first dissections of human cadavers. He made precise drawings of human skeletal muscles and the way they were attached to skeleton, and his drawings of anatomical details, having an almost modern value, are popular in our time for illustration of medical books. He studied nature and designed engineering artifacts modeling it. His interest in biological realm was broad: physiology of blood circulation, biological vision, muscular and skeletal function, to mention but a few topics. Da Vinci was a scientist, more precisely a natural phylosopher, with an immensly broad interest in nature, art and engineering, and with many technical ideas and concepts ahead of his time [6]. He wrote: „On the human body", O'Malley and Saunders, Henry Schuman, New York, 1952. Since that time, human anatomy has been roughly known for purposes of gross body movement analysis. Andreas Vesalius (1514–1564), a Belgian anatomist, added to Galen's work, correcting and upgrading it („On the structure of the human body") [3].

Galileo Galilei (1564–1642), an Italian physicist, astronomer, and mathematician, introduced an experimental method. He investigated by using instruments of that time, such as optical microscope and telescope. So, both analytical and experimental knowledge have been advancing. Due to his insights into the fields such as allometry, through its relevance to the built of certain body parts, Galilei may even be considered a founder of biomechanics according to some sources [7]. He combined, for the first time, a deductive reasoning with experimental observation.

Giovanni Alfonso Borelli (1608–1679), (Fig. 2.1) an Italian physician, physiologist and astronomer, professor of mathematics, born in Naples, applied Galileo's mechanicistic view to biological systems, to man and to animals, primarily to static

A)

B)

Fig. 2.1 **A)** Giovanni Alfonso Borelli (1608-1679) https://commons.wikimedia.org/wiki/File:Giovanni_Alfonso_Borelli.jpg (public domain). **B)** From the Borelli's book „De motu animalium“ (1680) https://collections.nlm.nih.gov/catalog/nlm:nlmuid-101448301-img (public domain)

body postures. His work is „De motu animalium ex principio mechanico statico" 1680. („On the movement of animals"), bearing the same title as Aristotle's book and printed after his death. The book treats movements and displacements of the limbs of man and animals. Borelli pursued iatrophysics (or iatromechanics), a medical theory which viewed human organism as reduced essentially to mechanical function. He had intuitive knowledge on the action of forces on the body, primarily in static situations. Analysing correctly, in principle, mechanics of static body postures, Borelli failed to take into account the influence of force of inertia. Not performing anatomical sections himself, he relied on experimental anatomical knowledge acquired by others. Borelli is considered a father of modern biomechanics. The most prestigious honor given by the American Society of Biomechanics (ASB) is „The Borelli Award".

Isaac Newton (1642–1727) was an English physicist, mathematician, astronomer, and natural philosopher. He synthesized knowledge on mechanics of the time and formulated laws of classical mechanics which are valid as long as respective motion of bodies happens at a relative speed much lower than the speed of light. („Philosophiae naturalis principia mathematica", 1686). His three laws of mechanics form the basis for the biomechanics of locomotion. Newton served as president of the Royal Society from (1703 1727).

Herman Boerhaave (1668–1738), the Duchman, of Leiden, a physician, chemist, and botanist, followed the work of Newton and was the first to consider dynamics, i.e. the kinetics of movement, taking inertial influences into account. In 1703. he gave a speech at the University of Leiden entitled „On the use of mechanical method in medicine" in which he foresaw future events in the field of biomechanics that would take place two hundered years later with the development of modern data acquisition and processing systems [8]. In time to follow, many a scientist at the side of exact, natural sciences further contributed to mathematical and physical knowledge relevant to mechanics of moving bodies: Leonhard Euler (1707–1783), Jean le Rond d'Alembert (1717–1783), Joseph-Louis Lagrange (1736–1813), to name but a few.

Tackled by scientific curiosity, a number of natural scientists, or natural phylosophers, studied and experimented with body tissues, which signalled the beginnings of the fields such as muscle physiology, neurophysiology and biochemistry [8]. Francesco Redi (1626–1698), an Italian physician, natural scientist and poet, has been the first to recognize the connection between muscle and generation of electricity. He documented, in 1666., a very specialized muscular tissue in electric fish. Jan Swammerdam (1637–1680), a Duch zoologist (anatomist and biologist), discovered that hitting the nerve innervating m. *gastrocnemius* in frog induces muscular contraction. According to some sources [9], Swammerdam did the first experiments in electrical stimulation (143 years prior to Luigi Galvani, to be mentioned later). Alessandro Volta (1745–1827), in Italy, invented the device which produced electricity (electrical battery, galvanic cell), that could be used for stimulation of muscle. Luigi Galvani (1737–1798), an Italian physicist and physician, showed that electrical stimulation of muscle tissue produces contraction and force, on frog m. gastrocnemius (1791). He is considered a father of neurophysiology. Due to limited instrumentation, his work was completely accepted only 40 years later. Carlo Mateucci (1811–1868), in Italy, proved by using galvanometer that electricity is connected with muscular contraction (1838). (An ever greater impact of new technical devices can be witnessed.) Mateucci demonstrated the existence of action potential in the frog muscle, and his experiments were at the beginning of bioelectrochemistry. Emil Du Bois Reymond (1818–1896), in France, postulated the existence of a potential difference between the interior and the exterior of a muscle cell and, consequently, of a membrane potential; he discovered the nerve impulse, the basic mechanism of information transmission in the nervous system, and was the first to detect electrical manifestation of voluntary muscular contraction (1848).

Wilhelm (1804–1891), a physicist, and Eduard (1806–1871), a physician, Weber, from Göttingen, Germany, studied human gait and modeled it in their book: „Die Mechanik der menschlichen Gehwerktzeuge. Eine anatomisch-physiologische Untersuchung“, Dietrische Buchhandlung, Göttingen, 1836. The book contains anatomical discussions of the joints used in walking and running, simple measurements made on living subjects, and a mathematical theory relating the length and duration of a step to anatomical parameters. They advocated the principle of least muscular effort, assuming that gait occurs automatically taking advantage only of force of inertia. Although their belief is disputed today—modern techniques exist for measurement and assessment of dynamic muscle action—there is indeed a tendency for metabolic energy expenditure needed for muscular work to be minimized during normal ambulation.

Two contemporaries, Eadweard Muybridge (1830–1904), a British photographer (real name Muggeridge), and Étienne Jules Marey (1830–1904) (Fig. 2.2), a French physician and physiologist, happen to be rather influential for the field of human motion study. Their parallel lifes marked an important step in measurement of kinematics of locomotion.

Marey, being a renowned physiologist gifted with engineering talents, begun by constructing various technical devices to measure and record physiological and locomotor phenomena. He was a thorough explorer: he approached measuring human and animal movement by constructing original apparatus that were attached to the body and produced recordings of physiological functions (like blood pulse) or movement (like movements of bird’s wings!). When detecting contacts between the foot and the ground during locomotion he took advantage of a hydraulic principle to transmit the detected change of pressure to the recording device worn by the subject (Fig. 2.2): rubber tubes connected air chambers in the shoes to the kymograph, a hand held recording drum (pen recorder). The device on the runner’s head is an accelerometer. It is illustrative to quote Marey’s description of the Station Physiologique in the Park de Princes at the place of today’s Roland Gaross in Paris, which was in fact the first

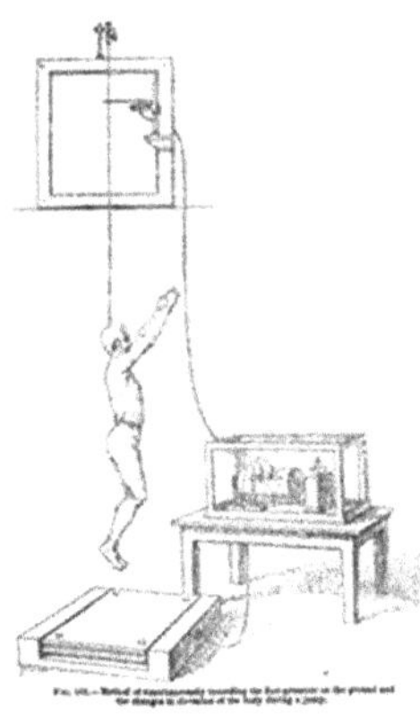

Fig. 2.2 Étienne-Jules Marey (1830–1904) (https://upload.wikimedia.org/wikipedia/commons/b/b4/Etienne-Jules_Marey._Photograph_by_Nadar._Wellcome_V0026815.jpg) (CC BY 4.0) and some of his apparatus in the pre-chronophotography era (https://books.google.hr/books?hl=hr&lr=&id=nCYmfOC-ya8C&oi=fnd&pg=PR2&dq=%C3%89tienne-Jules+Marey&ots=Tvs57pKzVX&sig=OMhwY4FuoIgkLiYMZLG_3JsgRWY&redir_esc=y#v=onepage&q&f=false) (public doman) (https://books.google.hr/books?id=oubaAAAAMAAJ&printsec=frontcover&dq=%C3%89tienne-Jules+Marey&hl=hr&sa=X&redir_esc=y#v=twopage&q&f=false) (public domain)

physiological-biomechanical laboratory: "*Registering dynamometers, spirometers, pedometers, and various apparatus for the measurement of objects under observation are devoted to the study of human locomotion. In addition, pneumographs, sphygmographs, and cardiographs enable the investigator to study the effect of athletic exercises on the function of organic life, and to follow step by step the improvement under training.*" [10]. The same is also the aim of modern laboratory facilities for movement analysis and for physiology of activity (exercise physiology).

The invention of a photographic method in 1839. was an epochal event, a product of work of a many an inventor, the most significant among them being Louis Jacques Mandé Daguerre (1787–1852), a French painter: i.e. it is named daguerreotype after him. Besides static situations, the invention enabled also registering human or animal body in motion. Namely, a photographer Muybridge was engaged in making pictures of American national parks. Then, according to a legend, Leland Stanford, a governor of California and founder of Stanford University, an *amateur* of race horses, gave him an assignment to record running of a horse in order to deduce whether during running there is a moment when the animal lifts all four hoofs off the ground. Muybridge realized the task by arranging a series of cameras in a row and making an animal to trigger cameras in sequential order while running, realizing a so-called chronophotography (Fig. 2.3). (Baker [11], however, reports on Marey's earlier method to directly register

Fig. 2.3 The horse in motion, by E. Muybridge 1878 [12] (https://upload.wikimedia.org/wikipedia/commons/7/73/The_Horse_in_Motion.jpg) (public domain)

contacts between horse's hoofs and the ground which made this timing visible). Muybridge's works are: „The horse in motion" 1882, „Animal locomotion" 1887, „The human figure in motion", Dover New York 1955. He made thousands of high quality records of human and animal locomotion, which is witnessed by modern reproductions of his books and may still today be used as a reference and for visual movement analysis.

With the invention of the chronophotographic method by Muybridge, Marey soon adopted the novelty and begun to use it in his studies of human and animal movement. For that purpose he constructed cameras such as a mobile camera vehicle at the Station Physiologique (Fig. 2.4) and a photographic gun (Fig. 2.5), this later device having a rotating shutter which, by providing sequential exposures of a film negative, could record movement at a speed of 12/s [10]. The photographic gun device is a precursor of sort to the the technique of cinematography, introduced in 1895. by Lumière brothers. Marey's pupils and collaborators in his work on measuring motion deserve to be mentined: Gaston Carlet (1849–1892), Karl Hermann Vierordt (1853–1944), and Georges Demeny (1850–1918). A comprehensive reference on Marey is: Marta Braun (1994) „Picturing time: the work of Étienne-Jules Marey (1830–1904)", University of Chicago Press.

Fig. 2.4 Mobile camera at the Station Physiologique, c. 1883 (https://upload.wikimedia.org/wikipedia/commons/2/2e/Marey_La_m%C3%A9th._graphique_0744.jpg) (public domain)

Two Germans, anatomist Wilhelm Braune (1830–1892) and physicist Otto Fischer (1861–1917) developed an approach whereby they measured human movement with two cameras, using chrono-photography, and combined two projections of an image into a three dimensional (3D) representation, i.e. they realized spatial measurement, a stereometry. They used special Geissler light bulbs controlled by Rhumkorff coils as active markers, attached at body prominences. Further, for the first time, they realized an inverse dynamics approach. For this they needed to dissect human cadavers in order to estimate relative masses of particular body segments and calculate their inertial properties. It took them, however, 6 to 8 h for measurement of gait, while calculations lasted several months [8]. They proved that during walking muscles are active, contrary to the former thesis by Weber brothers [13]. Their work marked the beginning of both analytical photogrammetry and inverse dynamics to be applied in the field of human motion study. Their famous book is „Der Gang des Menschen" (1895) (translated into English by Maquet and Furlong, The human gait, Springer Verlag, Berlin, 1987) (Fig. 2.6).

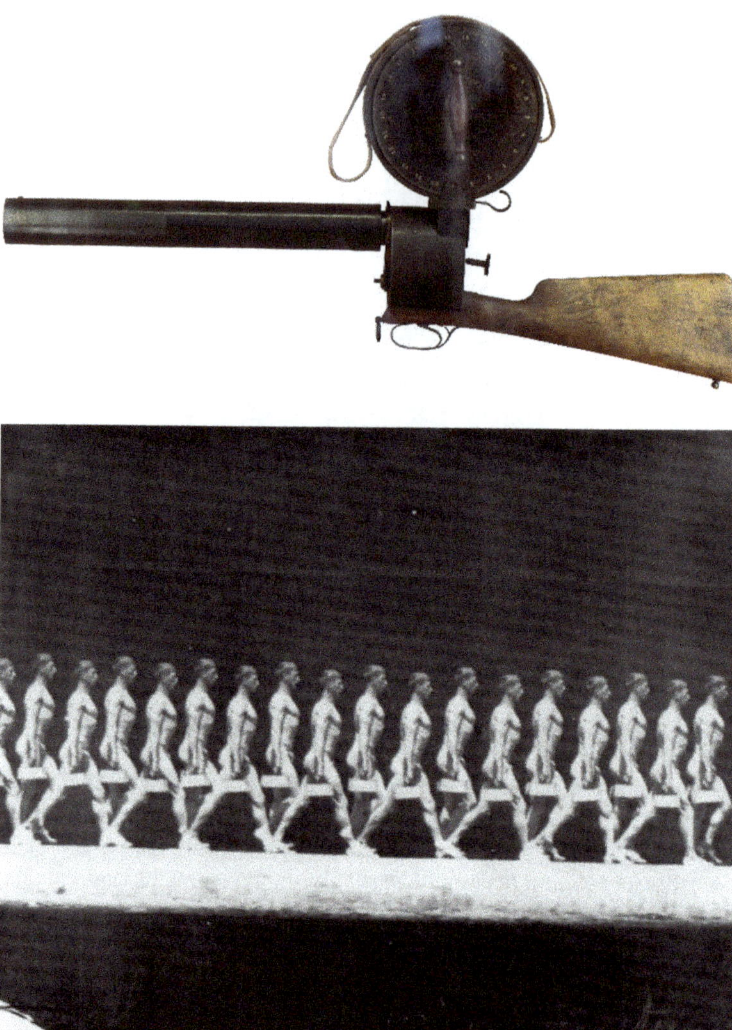

Fig. 2.5 Photographic gun by Marey (1895) (https://upload.wikimedia.org/wikipedia/commons/b/bc/Chronophotographic_gun-CnAM_16955-IMG_5275-white.jpg) (CC BY-SA 3.0) and record of a walking man carrying a weight (https://hyperallergic.com/197464/the-scientist-who-shot-his-photos-with-a-gun-and-inspired-futurism/) (public domain)

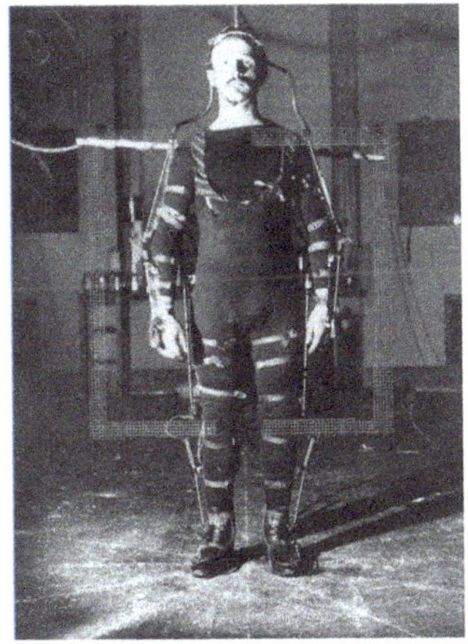

Fig. 2.6 From measurements by Braune and Fischer (Woltring [14]. With permission of the author's family)

Fig. 2.7 „A nude descending a staircase" by Marcel Duchamp (1912). Oil on canvas. Philadelphia Museum of Art. Influences on art: elements of fragmentation and synthesis belong to cubism while movement and dynamics to futurism https://philamuseum.org/collection/object/51449 © Association Marcel Duchamp/ADAGP, Paris 2021. With permission

Entering the twentieth century, new technologies of recording movement using high-speed photography have also influenced arts, as nicely exemplified, for instance, in the art of painting by pictures such as „A nude descenting a staircase" by Marcel Duchamp (Fig. 2.7).

In studying energetics of locomotion, ever more technical devices and instrumentation were introduced into the use whereby an intention was to subject a human body to physical load and to record its reaction. One such attempt is shown in Fig. 2.8,

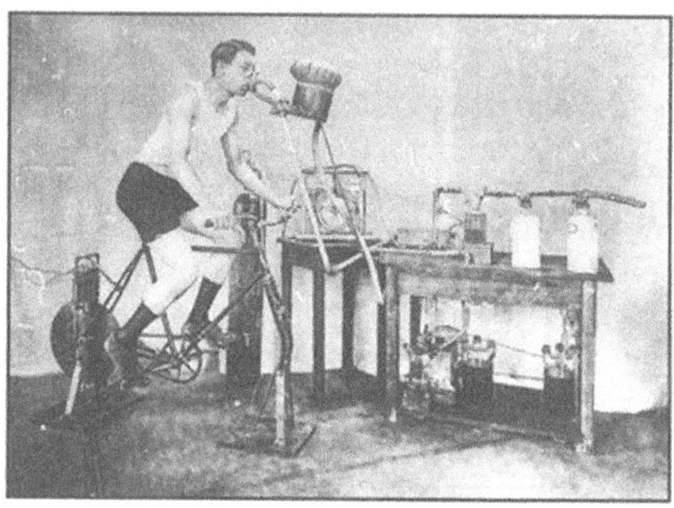

Fig. 2.8 Bicycle ergometry with examination of gas metabolism, carried out by Atwater and Benedict around 1900 in the US (from Hollmann and Prinz [15]. Reproduced by permission of Springer Nature)

refering to the work of Francis G. Benedict (1879–1957), biochemist and physiologist, in USA, where cycle ergometer provided mechanical loading of the body while pulmonary output was measured. Oxygen consumption during light, moderate, and severe exercise were measured.

Another possibility of loading is by a treadmill. Here, compared to a cycle ergometer case, the subject during exercise bears his whole body weight, a factor working in favor to light compared with heavy-weight subjects.

Meanwhile, in Germany, similar attempts took place.

Robert Herbst (1890–1962), a medical doctor, was active as lecturer at the „Deutsche Hochschule für Leibesübungen" (DHfL) (The German College for Physical Education, established in 1920 in Berlin; a forerunner of „Deutsche Sporthochschule Köln" (DSHS) (The German Sport University Cologne) established in 1947). He worked on maximal oxygen uptake measurements, having a general interest in how various exercise forms affect bodily functions.

Another interesting attempt to examine breathing process when running is pictured in Fig. 2.9.

Fig. 2.9 Examination of the gas metabolism during cycling in connection with the Douglas bag, performed by Kost in 1928 in Berlin (from Herbst [16]. With permission)

In Russia the school of biometrics developed, led by Nikolaj A. Bernstein (1896–1966), a physiologist, working in the Russian All-Union Institute of Experimental Medicine in Moscow, who devised stereometric kinematic recording equipment with mirrors. Based on his physiological and kinesiological research he conceptualized a hierarchical model of the nervous system in control of motorics. The modern field of motor control is by a significant amount influenced by bernsteinian school of thought [17].

Herbert Elftman (1902–1988), in USA, constructed a force platform using springs („The measurement of the external forces in walking“, Science, 88, 152–153, 1938. Figure 2.10), a kind of a successor to the Marey's and Carlet's device. Robert Plato Schwartz (1892–1965), medical doctor at the University of Rochester, Minnesota, who led the Myodinamics Laboratory established in 1926 is, according to Brand [18], probably the first medical clinician who developed locomotion measurement methods for clinical, and not primarily research applications in the 1930s, and systematically collected and analysed gait measurement data. He used a device with air chambers similar to Carlet's [8].

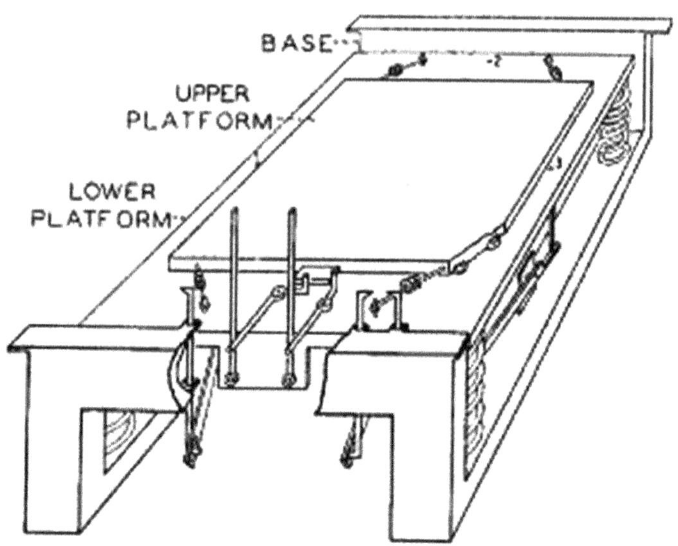

Fig. 2.10 Force platform using springs by Elftman (Science, 88:152–153, 1938) (With permission)

Herbert Henri Jasper (1906–1999), a Canadian neurologist, constructed first electromyograph device (term electromyography was coined by Marey) (1942–44) at the McGill University (Montreal Neurological Institute). He made unipolar needle electrode for application in epilepsy.

After the World War II, in USA at University of California in San Francisco, a multidisciplinary group of medical doctors and engineers was formed with the aim to supply war injured soldiers with prostheses of extremities. In order to accomplish the task it was necessary to thoroughly study normal gait as a fundamental biological phenomenon, to provide a basis for approaching pathological situations. The group, known as the Berkeley group, was led by Verne Inman (1905–1980) and Howard Eberhart (1906–1993). They authored: „Human walking", which in our days is in its 3rd edition [19] and constitutes a premier reference for the field. Their research revealed six, a so-called determinants of gait, describing a kinematic model of gait. The issue of determinants of gait has been discussed in [20]. Also, in [19], DS Childress and SA Gard provide minor modification (pp. 19–21) to those from the original paper (JB Saunders, VT Inman, HD Eberhard (1953) „Major determinants in normal and pathological gait". J Bone Joint Surg 35:-A:543) but the concept remained essentially unchanged however. Successors to the Berkeley group were Jacqueline Perry (1918–2013), combining expertease in physical education, physical therapy, and orthopedic surgery, of Downey, California, and David Sutherland (1923–2006), orthopedic surgeon, of San Diego, California, who were developing clinical applications, including a standardized observational analysis protocol for human gait. Mulitchannel EMG was also introduced into clinical measurements. In Europe, the field has been advanced by Jürg Baumann (1926–2000), a Swiss ortopedist and neuro-ortopedist.

Technique of high-speed photography was perfected; a notable expert Harold E. Edgerton (1903–1990) of MIT is to be mentioned. Around 1960es, a transition from high-speed photography to automated electronic sensor based measurement of kinematics of locomotion took place [21]. As regards measurement of kinetic quantities, the first professional grade six-component force plate that appeared on the market was a result of joint venture between Jürg Wartenweiler (1915–1796), a Swiss educator, of the ETH Zürich and a private company [1]. In USA, first force platforms in modern sense appeared by beginning of 1970es. Worth mentioning is the work in Bioengineering Unit, University of Stratchylde, Glasgow, UK, where biomechanics of gait has been studied and first hip prosthesis developed. It is a premier example of a biomedical engineering prosthesis, prior to a pacemaker [22].

To bring this historical review to an end, three last decades of the 20th, and the beginning of the twenty-first century are looked upon, characterized by development and standardization of biomechanical and physiological measurement instrumentation and methods. A few selected examples of modern locomotion biomechanics research and practice are mentioned here, the selection being rather subjective, a personal preference of the author of this Chapter, while an objective historical overview will only be possible from a greater time distance.

James Gage, an American orthopedic surgeon and biomechanist devised a first automated motion analysis laboratory in Newington Children's Hospital, Newington, CT, USA, in 1976. He developed clinical gait analysis dominantly to improve diagnostics and treatment of cerebral palsy patients, where he organized a multidiciplinary team work [23]. He is now with Gillette Children's Specialty Healthcare, Saint Paul, Minnesota; The James R. Gage Center for Gait and Motion Analysis is named after him. Richard Baker pursued significant work on clinical gait analysis first in the Royal Children's Hospital in Melbourne, Australia, and now at the University of Salford, Manchester, UK [11]. Baker is holding World's first Professorship of Clinical Gait Analysis.

Skeletal muscle was researched the most during the twentieth century at the microstructural, biophysical, biochemical, biomechanical, as well as at the control level. It has been enabled by technological and methodological advancements such as the invention of electron microscopy, development of electrophysiological recording techniques, pursuing the concept of cybernetics and systems analysis, etc. Basic muscle contraction mechanism was elucidated and mathematically modeled: see, for instance the explanation of Huxley models of muscle contraction by McMahon (20, front cover shown in Fig. 1.1, Chap. 1).

Fig. 2.11 Peter Greene and Thomas McMahon shown working with a spring-dashpot mechanical leg at Harvard University 200-m indoor "tuned" track (Reproduced by permission of the authors: Greene and Medved [24]. Creative commons attribution license: http://creativecommons.org/licenses/by/4.0/)

Thomas A. McMahon (1943–1999), a physics engineer and professor of applied mechanics and biology at Harvard, encompassed the field ranging from biophysical and bioengineering aspects of skeletal muscle function to whole body locomotion of men and animals, implying allometric aspects as well. He was instrumental, together with his collaborator Peter R. Greene, in designing a "fast running track" at Harvard University that increased running speed, cutting simultaneously number of injuries (Fig. 2.11).

The Harvard Track was manufactured by Tony DiNatale at Championship Sports Floors in Florida. It was mechanically "tuned" to biomechanical characteristics of a runner. Similar indoor tracks were also installed in Madison Square Garden in New York, at Yale University, the Meadowlands Sports Complex, while the outdoor experimental track was installed at Loughborough University, UK. It has proven quite durable, lasting now 45 years; the upper poly-urethane surface was recently re-surfaced in 2018. (Most tracks of this nature, at best, are guaranteed for five years.)

John Basmajian (1921–2008), anatomist (Fig. 2.12), and Carlo De Luca (1943–2016), electrical engineer, have comprehensively covered the method of surface electromyography (sEMG) [26]. It included mathematical modeling of signal formation, and established sEMG as an important tool in grasping human movement and locomotion by providing empirical evidence on bioelectric activity of muscle tissue and quantitative measure of neuromuscular system function.

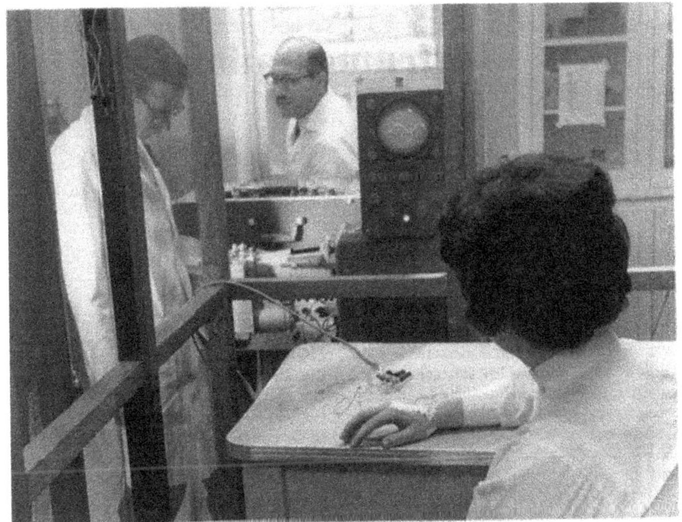

Fig. 2.12 Surface EMG recording of first *dorsal interossei* conducted in a copper-screened room. John Basmajian, the father of sEMG, is shown in the middle. Courtesy of John Basmajian (from Cram et al. [25]. With permission by Nancy Basmajian, daughter of J.V. Basmajian)

Recent progress in the field of biomechanics and neural control of movement is characterized with the following diverse but mutually interdependant issues such as; widespread development and proliferation of biomedical engineering as an academic field; focus on research of the manner in which the brain controls movement; extraction possibility of neural representations of movement through implementing multi-electrode arrays; neuro-musculo-skeletal system control properties elucidation; cortical control of neuroprostheses, to name but a few areas [27]. Notable are contributions by Merletti and Farina [28]. Spanning areas from neural control of movement to biomechanical movement mechanisms, a field called neuromechanics was introduced, as evidenced, for example, by the title of the book by Roger Enoka, a kinesiologist of New Zealand, active in USA [29].

Many a modern scientist contributed to development of quantitative methodology for biomechanical analysis of movement, by combining measurement with mechanical modeling through inverse dynamics and development of appropriate signal elaboration and processing methods and algorithms: Woltring [13, 14], Lanshammar [30, 31], Cappozzo [32, 33], and Vaughn [34] are examples.

The Stanford bioengineering school led by Felix Zajac is personalized by his student Scott Delp, who transformed the field of movement science by creating highly accurate computer models of musculoskeletal structures providing them to researchers worldwide using a software system (SIMM and OpenSim) that he and his team developed [35–38]. His work draws on computational mechanics, biomedical imaging, and neuromuscular biology. Faithfull simulation of human and animal motion—focused in application to particular body parts—is thus enabled. Improvements of treatments for individuals with physical disabilities are attempted. Delp's software has become the platform for an international collaboration.

In the introductory Chap. 1, trends in twentieth century electronics and automation impacting the field of human motion measurement and analysis were identified (Fig. 1.3). Nigg [39] has resumed well historical development of biomechanics in the context of global scientific and cultural development in World history. He made an effort to evaluate impact of the several scientific fields on development of biomechanics: mechanics, mathematics, anatomy, muscle, and locomotion (Fig. 2.13).

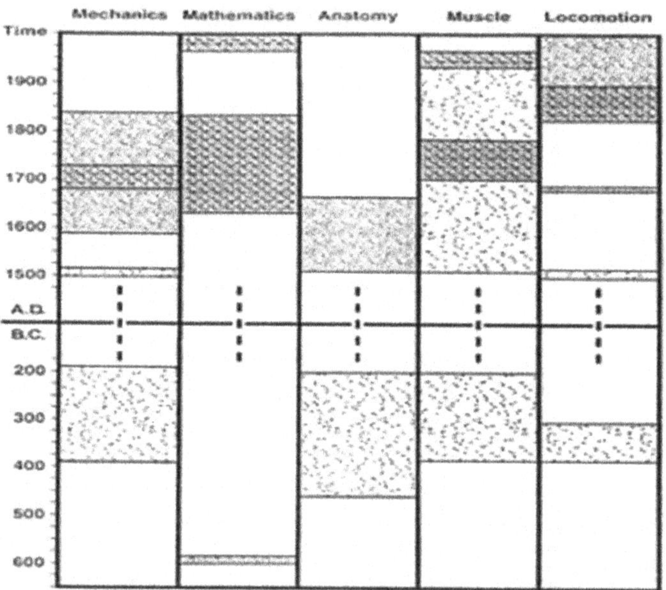

Fig. 2.13 Schematic illustration of the time periods with significant developments in the areas of mechanics, mathematics, anatomy, muscle, and locomotion research as they relate to biomechanics. Dark shades indicate the greatest development, medium shades indicate medium development, and light shades indicate light or no development. Note: greatest, medium, and light levels have been set arbitrarily (from Nigg [39]. With permission)

2.2 Human Locomotion Study: Elements of Current Research Methodology

It is clear that actual research methodology in a certain field—and the field of human locomotion study is not an exception—can't be adequately covered in a small subchapter such as the present one. However, there are some elements of methodology accepted, in a way, as a general standard, and these elements are shortly depicted here.

Multiple Rigid Body Modeling Paradigm

Biomechanical methodology for studying human movement resides on a multiple rigid body modeling paradigm in the representation of human body. Body segments are presumed to be of a regular geometrical shape and rigid, and interconnected by joints, so that a kinematic chain is formed as a relevant description of movement of the body as a whole (gross body movement) (Fig. 2.14).

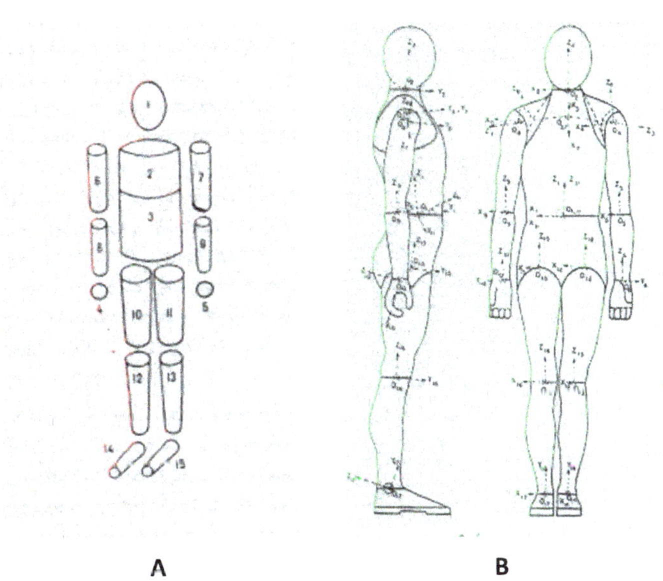

Fig. 2.14 Hanavan (**A**) and Hatze (**B**) models, first one being particularly suitable for inverse dynamics (**A**—Reproduced from Hanavan [40]; **B**—Reproduced from Hatze [39]. With permissions)

At this level of abstraction, and simplification, laws of classical mechanics can be applied to the system (equilibrium of forces and moments), whereby experimentally obtained, measured kinematic data are combined with inertial properties of body segments, taking into account possible external acting forces and moments (couples) which are prone to measurement; ground reaction typically. Mathematical calculation of internal resultant (net) forces and moments (couples) acting in virtual joint centers is thus enabled, the procedure called inverse dynamics. To accomplish the task, a 3D measurement of kinematics of the body in motion is to be provided, so that, consequently, relevant Newtonian equations can be implemented on a system.

A detailed elaboration of the subject matter describing the approach is available in representative literature [31–36]. It is the basis of modern biomechanics of human movement and locomotion, in sportive, medical, ergonomic and other applications.

Surface Electromyography and Neuro-Muscular Modeling

Skeletal muscle is a system characterized by mechanical, thermal and electrical energy outputs. Mechanical action of skeletal muscle as a whole is succinctly described well by the „tension-length“ and „force–velocity“ curves (Fig. 2.15), its model including active, elastic and viscous components (Fig. 16).

EMG-supplied information is related to the fundamental muscle function as a metabolically nurtured force generator. The active component, namely, is the one representing genuine feature of muscular tissue to use metabolic energy and mechanically contract, and this component is correlated with electrical events; being manifested ultimately as electromyographic signals (provided adequate detection and recording). Contributions of elastic and viscous components of the model to the muscle force, on the contrary, are not „visible“ in the EMG.

Up to the 1980s, theoretical and experimental basis was made available for creating faithful mathematical models of the muscle–tendon complex [35], based on which—as already mentioned—computer-supported quantitative and graphics-based solutions to simulate action of the neuro-musculo-skeletal system were realized [36]

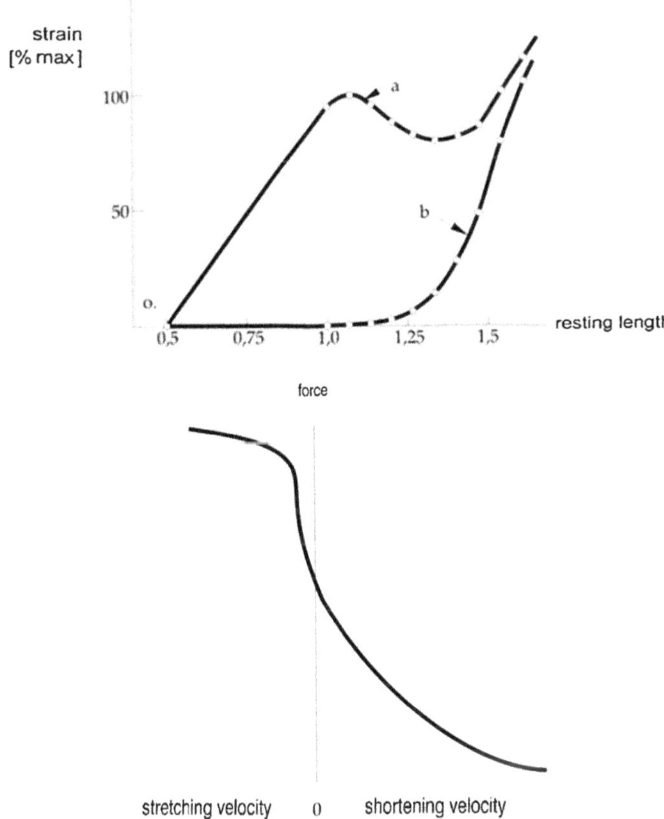

Fig. 2.15 Tension-length and force-velocity dependencies for skeletal muscle (from Medved [8]. Reproduced by permission of Taylor & Francis Group)

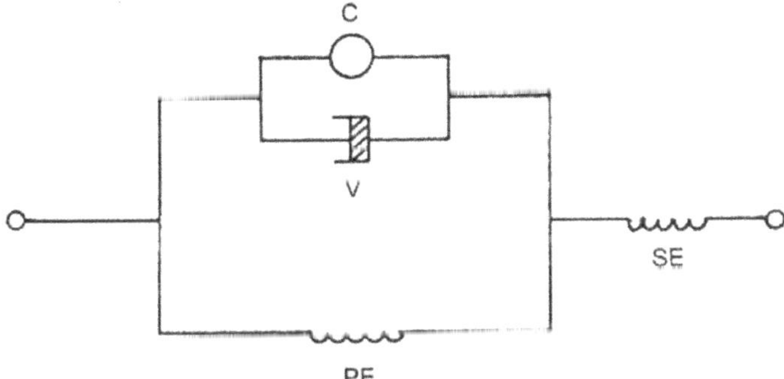

Fig. 2.16 Equivalent mechanical scheme of skeletal muscle. The symbols denote: C, active contractile element; SE, serial elastic element; V and PE, parallel viscious-elastic (attenuating) element (from Medved [8]. Reproduced by permission of Taylor & Francis Group)

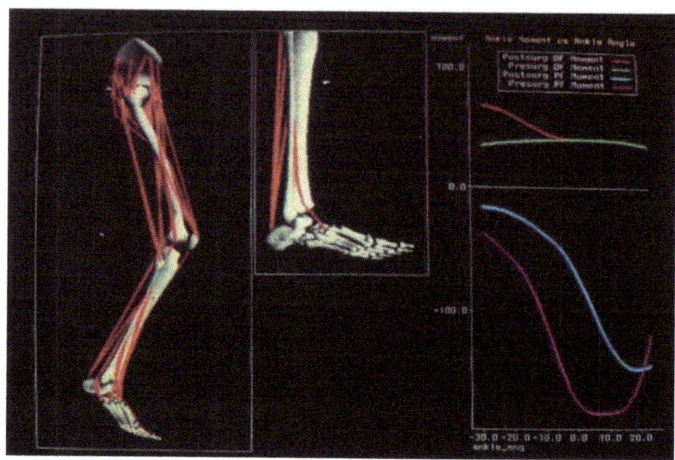

Fig. 2.17 Display from a simulated surgery in which the Achilles tendon was lengthened (1 cm) and the tendon of tibialis posterior was transferred to the base of the third metatarsal. The left postsurgery musculo-skeletal geometry. The plot on the right shows presurgery and postsurgery ankle plantarflexion and dorsiflexion moments (in N-m) versus ankle angle. Positive (negative) ankle angles and moments indicate dorsiflexion (plantarflexion). Postsurgery dorsiflexion moment is increased significantly, but only in the range of ankle plantarflexion (cf. red and green lines). Postsurgery plantarflexion moment is greatly decreased (cf. purple and blue lines) (from Delp et al. [36]. With permission)

(Fig. 2.17). This approach has been implemented in research and in clinical applications [37, 38]. It adds to the classical inverse dynamics approach, enabling, in effect, further sub-division of calculated resultant (net) forces and moments in the joints into their components corresponding to particular muscles (muscle–tendon complexes), resulting with a detailed and realistic biomechanical modeling and simulation possibilities of complex neuro-musculo-skeletal structures.

Winters [42] has presented a conceptually useful block schematics of neural control of movement (Fig. 2.18). Basic control and sensory pathways in a system are depicted in a simplified manner. A block schematics of this kind provides a general framework for modern efforts to describe, quantitatively assess, and explain human locomotion.

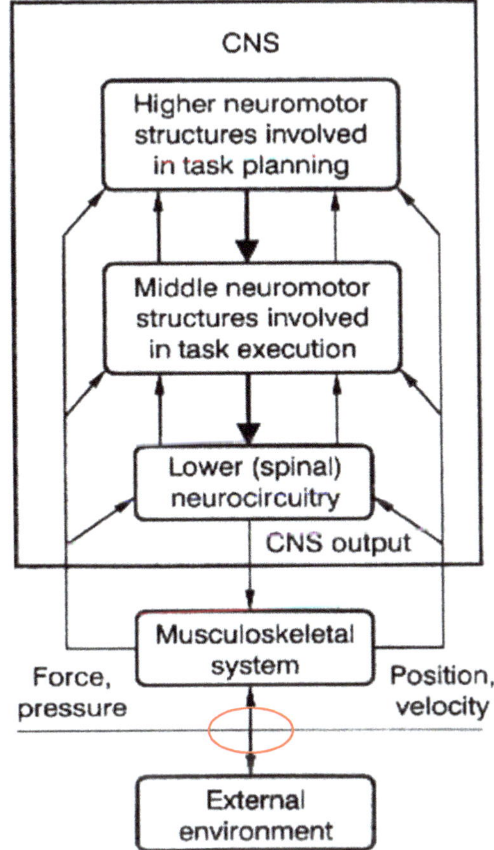

Fig. 2.18 Simplified representation of information flow in the neuro-musculo-skeletal system. The red elipse denotes the site pertaining to biomechanical aspects of the „body-environment“ system (contact of the body with the environment) (modified from Winters [42]. With permission)

Energetics of Locomotion and Exercise Physiology Approach

In addition to the approach of classical mechanics in characterizing energetics of locomotion—a dominant orientation pursued in this book—medico-physiological aspects of locomotion are also relevant. They are the subject of appropriate measurements, testing, and diagnostics of the cardio-pulmonary function of an organism in course of performing movement and locomotion: spontaneous or under load. Metabolic aspects of the physiology of activity are primarily considered. It is refered to standard methods and techniques of spiroergometry and cardiac function testing and evaluation (stress ECG) that are to be performed [2, 43, 44].

References

1. Cappozzo A (2018) Observing and revealing the hidden structure of the human form in motion throughout the centuries (Wartenweiler Memorial Lecture at ISB Symposium in Glasgow 2015). In: Müller B, Wolf SI (eds) Handbook of human motion. Springer, part of Springer Nature, pp 3–15
2. Tipton CM (ed) (2003) Exercise physiology: people and ideas. Oxford University Press
3. Martin RB (Oct 23, 1999) A genealogy of biomechanics. Presidential Lecture presented at the 23rd annual conference of the American society of biomechanics. University of Pittsburgh, Pittsburgh PA
4. Goss CM (1968) On movement of muscles by Galen of Pergamon. Am J Anat 123(1):1–25
5. Seireg A, Arvikar R (1989) Biomechanical analysis of the musculoskeletal structure for medicine and sports. Hemisphere, New York
6. Capra F (2007) The science of Leonardo: inside the mind of the great genious of the Renaissance. Doubleday, New York
7. Ascenzi A (1993) Biomechanics and Galileo Galilei. J Biomech 26(2):95–100
8. Medved V (2001) Measurement of human locomotion. CRC Press Inc., Boca Raton, Fl
9. Clarys JP (1994) Electrology and localized electrization revisited. JEMG Kinesiol 4:5–14
10. Marey EJ (1895) Movement. New York, Appleton (Reprint edition 1972 by Arno Press, Inc.)
11. Baker R (2007) The history of gait analysis before the advent of modern computers. Gait Posture 26(3):331–342
12. Ancillao A (2016) Analysis and measurement of human motion: modern measurement protocols and clinical considerations. J Robot Meg Eng Resr 1(4):31–37
13. Woltring HJ (1992) One hundered years of photogrammetry in biolocomotion. In: Proceedings of the symposium on biolocomotion: a century of research using moving pictures. Formia
14. Woltring HJ (1977) Marey revisited: prospect and retrospect, in measurement and control of human movement. Dissertation. Nijmegen, Katholieke Universiteit, pp 175–199
15. Hollmann W, Prinz JP (1997) Ergospirometry and its history. Sports Medicine 23:93–105
16. Herbst R (1928) Determining maximal oxygen consumption in healthy persons [in German). Dtsch Arch Klin Med 162:33
17. Whiting HTA (1984) Human motor actions: Bernstein reassessed. Elsevier
18. Brand RA (1992) Assessment of musculoskeletal disorders by locomotion analysis: a critical historical and epistemological review. In: Tosi V, Cappozzo A (eds) Proceedings of the symposium on biolocomotion: a century of research using moving pictures. Formia
19. Rose J, Gamble JG (eds) (2006) Human walking. 3rd edn. Lippincott Williams & Wilkins, Philadelphia, Pa
20. McMahon TA (1984) Muscles, reflexes, and locomotion. Princeton University Press, Princeton, NJ
21. Furnée EH (1989) TV/Computer motion analysis systems: the first two decades. Dissertation. Delft University of Technology
22. Paul JP (1966) Biomechanics of the hip joint and its clinical relevance. Proc Roy Soc Med 59:943–949
23. Gage JR, Schwartz MH, Koop SE, Novacheck TF (2009) The identification and treatment of gait problems in cerebral palsy, 2nd edn. McKeith Press, London
24. Greene PR, Medved V (2019) Compliant surface dynamics, the Harvard tuned track. Biol Eng Med 4. https://doi.org/10.15761/BEM.1000172
25. Cram JR, Kasman GS, Holtz J (1998) Introduction to surface electromyography. Aspen Publishers Inc., Gaithersburg
26. Basmajian JV, De Luca CJ (1985). Muscles alive: their functions revealed by electromyography. Fifth edition. Williams & Wilkins, Baltimore, Md
27. Nordin AD, Rymer WZ, Biewener AA, Schwartz AB, Chen D, Horak FB (2017) Biomechanics and neural control of movement, 20 years later: what have we learned and what has changed? J NeuroEng and Rehab 14:91

28. Merletti R, Farina D (eds) (2016) Surface electromyography: physiology, engineering, and applications. 1st edn. IEEE, Inc. Wiley, Inc.
29. Enoka R (2008) Neuromechanics of human movement, 4th edn. Human Kinetics, Chicago
30. Lanshammar H (1982) On practical evaluation of differentiation techniques for human gait analysis. J Biomech 15(2):99–105
31. Lanshammar H (1982) On precision limits for derivatives numerically calculated from noisy data. J Biomech 15(6):459–470
32. Cappozzo A (1984) Gait analysis methodology. Hum Mov Sci 3:27–50
33. Cappozzo A (1985) Experimental techniques, data acquisition and reduction. In: Berme N, Engin AE, Correia da Silva KM, (eds) Biomechanics of normal and pathological human articulating Joints. The Hague, The Netherlands, Martinuus Nijhoff, pp 53–81
34. Vaughan CL, Davis BL, O'Connor JC (1999) Dynamics of human gait. Kiboho Publishers, Cape Town
35. Zajac FE (1989) Muscle and tendon: properties, models, scaling, and application to biomechanics and motor control. CRC Crit Rev BME 17:359–411
36. Delp SL, Loan JP, Hoy MG, Zajac FE, Topp EL, Rosen JM (1990) An interactive, graphic-based model of the lower extremity to study orthopaedic surgical procedures. IEEE Trans BME 37:757–766
37. Delp SL, Anderson FC, Arnold AS, Loan P, Habib A, John CT, Guendelman E, Thelen DG (2007) OpenSim: open source software to create and analyze dynamic simulations of movement. IEEE Trans BME 54:1940–1950
38. Seth A, Hicks JL, Uchida TK, Habib A, Dembia CL, Dunne JJ et al (2018) OpenSim: simulating musculoskeletal dynamics and neuromuscular control to study human and animal movement. PLoS Comput Biol 14(7):e1006223. https://doi.org/10.1371/journal.pcbi.1006223
39. Nigg BM (1994) Selected historical highlights. In: Nigg B, Herzog W (eds) Biomechanics of the musculo-skeletal system. Wiley, New York, pp 3–35
40. Hanavan EP (1964) A mathematical model of the human body, aerospace medical research laboratory, Wright-Petterson Air Force Base, Ohio, USA
41. Hatze H (1980) A mathematical model for the computational determination of parameter values of anthropometric segments. J Biomech 13:833–843
42. Winters JM (1995) Concepts in neuromuscular modeling. In: Allard P, Stokes IAF, Bianchi J-P (eds) Three-dimensional analysis of human movement. Champaign Il, Human Kinetics, pp 257–292
43. Medved R (ed) (1987) Sportska medicina (Sports medicine). JUMENA, Zagreb
44. Hollmann W, Strüder HK (2009) Sportmedizin: Grundlagen für körperliche Aktivität, Training und Präventivmedizin 5. Auflage. Schattauer Verlag

Chapter 3
On Evolution and Development of Human Gait

Marija Rakovac

Abstract Human bipedal locomotion is a very complex behavior. The evolution of upright posture and obligate bipedal gait—a central event in the evolution of humankind—took several millions of years and its incentives are still (and surely will be in the future) a subject of scientific debate. Some of the evolutionary milestones are still recognizable in the development of mature gait in toddlers. To provide an overview on evolution and development of human gait, this chapter starts off with a short description of modern human gait modalities—walking, running, and, (arguably) skipping. The broad topic of evolution of human locomotion is first approached by a comparison of the musculoskeletal anatomy of the humans and extant apes. The positive and negative aspects of bipedalism are then tackled, followed by an overview of existing theories on human evolution and theories on evolution of human bipedalism. The chapter finishes off by describing the development of mature gait in toddlers, an amazing chapter in human life in which we could argue that 'ontogeny recapitulates phylogeny'—the development of the species is reflected in the development of a single person.

3.1 Human Gait Modalities—Walking and Running Pattern in Modern Humans

Upright bipedal locomotion is characteristic for humans and our immediate ancestors, while other primates mainly use quadrupedal modes of locomotion [1]. Locomotion can be defined as "the translation of the center of mass through space in a way to require the least energy expenditure" [2]. In modern humans, locomotion is characterized by two distinct gait modalities—walking and running.

M. Rakovac (✉)
Faculty of Kinesiology, University of Zagreb, Horvaćanski zavoj 15, 10000 Zagreb, Croatia
e-mail: marija.rakovac@kif.hr

V. Medved (ed.), *Measurement and Analysis of Human Locomotion*,
Series in Biomedical Engineering, https://doi.org/10.1007/978-3-030-79685-3_3

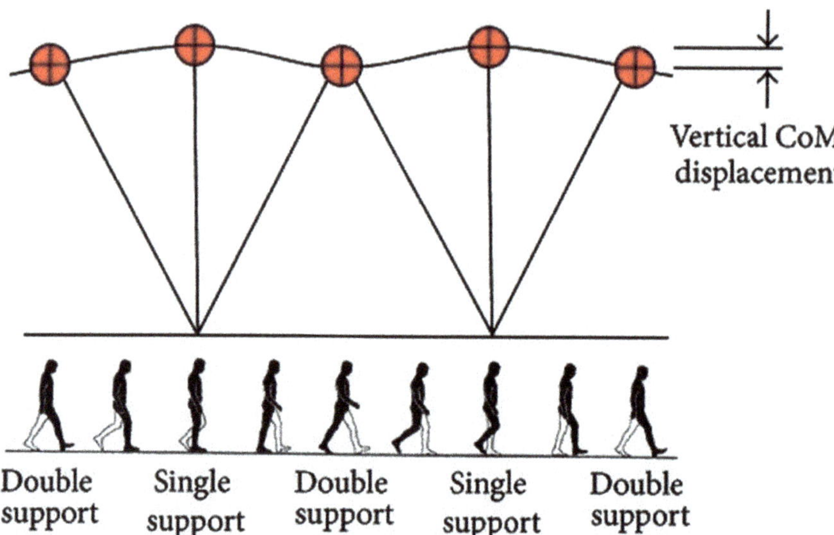

Fig. 3.1 Inverted pendulum model of walking. Center of body mass (CoM) (from [7], Lobet S, Detrembleur C, Massaad F, Hermans C (2013) Three-dimensional gait analysis can shed new light on walking in patients with haemophilia. ScientificWorldJournal 2013:284358. An open access article distributed under the Creative Commons Attribution License)

Walking is a gait used at relatively lower speeds [3]. It is a gait in which the body's center of mass vaults over stiff lower limbs (stiff postures of the hip and knee joints). Walking can be described by an inverted pendulum model. The center of the pendulum represents the body's center of gravity (situated in front of the second sacral vertebra) which vaults up and over two extended lower limbs in an arc [4, 5; Fig. 3.1], with a vertical displacement of around 5 cm [6].

Walking gait cycle is characterized by two phases or periods: (1) a stance phase, during which the foot is in contact with the ground (comprising 62% of gait cycle) and (2) a swing phase, in which the foot is in the air (38% of the gait cycle) [6]. In its vertical displacement, the center of mass is at its lowest point at heel-strike (beginning of the stance phase) and at its highest point at midstance [6]. Walking gait provides an effective exchange of gravitational potential and kinetic energy. Namely, during the first half of the stance phase kinetic energy is transformed into gravitational potential energy (highest at midstance), which is then partially recovered during the second half of the stance phase, when the center of mass falls forward/downward (around 65–70% of gravitational energy at the stance phase is transformed into kinetic energy during the propulsive phase) [4, 5]. Unlike running, walking is characterized by a double support period, in which both feet are in contact with the ground.

Running is used at higher speeds. This gait modality can best be described by a spring-mass model [8]. It is a bouncing gait, in which bouncing minimizes energy expenditure. It is characterized by relatively flexed (compliant) postures of the lower limb joints—hip and knee. There is no double support period, meaning that in no time point two feet simultaneously touch the ground. There is, on the other hand, an entirely airborne "flight" phase, not found in walking. During this gait, the gravitational potential and kinetic energy are in phase [6].

As the leg touches the ground, both kinetic and gravitational potential energy are stored as elastic energy by muscles, tendons, and ligaments of the lower leg and the sole of the foot. They act as springs and reutilize the elastic energy during the propulsive phase—second half of the stance phase [5, 6]. The leg function during running is thus similar to a spring repeatedly compressed (by body mass) and decompressed (release of the elastic energy during the propulsive phase) [9, 10]. However, the spring-mass model is not entirely adequate for description of high-speed sprint running [11].

Walking and running are the two most common modes of human locomotion (Fig. 3.2). Since they are inherently different, the transition from walking to running at higher speeds is abrupt and clearly recognizable [5]. Increase in speed is associated with changes in gait cycle length and duration, duration of stance and swing phase, higher intensity of muscle activation, higher range of motion in all lower limb joints [5, 12]. The average preferred speed of transition between walking and running is around 7.2 km/h (2.2 m/s) [8, 13, 14]. It is characterized by a decrease in ground contact time by 35 and a 50% increase in peak ground reaction force [5].

A "third locomotion paradigm" in humans is related to skipping [8, 15]. It is a high-speed gait, mainly used by children. It can be unilateral or bilateral. It consists of a stance and a swing phase with one foot, then a double stance phase, and a stance and a swing phase with the other foot [15]. Its longer flight phase differentiates it from walking, while the double support phase differentiates it from pure running [15]. It combines pendulum- and spring-mass mechanisms [15].

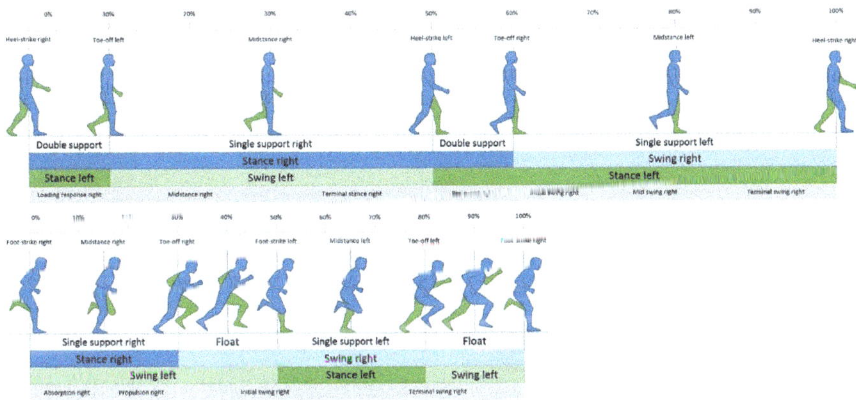

Fig. 3.2 Walk and run cycle (https://commons.wikimedia.org/wiki/File:Walk_and_run_cycle.jpg; The file is licensed under the Creative Commons Attribution-Share Alike 4.0 International license.)

3.2 Musculoskeletal Anatomy of Modern Humans and Extant Apes

To start the discussion on the origin and development of human bipedal locomotion, let us first compare locomotion behavior and musculoskeletal anatomy of modern humans and extant apes. As we know, modern humans share many features with apes. Apes and humans both belong to the order *Primates*, superfamily *Hominoidea*, which is split into two families—*Hylobatidae* (lesser apes—gibbons and siamangs) and *Hominidae* [16] (Fig. 3.3). *Hominidae* family consists of hominids—great apes and humans. It is split into three subfamilies—*Ponginae* (orangutans), *Gorillinae* (gorillas) and ***Homininae*** (chimpanzees and bonobos (genus *Pan*), and humans (genus *Homo*)) [16]. So, humans' closest relatives among primates are the African chimpanzees and bonobos.

When one explores the **locomotor behavior of extant apes**, the first impression is its **diversity** [3, 8]. Gibbons, orangutans, chimpanzees, and gorillas use several modes of both arboreal and terrestrial locomotion:

(1) ***brachiation***—an arm-swinging, forelimb suspensory arboreal locomotion [8]. Gibbons and siamangs use this type of forelimb locomotion to effectively move through canopies [17].
(2) ***quadrumanous climbing***—when climbing the trees, apes use both forelimbs (mainly for suspension) and hindlimbs (mainly for propulsion). This type of locomotion is especially studied within the context of evolution of human bipedal gait due to (electromyographic) similarities in hindlimb muscle activation between these two gaits, which is more similar than between human bipedalism and ape terrestrial locomotion (either quadrupedal or bipedal) [3]. This will be further discussed later in the chapter.
(3) ***quadrupedal knuckle- or fist-walking***—a terrestrial four-limb locomotion mode, in which apes rely on forelimbs for support. A number of musculo

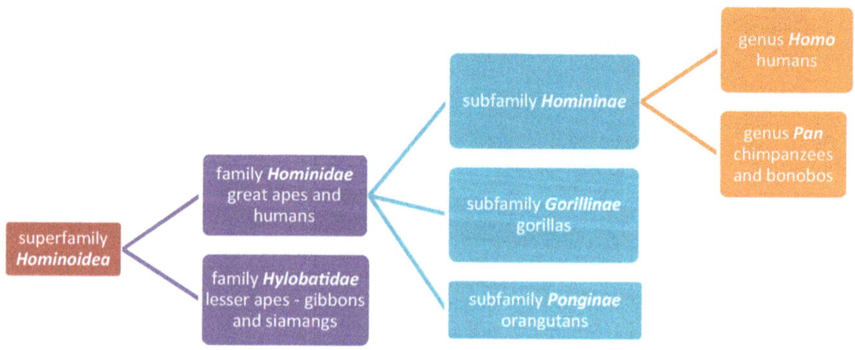

Fig. 3.3 Classification of the Primate superfamily *Hominoidea*. *Note* chimpanzees, gorillas, orangutans are classified as great apes

skeletal adaptations of the trunk and forelimbs allow them to perform this mode of locomotion with ease [18].

(4) ***bipedal locomotion*** [3, 8]—bipedal locomotion is used in different extent by different apes. E.g., gibbons show prolonged bipedalism, which makes up to 10% of their locomotor behavior. They employ a bouncing, running-like gait (although generally without a flight phase) and can attain speeds of 3.5 m/s [8]. However, in their bipedal locomotion apes cannot sustain the upright posture (even when walking on two feet the trunk always remains somewhat inclined) and their lower limb joints are flexed. The exception are orangutans. Remarkably, although not being the closest relatives to humans, orangutans employ a stiffed-legged, upright bipedal gait, otherwise characteristic for humans [19].

Let us now compare the main features of musculoskeletal anatomy of **humans** and extant apes (Fig. 3.4), including their closest relatives—chimpanzees. The most pronounced differences and humans' adaptations to upright stance and bipedal gait can be noticed in the anatomical regions of trunk and pelvis, femur and foot (Table 3.1).

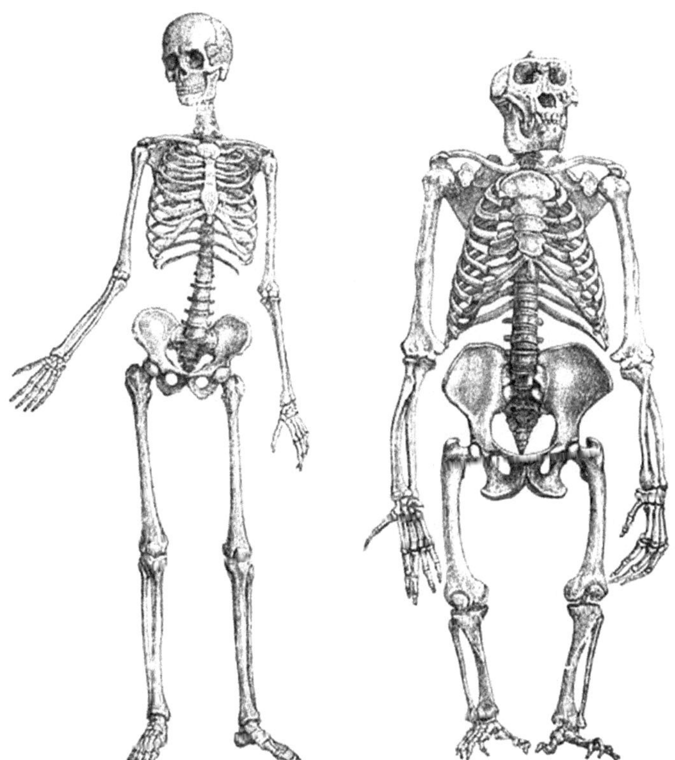

Fig. 3.4 The human and gorilla skeleton [https://commons.wikimedia.org/wiki/File:Primatenskelett-drawing.jpg (Public domain)]

Table 3.1 Some of the pronounced differences between the human and chimpanzee skeleton

Humans	Chimpanzees
Trunk	
– S-shaped vertebral column	– Bow-shaped vertebral column
– Bowl-shaped pelvis (and ilium)	– Long pelvis, flat ilium
Leg	
– Inward angled femurs	– Vertical femur orientation
Foot	
– Hallux aligned with other toes – Large and rigid tarsal bones – Longitudinal and transverse arches – Longer legs, powerful muscles, spring-like tendons	– Opposable hallux – Smaller tarsal bones
Arms	
– Shorter	– Longer and stronger

Regarding the trunk and pelvic region, humans have developed the S-shaped vertebral column, with four compensatory curves, designed for pressure transfer and amortization. They allow for the weight to be transferred roughly through the centers of the main weight-bearing joints (hips, knees, ankles), minimizing the muscle activity needed for maintenance of the upright posture. The center of mass is placed over the feet [1]. There are two anteriorly convex spinal curves (cervical and lumbar lordosis) and two curves convex to the posterior (thoracic and sacral kyphosis). During the intrauterine development, the human fetus displays only a primary kyphotic curve along the spine. When the child is born, during the early motor development, the secondary (lordotic) curves develop. The cervical lordosis develops when the child starts to raise the head, and the lumbar lordosis develops when the child starts to stand and walk. The chimpanzees, on the other hand, have a bow-shaped vertebral column without the compensatory curves.

Upright gait on extended lower limbs requires lumbar lordosis, a crucial initial adaptation in the evolution of human morphology [20, 21]. Unlike the African apes, humans have a long flexible lumbar spine with pronounced lordosis which permits the vertical placement of the upper body above the large lower limb joints, and its extended upright position during locomotion [21]. Apes' lumbar spine is very stiff, with only about two potentially mobile lumbar vertebrae [22].

There is also a difference in the number of vertebrae in different vertebral regions. Humans, chimpanzees, and orangutans differ in the number of vertebrae per region [23]. Humans have 12 thoracic and 5 lumbar vertebrae, orangutans have 12 thoracic and 4 lumbar vertebrae, while in chimpanzees there are 13 thoracic and 3 to 4 lumbar vertebrae [23]. When in the upright stance, their weight is transferred in front of the hip joints, the trunk is inclined, meaning the trunk (back) muscles need to be more active (and expend more energy) to ensure the bipedal gait. Further caudally, humans have a bowl-shaped pelvis and ilium bone, while the pelvis of the chimpanzees is long. The ilium bone is flat and much longer in apes than in humans [1]. Extant apes have very narrow sacra—it is related with their lower trunk being more rigid during suspension due to their lower part of lumbar spine being "trapped" between pelvic bones [20]. In humans, the sacrum became wider [1].

There is a difference in the morphology of the hip joint between humans and great apes [24]. In chimpanzees and gorillas femoral head is relatively small and femoral neck is short [24]. In chimpanzees acetabulum faces laterally, in gorillas it is deeper than in other apes, making the joint less mobile. In orangutans, the hip mobility is greater, with larger femoral head and longer femoral neck [24]. Human hip is adapted to bearing an increased load on only two limbs by a relatively large femoral head and a deep, anteriorly facing acetabulum [24]. In humans there is a long femoral neck, femur is inward angled, and there is a valgus angle at the knee, due to the increased bi-acetabular diameter [24], while in chimpanzees the femur is vertically oriented. Furthermore, humans have a developed collo-diaphyseal angle and femoral antetorsion, putting the knee in the central position and enabling the weight transfer to the foot directly through the ankle, distributing the weight onto the whole foot, while in chimpanzee the weight is transferred to the lateral border of the foot [3].

The most pronounced differences can be described in the structure of the foot. In humans, the foot is adapted to bipedal gait, and it has mainly lost its grasping function. The hallux is aligned with other toes, tarsal bones are large and rigid because they have to transfer the weight of the body to the ground. Longitudinal and transverse foot arches are well-developed and used for pressure transfer and elastic energy use. Chimpanzees have an opposable hallux adapted for branch grasping and their tarsal bones are smaller. The foot arches are not developed. Overall, the human legs are longer, muscles are powerful, with spring-like tendons not found in apes [8]. The upper limbs are shorter in humans, while they are longer and stronger in chimpanzees.

The skeletal characteristics found in all apes are the stiff lumbar spine, broad flattened ribcage, broad pelvis and long forelimbs [8].

Bipedal gait also transformed the roles and function of the muscle groups of the pelvis and lower leg [1]. E.g., gluteal muscles went through prominent adaptations. In the chimpanzee the *gluteus maximus* is a relatively minor muscle, while in the humans it became the largest muscle in the body, stabilizing the trunk and preventing it from flexing forward during upright locomotion [1].

Also, *gluteus medius* and *minimus* (anterior gluteals) changed their position and function with the development of bipedal gait [1]. As the pelvic bones reoriented in humans, the anterior gluteals assumed a lateral, abductor position. Their function changed from propulsive muscles (in apes) to muscles stabilizing the body weight against the weight-bearing leg during walking [1].

3.3 Bipedal Walking in Modern Humans and Extant Apes

Obviously, there are clearly visible and biomechanically measurable differences in bipedal walking of humans and apes. While walking, humans' center of gravity is placed directly over their feet, the characteristic of the upright position. Humans have well-developed pelvic (gluteal) abductor muscles. The *gluteus medius* and *minimus* muscles are indispensable for alignment of the pelvis during the stance phase of the gait, otherwise the only possible gait is a wobbling, energy-inefficient one. The body weight is distributed across the whole foot, the arches act to take and distribute the pressure, and there is a harmonious counter-rotation in the trunk (mainly chest) area [3].

On the other hand, apes, when walking bipedally, show different characteristics. Their hip- and knee joints are bent when standing upright. They put their feet wide apart, having a wide base of support. Due to the weak pelvic muscles they move their body (pelvis) side to side, showing a pronounced wobbling gait. The weight is distributed only on the outer part of the foot. The arches are not developed, meaning there is a higher energy expenditure and inadequate pressure distribution. The counter-rotation takes place in the hip, not in the trunk [3].

3.4 The Advantages and Negative Aspects of Bipedalism

In evolutionary terms, there were some clear advantages of the acquisition of obligate bipedalism, but as this process is still ongoing today, we are witnessing some negative aspects, too. The relatively rapid evolution of bipedal stance and gait continues to affect health of modern humans [23].

The main advantages of acquired bipedalism were the more efficient locomotion, increased body height which was especially useful for hunting, raised head which increased the field of view, useful for spotting both prey and danger. The hands were freed to perform other tasks, such as hunting, manipulating objects, and care for offspring. The thermoregulation became more efficient as, by acquiring upright stance, less surface of the skin was exposed to direct sunlight (as opposed to the dorsal side of the body in the horizontal body position during quadrupedal gait), and more skin surface was at disposal for (mainly) convective heat loss [25].

However, our bipedal ancestors were deprived of speed and agility, which came as a shortcoming [1]. Upright walking brought about challenges such as increased instability due to the reduced area of support, adaptation to load distribution on two lower limbs and a more complex control of the upright stance [26].

The negative aspects are present still in nowadays humans. The upright stance brought about changes in the form of the pelvis, which lead to a narrower birth canal and a more difficult labor in females (one of the most painful among animal species) [1], and humans still have problems with low back, knee problems, hernias, and fallen arches.

The evolution of morphology and function of human low back and pelvis are unique among mammals [22]. Arguably, anatomical adaptations to upright stance and bipedal walking, which preceded all other human characteristics, such as tool making and brain development, can explain why humans suffer more degenerative and traumatic pathologies of the spine than other primates [22, 23]. Human lumbar spine is elongated and lordotic and stabilized by a rather small *erector spinae* muscle [22]. As such, it is prone to several injuries and degenerative changes such as [22]: (1) idiopathic scoliosis, found only in humans; (2) spondylolysis (and spondylolisthesis), (3) femoral neck fracture, which can be related to the changes in bone mineral density, amount of cortical bone and distribution of trabecular bone brought about by the upright locomotion; and (4) pathologies of the sacroiliac joint, as a result of its higher loading [22].

It was found that herniation of intervertebral disc happens more often in persons whose vertebrae "are towards the ancestral end of the range of shape variation" among humans—meaning they had larger and rounder bodies, larger neural foramen and shorter, narrower pedicles providing less support for the spine [23].

3.5 Timepoints in the Human Evolution Related to the Development of Bipedalism

Some fossil primates with adaptations for orthograde posture date back to the period of **Miocene** (more than 7 million years ago) [25] (Fig. 3.5). There is an ongoing debate around the last common ancestor (LCA) of humans and apes, and the evidence suggests that already the LCA of panids and hominids did not possess morphological characteristics mainly adapted for arboreal suspension and vertical climbing present in extant apes (e.g., narrow elongated sacrum, elongated thorax, arrangement of the hip musculature, decreased number of lumbar vertebrae) [20].

The earliest evidence of the hominid clade dates back to **Genus Ardipithecus** (more than 5 million years ago) [25]. As already mentioned, adaptations of the lumbar spine and pelvic girdle were the initial for the development of postcranial morphology of humans [21]. In the evolution of the morphology of hominin pelvis, two broad phases are described: the development of bipedal pelvic morphology (earlier or by the mid-Pliocene), and later morphologic changes required for the enlargement of the birth canal due to the increase in brain size in the genus Homo (Pleistocene) [27]. There is a scientific "obstetrical dilemma" questioning concurrent needs of birth canal enlargement and factors associated with bipedal gait [28]. The pelvis of Ardipithecus ramidus has been used as the model for discussions on LCA pelvis [29]. Ardipithecus ramidus had a long, caudally oriented ischium not adapted for extended-hip bipedal gait. Shortening of the ischia allowed a functional hip extension during bipedal gait. Ischia of the modern humans are the shortest among primates [30].

Genus Australopithecus (around 3.8 million years ago) were the ones that habitually walked, showing anatomy adapted both for arboreal locomotion and for bipedal

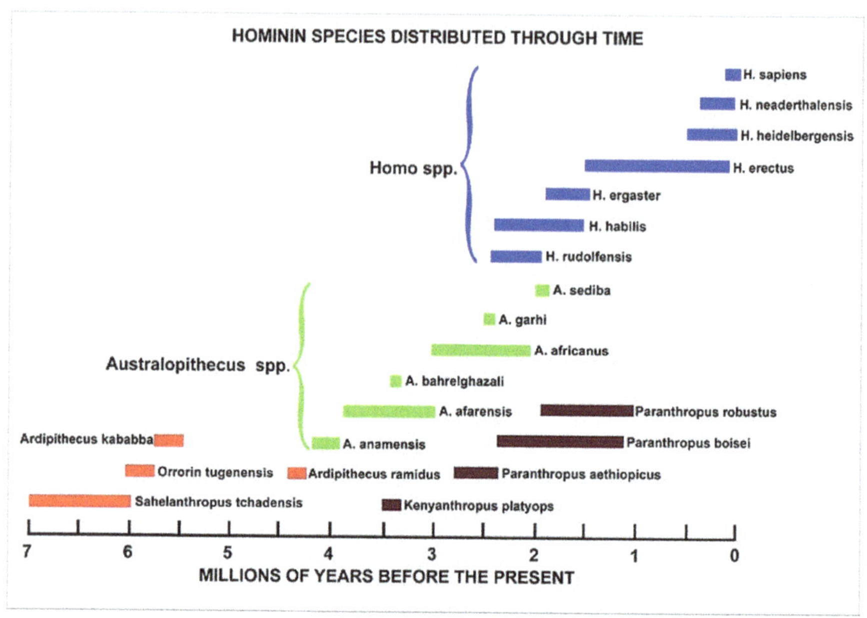

Fig. 3.5 Hominin evolution timeline (https://commons.wikimedia.org/wiki/File:Hominin_evolution.jpg The file is licensed under the Creative Commons Attribution-Share Alike 4.0 International license)

gait. *Australopithecus afarensis* (Lucy, AL-288-1) was discovered in 1974 [1]. She employed a fully upright, bipedal gait. She probably had six lumbar vertebrae (characteristic also for later genus Homo) and a developed lumbar lordosis [21, 22], although there was no progressive increase in the vertebral size along the lumbar region, present in modern humans [22]. Her pelvis had distinctive characteristics of a biped—short and bent ilia, hip muscles (gluteus maximus, abductor muscles, iliopsoas and quadriceps) size and arrangement already similar to modern humans [1] (Fig. 3.6).

However, the birth canal was wide but short from front to back. As human ancestors developed a larger brain, the form of the pelvic opening had to conform. However, the concurrent needs for a mechanically efficient bipedal hip joint and an enlarged birth canal have not been adequately met, making the human birth process, as already mentioned, one of the most difficult among animals [1].

Although Lucy combined bipedal gait and arboreal four-limb locomotion, her femoral neck was already primarily adapted for bipedal walking [1]. Femur and tibia were angled at the knee, which was already adapted for weight transfer during complete extension to the foot placed directly under the center of body mass during one-legged stance [1]. However, femur and tibia are longer in modern humans than in A. afarensis, pointing to the elongation of the lower extremity during Pleistocene [21]. Lucy's foot arch was developed and great toe was parallel to other toes and

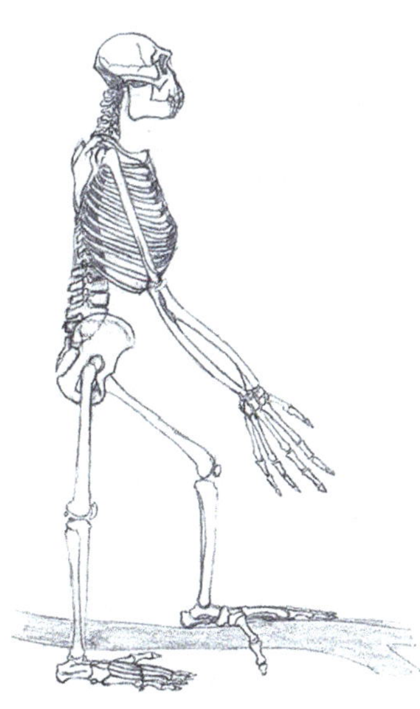

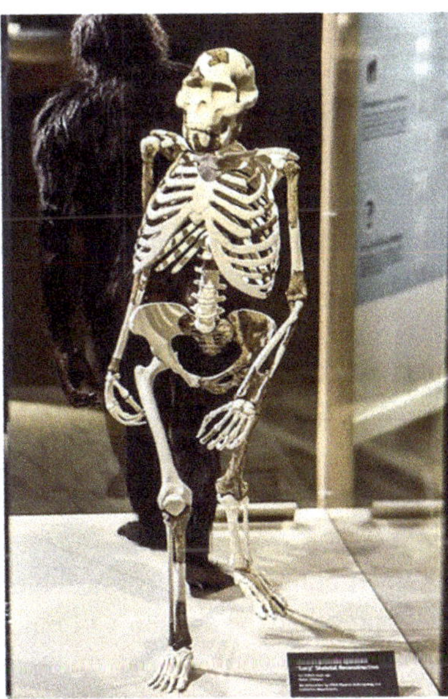

Fig. 3.6 Ardipithecus (right) and Australopithecus afarensis (Lucy, AL-288-1) (right) (https://commons.wikimedia.org/wiki/File:Ardipithecus_Gesamt1.jpg. The file is licensed under the Creative Commons Attribution-Share Alike 3.0 Unported license), https://commons.wikimedia.org/wiki/File:Lucy_(30709431042).jpg. The file is licensed under the Creative Commons Attribution 2.0 Generic license)

non-opposable [1]. The foot became a propulsive lever for bipedal terrestrial locomotion. The upper limb, including fingers, was shorter than in apes [1]. Although Australopithecines were bipeds, their body size (mass and height) were small, and their body proportions were still similar to apes (relatively long arms and shortened legs) [31].

Modern human-like body shape was a trait of the **Genus Homo** (Homo Erectus, Africa, around 1.8 million years ago). Genus Homo represented a *'major adaptive shift'* in evolution [31]. Both brain and body size enlarged markedly and body proportions were comparable to modern humans (shortened arms and longer legs) [31]. Brain size of early Homo erectus was 800–900 cm^3 [31]. It was a major developmental step, since the brain size of the australopithecines evolved from 438 to only 530 cm^3 (brain size comparable to extant apes) in over 2 million years [31].

3.6 Theories Behind the Evolution of Bipedalism

Obviously, we cannot draw certain conclusions about the evolution of bipedalism. There are many theories trying to explain the onset of upright stance and bipedal gait. The still ongoing scientific debates [23, 32] include:

(1) the locomotor repertoire of our ancestors that actually preceded bipedalism.
(2) the ecological/behavioral reasons that influenced the evolution of bipedalism.
(3) the timing of the appearance of bipedalism.
(4) the degree to which certain hominin taxa were indeed obligate bipeds and in what extent their bipedalism resembled that of the modern human.

A number of theories on the evolution of bipedalism were put forward in the last 100 years [32]:

(1) There are theories based on anatomical differences between modern humans and extant great apes. They consider humans' and apes' **postcranial morphology** related to locomotion as well as their **locomotor behavior**.
(2) Since the 1960s there are theories based on fossil material discovered during that period.

Since it is widely accepted that humans' closest relatives are chimpanzees, and bonobos, their common ancestor is debated in terms of the locomotor behavior that might have preceded the evolution of bipedal gat [23]. The most frequent hypotheses on the locomotor behavior that might have preceded bipedalism claim: (1) the common ancestor was a knuckle-walker (as extant apes—chimpanzees, bonobos, and gorillas), (2) the common ancestor was arboreal quadrumanous climber (like orangutans) [23].

Up to the 1940s the dominant theory claimed an arboreal origin of bipedalism [8, 32]. The theory postulated the bipedalism developed from a brachiating arboreal ancestor, belonging to the 'hylobatians' (small-bodied orthograde primates, similar to gibbons). It was postulated the hylobatians later evolved into larger-bodied apes. Their locomotor behavior went through an arboreal vertical climbing phase, then through a terrestrial knucklewalking phase and finally evolved into obligate bipedalism [8, 32].

For the last 60 years the dominant theory described a knucklewalking quadrupedalism model [8, 19]. This theory is based on the close relationship between *hominins* (humans and ancestors) AND *panins* (bonobos and chimpanzees). **The knucklewalking quadrupedalism model** [8, 19] assumes there was a common chimpanzee-like ancestor of hominins and panins. The theory claims bipedalism developed from terrestrial, knucklewalking quadrupedal ancestor. The explanation is based on a similarity between the heel-strike plantigrady in African (and Asian) great apes, knucklewalking quadrupedalism, and adaptations observed in the hominin foot—the heel makes the first contact with the ground and the weight is transferred through the whole plantar surface of the foot.

However, there are some strong criticisms of the knucklewalking quadrupedalism model [8, 19]. Namely, the kinematics of knucklewalking is rather dissimilar to modern human bipedal gait and there is not enough evidence to explain the evolution of flexed hindlimb postures in knucklewalking into extended hindlimbs during upright walking in humans.

During the 1980s Prost and Fleagle proposed the **vertical climbing hypothesis** [8, 19]. Prost [33] argued that in terms of kinematics, the most similar locomotor behavior of African apes to modern human locomotion was their vertical climbing. Fleagle [34] found significant similarities in EMG activity of hip, gluteal, and thigh muscles between African apes' vertical climbing and human walking, arguing that 'vertical climbing would pre-adapt the hindlimb musculature for human-like bipedal walking'. Crompton et al. [19] argue that compressive orthograde in arboreal locomotion may be 'the oldest crown-hominoid locomotor adaptation'.

Since the 1970s, many ape fossils from the Miocene epoch (13–5.3 million years ago) have been also discovered in Europe [35]. A very recent discovery of the well-preserved 11.62-million-year-old fossil ape Danuvius guggenmosi (Allgäu region of Bavaria) provides evidence for a new form of evolutionary behavior comprising equal contributions of fore- and hindlimbs—**extended limb clambering** [35]. The Danuvius can be considered a model for the common ancestor of humans and great apes, being a great ape with combined morphological adaptations of bipeds and arboreal apes (broad thorax, elongated lumbar spine, extended hip and knee joints, elongated and extended forelimbs) [35]. This discovery sheds a new light on the evolution of bipedalism in humans, providing evidence—and for first time from a fully preserved fossil—that both human bipedal gait and ape suspensory behavior evolved from a newly suggested arboreal locomotion characteristic for the common ancestor from the middle Miocene [35]. Extended limb clambering combines characteristics of bipedalism and orthograde suspension—the flat and adducted foot, opposable grasping hallux, habitually extended hip and knees, the elbow with a full range of motion, mobile wrists and hands with curved phalanges [35]. The authors thus propose extended limb clambering as a precursor behavior to obligate bipedalism.

There is also an interesting hypothesis regarding the development of human locomotion, and that is the impact of **endurance running (ER) in human evolution** [36]. Running was rarely considered as locomotor behavior that strongly influenced human evolution since humans are poor sprinters (10.2 m/s for less than 15 s) [36]. However, ER is a unique human trait among primates. It is also not common even among quadrupedal mammals, except for cursorial hunters (e.g., dogs and hyenas) or horses [36]. ER is a trait of the genus Homo (about 2 million years ago) [36]. It is suggested the ER had a strong influence on the evolution of the modern human body form. The hypothesis is that Homo covered long distances by both walking and running. It can also be argued that some characteristics of the foot, unique to humans, such as spring-like longitudinal arch and short toes, were indeed adaptations to long distance running [37]. Also, running velocity might be one of the possible explanations of the elongation of the lower extremity during human evolution, once our ancestors already became bipedal, despite the clear disadvantages of the elongation, such as increased probability of injuries of hamstrings muscles, ankle and knee, common in modern humans [21].

Table 3.2 Characteristics of the human skeleton (first identified within the genus Homo) and their functional role that greatly benefited endurance running [arranged according to Bramble and Lieberman (36)]

Body region and skeleton feature	Functional role beneficial for endurance running performance
Head	
– Enlarged semicircular canals (anterior/posterior) – Balanced head – Nuchal ligament	– Stabilization
Trunk and Upper Extremity	
– Lowered shoulder girdle with wider shoulders – Shortened forearms – Narrow thorax, waist and pelvis	– Counter-rotation of trunk
– Stabilized sacroiliac joint – Larger surface area of the origins of mm. erector spinae and m. gluteus maximus	– Stabilization
Lower Extremity	
– Long Achilles tendon – Plantar arch	– Energy storage and shock absorption
– Close-packed calcaneocuboid joint	– Energy storage

Bramble and Lieberman (36) considered characteristics of the human skeleton instrumental to different functional demands of endurance running: energetics, strength, stabilization and thermoregulation. The authors identified features of human skeleton—all with the earliest evidence found in different representatives of the genus Homo—that significantly benefited endurance running (the features, together with their functional role related to endurance running, are presented in Table 3.2).

There were many different ecological/behavioral hypotheses behind the development of upright stance and bipedal walking put forward by different authors, many by now already outdated, summarized by Niemitz [25]. They include [25]:

- *The Watching Out Hypothesis*—upright posture enlarged the field of view and enabled our ancestors to have a better visual control of their surroundings to spot both danger and prey.
- *The Freeing of the Hands Hypothesis*—the upper limbs were free to perform actions other than locomotion.
- *The Throwing Hypothesis*—upright stance was instrumental in usage and development of weapons.
- *The Infant Carrying Hypothesis*—acquisition of upright stance enabled females to carry their offspring in their arms.
- *The Reaching for Food Hypothesis*—claims our savannah-dwelling ancestors were forced to pick high-growing fruit.
- *The Carrying Food or Provisioning Hypothesis*—bipedal stance freed the hands to carry food.

- *The Display Hypothesis*—bipedal threat display was used to resolve interpersonal conflicts.
- *The Orthograde Scrambling Hypothesis*—our ancestors employed this kind of upright locomotion when moving through canopies.
- *The Scavenging Hypothesis*—puts forward the role of scavenging in the human evolution, but the theory actually refers to later historical periods, and is now not considered in relation to evolution of bipedalism
- *The Aquatic Ancestor Hypothesis*—explores a hypothesis of an aquatic ancestor, considering the similarity of some of the traits of modern humans and some aquatic mammals.
- *The Thermoregulation Hypothesis*—already mentioned in the Sect. 3.4., the thermoregulatory advantages of bipedalism in terms of alleviating the effects of solar exposure [25].

The **Ecological/behavioral hypotheses** [25] until the 1990s presumed bipedalism developed in savannah habitat, with Africa being the most likely part of the world for evolution of bipedality [20]. The 1990s discoveries related the development of bipedalism with a more forested landscape [25]. Ardipithecus arguably inhabited a forested landscape [20]. The newer hypotheses consider the proximity of water to the forests and suggest our ancestors first went through a wading phase which preadapted them to bipedal gait [25].

On the other hand, it is presumable that the onset of bipedality was probably not just a reaction to the environmental changes in habitats, but it also represented a part of larger changes in social structure [20]. Bipedality was certainly associated with several evolutionary important socio-behavioral adaptations, including nuclear family of monogamous parents who both cared for their offspring [1]. Males were in charge of providing high-energy food, while the females were freer to care for children [1]. Recent studies report a unique human neurochemical profile (elevated serotonin, neuropeptide Y, and striatal dopamine, lowered acetylcholine), distinctly different than in other primates. This has been linked with social sensitivity, altruism and empathy [38]. The neurochemical changes are hypothesized to have either preceded or happened parallel with the development of bipedality and promoted lifestyle and the mentioned social changes such as monogamy, cooperation, acquisition of language [38].

3.7 Development of Gait in Toddlers (The Development of Mature Gait)

Unlike the offspring of different animals, human babies take time to develop a bipedal stance and gait. The reason is the complex development of the nervous system and motor coordination. If we consider early human motor development, there are several distinct milestones. At around 6 months the baby starts sitting, crawling takes place at 9 months, walking with support around 1 year of age, walking without support at 15 months, and running roughly at 18 months [39].

We could argue that in the **start of independent walking** the '***ontogeny*recapitulates *phylogeny***'. During the development of mature gait toddlers go through stages which, we might argue, resemble the characteristics of walk in extant apes. Toddlers step with a wide base, their hips and knees are hyperflexed, their feet are externally rotated, there is no heel strike, arms are abducted in a 'high guard' position, elbows are flexed, there is a *staccato* pattern of movement and lower limbs move *en bloc* to lower the number of degrees of freedom, i.e., the segments which need to be coordinated [39, 40]. Their gait is characterized by a low average walking speed, high cadence, their step length is short, and they display a prolonged double stance phase [40] (Fig. 3.7).

The mentioned characteristics of the toddler's gait are mainly attributed to insufficient muscle force, immature balance and movement control [40]. With walking experience, the gait matures—the base width diminishes, movements become smoother, there is a reciprocal arm-swing, the step length and walking speed increase [39]. Two phases of gait maturation were described. The first phase occurs in the first 5–6 months after the start of independent walking. This phase is characterized by rapid

Fig. 3.7 Baby walking—high-guard arm position, wide base of support (https://commons.wikimedia.org/wiki/File:Baby_walking.jpg The file is licensed under the Creative Commons Attribution-Share Alike 3.0 Unported license)

development of different gait characteristics [40]. The second phase encompasses further maturation of the gait and it lasts until the age of 8 [40, 41].

The basic characteristic of bipedal walking, energy-saving inverted pendulum model, is not efficiently implemented at the start of independent walking [42], due to the balance problems and immature movement control [43]. The percentage of energy recovery in toddlers is lower than in adult gait (by around 50%) since the changes of gravitational and kinetic energy and their phase relation are very irregular [42]. As the start of independent walking triggers the developmental changes, the pendulum mechanism evolves toward mature features within a few months of its onset [42]. Some argue that the pendulum mechanism is not an innate characteristic, since, in that case, one would expect it at the very beginning of the independent walking [42]. However, toddlers beginning walking do not implement it, and it actually requires development of complex neural control and efficient between-segment coordination pattern, which takes place during the first months of independent locomotion, both triggered and promoted by walking experience [42].

The inefficient energy saving pendulum system in newly-walking toddlers (they have low self-selected walking speed and large vertical oscillations of the body, during 25–50% of the gait cycle potential gravitational and kinetic energy are in phase, etc.) enables them to recover up to 40% of energy at optimal speed [43]. After approximately 3 months of independent walking, toddlers walk at a speed at which energy exchange is optimal for their length [43]. Parallel with the increase in energy efficiency of the pendulum mechanism, there is a decrease of variability of kinematic and kinetic parameters [42]. New walkers show decreased variability in step width and high variability in step length. As walking experience increases, step width variability increases and step length variability decreases. By 6 months of walking the two parameters are equal and continue to develop towards adult levels [44].

Progressive changes of gait kinematics and kinetics follow the neural maturation of central pathways (e.g., myelination of descending tracts) that are important for postural and locomotor control, and the improved cognitive capacity [42]. By the age of 2, walking becomes functional with speed around 0.8 stature/s and 60% of energy recovery [39, 42, 45]. Children above the age of 3 years are dynamically similar to adults, at least concerning the energetics of walking [43].

Beside the development of gait parameters, toddlers are faced with the need to maintain an upright head and trunk while moving forward in space [44]. The poor head coordination at the onset of independent locomotion is likely to correspond to immature gaze stabilization strategies or to a difficulty in motor coordination. The stabilization of the head is already accomplished within the first weeks of the start of independent walking, while the fine head coordination progressively develops even after the first year of independent walking [46].

The newly walking toddlers display large oscillations of the trunk and have a wide base of support, to counteract their instability [47]. Upper extremities are very important in improving balance. While adults move their arms contralaterally and synchronized with lower limbs, toddlers keep their arms high, with abducted and laterally rotated shoulders and flexed elbows—a 'high guard' position, which is present during the first months of walking [45, 48–50]. This arm position, characteristic for very early phase of gait development, is then replaced by reciprocal swinging of arms five to six months after the beginning of independent walking [49, 51].

Three characteristics regarding joint dynamics in immature walking were identified [40]: (1) balance problems most likely influence the dominant extending moments of hip and knee joints throughout stance, (2) there is a reduced complexity of the joint moment profiles because of the immature movement control, (3) practically no push-off forces are generated at the ankle, meaning there is no active push-off [40].

As for the muscle activity, at walking onset muscle activity is highly variable [49]. Walking experience makes variability and coactivation lower, but still inconsistent, probably due to many possible options of muscle combinations. Muscle activation patterns are thus the slowest to stabilize among the subsystems in gait development. Gradually, via exploration and selection, early applied muscle activations are gradually replaced by stable and efficient patterns [49].

3.8 Conclusion

The development of upright stance and bipedal locomotion was without a doubt a central event and one of the most significant adaptations in human evolution. The well-adapted and harmonious bipedal gaits of walking and running are unique to humans. The development of bipedal locomotion preceded and enabled the evolution of other human characteristics such as enlarged brains and tool making. The evolution of bipedal locomotion took several million years. Australopithecines (around 3.8 million years ago) were habitual walkers, while genus Homo (1.8 million years ago) displayed modern human body form. Many questions regarding the evolution of human gait, however, still remain unanswered, including the roles of interactions between genes, environmental influence, and natural selection [8, 25]. The complexity of maintaining upright stance and bipedal gait is evident in development of independent bipedal locomotion in toddlers.

References

1. Lovejoy CO (1988) Evolution of human walking. Sci Am 259(5):118–125
2. Saunders JB, Inman VT, Eberhart HD (1953) The major determinants in normal and pathological gait. J Bone Joint Surg Am 35-A(3):543–558
3. Schmitt D (2003) Insights into the evolution of human bipedalism from experimental studies of humans and other primates. J Exp Biol 206(Pt 9):1437–1448
4. Cavagna GA, Thys H, Zamboni A (1976) The sources of external work in level walking and running. J Physiol 262(3):639–657

5. Cappellini G, Ivanenko YP, Poppele RE, Lacquaniti F (2006) Motor patterns in human walking and running. J Neurophysiol 95(6):3426–3437
6. Ayyappa E (1997) Normal human locomotion, Part 1: Basic concepts and terminology. J Prosthet Orthot 9(1):10–17
7. Lobet S, Detrembleur C, Massaad F, Hermans C (2013) Three-dimensional gait analysis can shed new light on walking in patients with haemophilia. ScientificWorldJournal 2013:284358
8. Crompton RH, Vereecke EE, Thorpe SK (2008) Locomotion and posture from the common hominoid ancestor to fully modern hominins, with special reference to the last common panin/hominin ancestor. J Anat 212(4):501–543
9. Cavagna GA, Franzetti P, Heglund NC, Willems P (1988) The determinants of the step frequency in running, trotting and hopping in man and other vertebrates. J Physiol 399:81–92
10. Ziliaskoudis C, Park SY, Lee SH (2019) Running economy—a comprehensive review for passive force generation. J Exerc Rehabil 15(5):640–646
11. Haugen T, McGhie D, Ettema G (2019) Sprint running: from fundamental mechanics to practice-a review. Eur J Appl Physiol 119(6):1273–1287
12. Dugan SA, Bhat KP (2005) Biomechanics and analysis of running gait. Phys Med Rehabil Clin N Am 16(3):603–621
13. Hreljac A (1993) Preferred and energetically optimal gait transition speeds in human locomotion. Med Sci Sports Exerc 25(10):1158–1162
14. Šentija D, Rakovac M, Babić V (2012) Anthropometric characteristics and gait transition speed in human locomotion. Hum Mov Sci 31(3):672–682
15. Minetti AE (1998) The biomechanics of skipping gaits: a third locomotion paradigm? Proc Biol Sci 265(1402):1227–1235
16. Encyclopædia Britannica (2020) Primate—mammal/Classification https://www.britannica.com/animal/primate-mammal/Classification. Accessed 6 Feb 2020
17. Usherwood JR, Bertram JE (2003) Understanding brachiation: insight from a collisional perspective. J Exp Biol 206(Pt 10):1631–1642
18. Simpson SW, Latimer B, Lovejoy CO (2018) Why do knuckle-walking African Apes knuckle-walk? Anat Rec 301(3):496–514
19. Crompton RH, Sellers WI, Thorpe SK (2010) Arboreality, terrestriality and bipedalism. Philos Trans R Soc Lond B Biol Sci 365(1556):3301–3314
20. Machnicki AL, Spurlock LB, Strier KB, Reno PL, Lovejoy CO (2016) First steps of bipedality in hominids: evidence from the atelid and proconsulid pelvis. PeerJ 4:e1521
21. Lovejoy CO (2005) The natural history of human gait and posture. Part 1. Spine and pelvis. Gait Posture 21(1):95–112
22. Lovejoy CO (2007) Evolution of the human lumbopelvic region and its relationship to some clinical deficits of the spine and pelvis. In: Vleeming A, Mooney V, Stoeckart R (eds) Movement, stability & lumbopelvic pain, 2nd edn. Churchill Livingstone Elsevier, London, pp 141–158
23. Plomp KA, Viðarsdóttir US, Weston DA, Dobney K (2015) Collard M (2015) The ancestral shape hypothesis: an evolutionary explanation for the occurrence of intervertebral disc herniation in humans. BMC Evol Biol 27(15):68
24. Georgiou L, Kivell TL, Pahr DH, Buck LT, Skinner MM (2019) Trabecular architecture of the great ape and human femoral head. J Anat 234(5):679–693
25. Niemitz C (2010) The evolution of the upright posture and gait–a review and a new synthesis. Naturwissenschaften 97(3):241–263
26. Maus HM, Lipfert SW, Gross M, Rummel J, Seyfarth A (2010) Upright human gait did not provide a major mechanical challenge for our ancestors. Nat Commun 1:70
27. Churchill SE, Vansickle C (2017) Pelvic morphology in Homo Erectus and Early Homo. Anat Rec 300(5):964–977

28. Ruff C (2017) Mechanical constraints on the hominin pelvis and the "Obstetrical Dilemma" . Anat Rec 300(5):946–955
29. Hammond AS, Almécija S (2017) Lower Ilium evolution in Apes and Hominins. Anat Rec 300(5):828–844
30. Lewton KL, Scott JE (2017) Ischial form as an indicator of bipedal kinematics in early hominins: a test using extant anthropoids. Anat Rec 300(5):845–858
31. Leonard WR (2010) Size counts: evolutionary perspectives on physical activity and body size from early hominids to modern humans. J Phys Act Health 7(Suppl 3):S284–S298
32. Harcourt-Smith WE, Aiello LC (2004) Fossils, feet and the evolution of human bipedal locomotion. J Anat 204(5):403–416
33. Prost JH (1980) Origin of bipedalism. Am J Phys Anthropol 52(2):175–189
34. Fleagle JG, Stern JT, Jungers WL, Susman RL, Vangor AK, Wells JP (1981) Climbing: a biomechanical link with brachiation and with bipedalism. Symp Zool Soc Lond 48:359–375
35. Böhme M, Spassov N, Fuss J, Tröscher A, Deane AS, Prieto J, Kirscher U, Lechner T, Begun DR (2019) A new Miocene ape and locomotion in the ancestor of great apes and humans. Nature 575(7783):489–493
36. Bramble DM, Lieberman DE (2004) Endurance running and the evolution of Homo. Nature 432(7015):345–352
37. Holowka NB, Lieberman DE (2018) Rethinking the evolution of the human foot: insights from experimental research. J Exp Biol 221(Pt 17). pii: jeb174425
38. Raghanti MA, Edler MK, Stephenson AR, Munger EL, Jacobs B, Hof PR, Sherwood CC, Holloway RL, Lovejoy CO (2018) A neurochemical hypothesis for the origin of hominids. Proc Natl Acad Sci U S A 115(6):E1108–E1116
39. Sutherland DH, Olshen R, Cooper L, Woo SL (1980) The development of mature gait. J Bone Joint Surg Am 62(3):336–353
40. Hallemans A, De Clercq D, Otten B, Aerts P (2005) 3D joint dynamics of walking in toddlers A cross-sectional study spanning the first rapid development phase of walking. Gait Posture 22(2):107–118
41. Kermoian R, Johanson ME, Butler EE, Skinner S (2006) Development of gait. In: Rose J, Gamble JG (eds) Human Walking, 3rd edn. Lippincott Williams & Wilkins, Philadelphia, pp 119–130
42. Ivanenko YP, Dominici N, Cappellini G, Dan B, Cheron G, Lacquaniti F (2004) Development of pendulum mechanism and kinematic coordination from the first unsupported steps in toddlers. J Exp Biol 207(Pt 21):3797–3810
43. Hallemans A, Aerts P, Otten B, De Deyn PP, De Clercq D (2004) Mechanical energy in toddler gait. A trade-off between economy and stability? J Exp Biol 207(Pt 14):2417–2431
44. Looper J, Chandler LS (2013) How do toddlers increase their gait velocity? Gait Posture 37(4):631–633
45. Iosa M, Fusco A, Morone G, Paolucci S (2014) Development and decline of upright gait stability. Front Aging Neurosci 6:14
46. Ledebt A, Wiener-Vacher S (1996) Head coordination in the sagittal plane in toddlers during walking: preliminary results. Brain Res Bull 40(5–6):371–373
47. Ivanenko YP, Dominici N, Lacquaniti F (2007) Development of independent walking in toddlers. Exerc Sport Sci Rev 35(2):67–73
48. Van de Walle P, Meyns P, Desloovere K, De Rijck J, Kenis J, Verbecque E, Van Criekinge T, Hallemans A (2018) Age-related changes in arm motion during typical gait. Gait Posture 66:51–57
49. Chang CL, Kubo M, Buzzi U, Ulrich B (2006) Early changes in muscle activation patterns of toddlers during walking. Infant Behav Dev 29(2):175–188

50. Ledebt A, Bril B (2000) Acquisition of upper body stability during walking in toddlers. Dev Psychobiol 36(4):311–324
51. Kubo M, Ulrich BD (2006) Early stage of walking: development of control in mediolateral and anteroposterior directions. J Mot Behav 38(3):229–237

Chapter 4
From the Archives of Zagreb School of Biomechanics: Measuring Biomechanical Properties of Lumbosacral Joint Specimens

Boris Božić

Abstract A specific part of human body is focused; a lumbar spine segment with surrounding tissues, biomechanics of which is crucial for maintaining the upward erect body posture and moving the body in the intended direction by means of walking, running and other locomotor modalities. The pelvic segment possesses a pivotal role in the walking process, with biomechanics of the lumbar spine being crucial for realizing dynamic spatial (3D) movements. Uprising of humans on lower legs during evolution has challenged this body part by subjecting it under big mechanical strain, as best witnessed, unfortunately, by neurosurgical praxis treating spinal injuries resulting from mechanical overloads, a consequence of sports training or occupational tasks. Zagreb school of biomechanics, appearing in second half of twentieth century, is introduced shortly. Approaches of anatomy, orthopedics and mechanical engineering were integrated. Initially viewed as essentialy a static construction, the spine was modeled subsequently as a dynamic mechanical structure. An archival example of in vitro type measurements of mechanical properties of lumbosacral joint speciments is presented in some detail, in the context of evaluating specific surgical interventions. Interlaminectomy has been found a procedure of choice in the treatment of disc hernias of the lumbar spine. Non-invasive in vivo measurements of spine kinematics during normal gait were performed later.

4.1 On Research of Spinal Biomechanics at the University of Zagreb

Human spine forms the base of the axial skeleton. It is located on the back of the trunk, and positioned centrally. It forms the base of the entire skeleton and allows for an upright posture. It consists of movable and fixed vertebrae. The movable spine consists of 7 neck, 12 dorsal, and 5 loins vertebrae. Fixed or static vertebrae are the sacrum, usually composed of five vertebrae, and the caneorcoccygeal bone, which

B. Božić (✉)
Department of Neurosurgery, Sestre milosrdnice University Hospital Center, Vinogradska cesta 29, 10000 Zagreb, Croatia

V. Medved (ed.), *Measurement and Analysis of Human Locomotion*, Series in Biomedical Engineering, https://doi.org/10.1007/978-3-030-79685-3_4

also typically has four or five fused vertebrae. Dynamic or movable vertebrae are dynamic in a biomechanical sense, while fixed vertebrae are static. Between the vertebrae there are intervertebral tiles—disci which usually account for the 23 of them, inter-joint extensions and arches of vertebrae. The spine is strengthened by a system of ligaments, muscles and connective tissue. This gives it firmness and stability in the static, while at the same time mobility in a dynamic sense. Due to its high flexibility, it enhances the dynamic strength of the muscles and transfers it to other structures of the movement system, to the appended or appendicular skeleton. These include the upper extremities—hands, connected to the spine via the scapula, and the lower extremities—legs, connected to the spine via the pelvic bones. The spine also has a very important protective effect of spinal cord from mechanical damage. Due to the upright posture and due to the force of gravity, the spine of a person is exposed to heavy loads. The special structure of the dynamic and static vertebrae mentioned above enables it to perform these seemingly conflicting tasks. In the biomechanical sense, the structure of the vertebrae that makes up a kinematic chain called the spine is very important. It consists of two types of bones. The part that cancels the dynamic load is called a spongy bone, which dynamically compensates for the centric, eccentric and torsional load of the spine. The part of the spine that cancels out static load is made of compact bone. These two bone structures are in a clear harmonic relation. Thanks to them, the human spine performs complex biomechanical functions described above. It should also be emphasized that with respect to the entire spine, the area of the lumbar spine in terms of anatomical structure, statics and mechanics is one of the most sensitive parts of the locomotor system of humans. This is the area where the load and motion of the whole body are transferred from the moving segments to the fixed base of the scarum fixed in the pelvic ring.

Looking into the dynamic behavior of the spine, due to the complexity of its geometry has been the subject of numerous researches by scientists from various scientific fields. Evans, King, Gibson, Hirsch, Brodetti, Nachemson [1] and Kazarian [2] stand out among those trying to solve these biomechanical puzzles. In Croatia, the research of spinal biomechanics began in the 1960s. One of the first studies was done at the Department of Anatomy at the School of Medicine, University of Zagreb (Fig. 4.1), led by Keros and his colleagues [3–6].

Following their research and within the same institution, Nikolić together with his associates continued the research [7].

The second center of research was the Orthopedic Clinic of the Zagreb School of Medicine, led by Ruszkowsky and the colleagues [8]. These were physicians who have in their original research—and especially according to Keros' approach—viewed spine primarily as a static and only then a dynamic rod. This is also illustrated by Keros' Ph.D. dissertation, where the spine was compared to a bamboo tree [9]. Opposite to them, a group of engineers has been formed within the Faculty of Mechanical Engineering and Naval Architecture in Zagreb, led by Muftić and his associates [10–15] (Fig. 4.2).

They made quite the opposite biomechanical claim, considering the spine primarily a dynamic rather than a static pole. They confirmed this with their research

Fig. 4.1 The University of Zagreb (since 1669). The Rectorate

Fig. 4.2 Professor emeritus Osman Muftić (1934–2010), doyen of human biomechanics in Croatia

on human speciments, using the so-called enlarged vertebral segment of the spine. Previously, the kinematic chain of two vertebrae, commonly known as the Junghans or Bewegungssegment was used as a rule [16].

Mutfić and the associates considered the kinematic chain consisting of only two vertebrae and the interverbal plate, the discus, to be inadequate. It shows biomechanical events at only one level, and spinal movements occur in at least two or more levels. Therefore, in their research they used an extended vertebral dynamic segment. Following the suggestion of Professor Muftić, Mijović [17], and afterwards Božić [18–24] based their research on the extended spinal segment described as above. They define it as a three-member dynamic segment, based on the model in accordance with human anatomical preparation, considering that it shows biomechanical events in the spine much more accurately. In their research, they used a fresh, not formalin-fixed, spine model as it used to be the case. (Under the term model it is ment an anatomical preparation, or a specimen, the term model being commonly used by Professor Muftić. We may use both terms in this text). They use a kinematic chain from the second, third and fourth lumbar vertebrae. According to research by Prof. Ivančić-Košuta of the Faculty of Kinesiology (former Faculty of Physical Education), conducted at the Department of Anatomy at the School of Medicine in Zagreb, in several parameters, the lumbar vertebrae are in a clear harmonic relations with each other, and the results obtained on one are easily comparable to the results obtained on the other vertebrae [25].

During the 1980s and later, there was a merge of medical doctors, engineers and kinesiologists studying spinal biomechanics. Following the research led by Muftić and the associates, especially Mijović and Božić, it was concluded that the spine is, first of all, a dynamic structure and only then a static one. These studies were subsequently confirmed in vivo on healthy subjects in the Biomechanics Laboratory at the Faculty of Kinesiology in Zagreb. The Laboratory was headed by professor Medved and the studies were performed by Kasović and Božić [26, 27]. These studies further confirmed the earlier thesis of the primarily dynamic and then static role of the spine. Biomechanical studies were performed using an automated 3D kinematic measurement system which recorded spinal movement in real time, that is, in vivo. (Spinal movement kinematics was estimated from body surface-obtained measurement data, using a marker-based detection technique; a measurement method described and presented in some detail in Chaps. 5 and 10). The volunteers were young healthy people who had no spinal issues. In spinal movements, the kinematic measurement system recorded deviations and, with its results, confirmed the thesis that the spine, in biomechanical terms is primarily a dynamic structure. The so-called physiological shift or instability of the lumbar spine, in the range from 0 (zero) to 1° was described. In this way, previous in vitro laboratory tests of the biomechanical properties of the spine on the extended human spinal dynamic segment, which were also in the range from 0 (zero) to 1°, were confirmed. Spinal biomechanics studies have also been conducted on patients who have undergone surgery for discus hernia of spine. That has proven that microdisectomy, that is, interlaminectomy, is the most viable procedure in spinal surgery. It does not impair the natural biomechanical features of the spine, nor does it increase the physiological instability described earlier, contributing to the immediate recovery of patients after such spinal surgery.

To illustrate in vitro laboratory research performed by the Zagreb biomechanics group, an archival example of research of biomechanical properties of lumbosacral joint speciments by Božić follows. It is a part of his dissertation, thesis advisor being Prof. Ivančić-Košuta [18].

4.2 Lumbar Spine Biomechanical Analysis on Intact and Operated Human Model

4.2.1 Introduction

Movements of the lumbar spine are very complicated and yet insufficiently described in the literature [28]. (The statement refers to the year of publication of this reference, 1985: comment by the editor). Vertebral shape is very complex and its normal function is hard to define. Through the lumbar spine load transfer, each vertebra is under a complex sum of forces, meaning that, apart from tensile and compressive forces, there are also flexion and twisting forces affecting the vertebra (Fig. 4.3). Each vertebra has three rotational and three translational degrees of free movement in relation to vicinal vertebra. The shape of articular plates as parts of facet joints has been described as a fragment of spherical surface that enables three rotational free movements, which has been examined by geometrical studies. The vertebral body is mainly made of a spongy bone, while the arch, spinous, and articular processes are vertebral parts made of a compact bone (Fig. 4.4). Terminal vertebral plates together with compact parts of vertebra (articular processes, joints, vertebral arch) are supportive parts during vertebral load. The spongy bone and the intervertebral

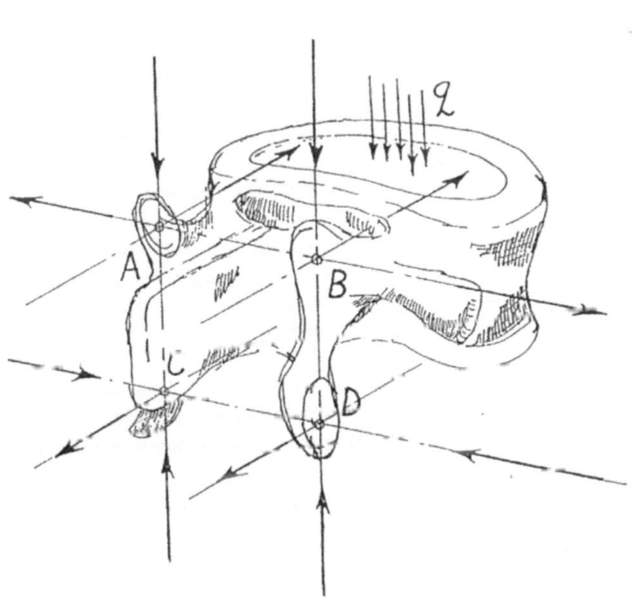

Fig. 4.3 Pressure transfer through the lumbar spine is a sum of complex forces that each vertebrae is exposed to

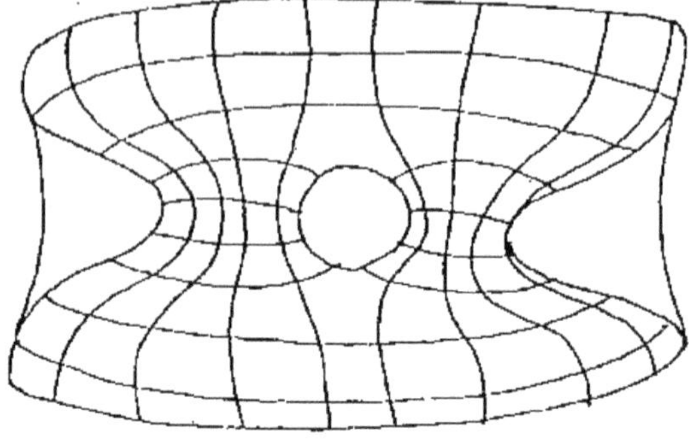

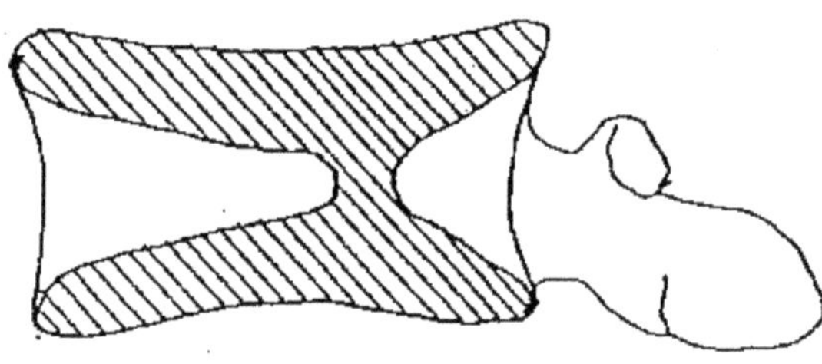

Fig. 4.4 The three-dimensional perspective of the density of a spongy and of a compact bone in the body of a lumbar vertebra

discs act as dynamic axial force absorbers. The four points of major load (two in the dorsal third of the vertebral body and one in each facet joint) form a trapeze (Fig. 4.5) whose diagonals and two axes of symmetry form a group of spindles around which the rotation of one vertebra in relation to the next is possible. This is how the relative vertebral motion is achieved. So far, several authors have studied the biomechanics of lumbar spine motion; Evans, King, Gibson, Hirsch, Brodetti, Nachemson, Schully [2, 29, 30]. Experiments were performed on a kinematic pair of two lumbar vertebrae with the corresponding intervertebral disc. This kinematic vertebral segment was named by Junghans as Bewegungssegment (Fig. 4.6). Dürrigl et al. named it vertebral dynamic segment, anatomically divided into the dorsal and the ventral part. The two adjacent vertebrae, intervertebral disc and anterior and posterior longitudinal ligaments make the ventral part. Vertebral arches with their processes and the corresponding ligaments make the dorsal part. The kinematic pair of two lumbar vertebrae with their intervertebral disc and corresponding ligaments show the biomechanical motions in one plane only, even though in the lumbar spine there are movements taking place in at least two or more planes [31]. Therefore we can conclude that this experimental model of lumbar spine does not enable a three-dimensional (3D) display of all the biomechanical events present in the lumbar spine.

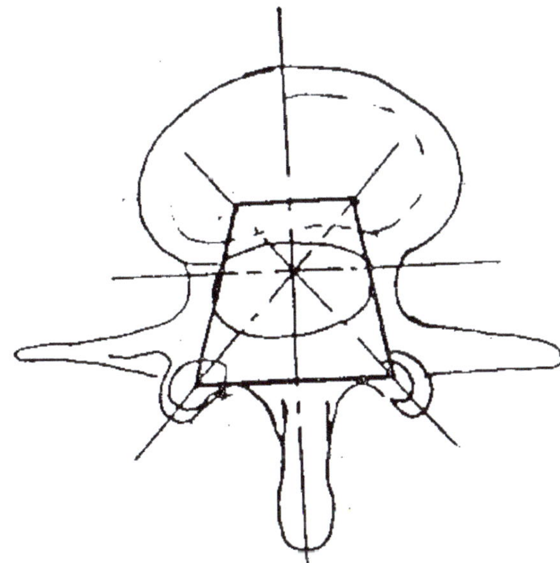

Fig. 4.5 The four points of major load in lumbar vertebrae (two in the dorsal third of the vertebral body and one in each facet joint) form of a trapeze

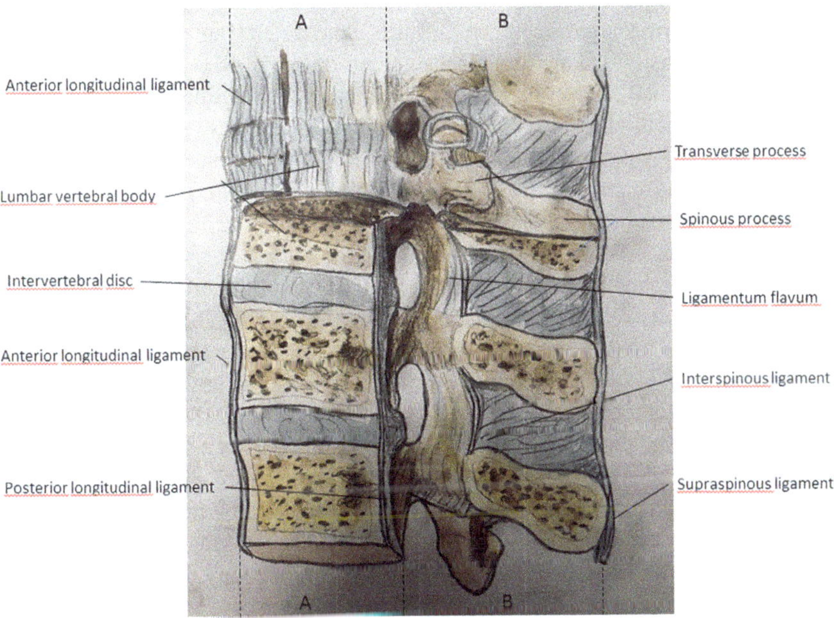

Fig. 4.6 Vertebral dynamical segment. **A** ventral part, **B** dorsal part

4.2.2 Materials and Methods

The extended vertebral dynamical segment of the lumbar spine has been used in our experiments to enable showing the real 3D movements present in the biomechanics of the lumbar spine (Fig. 4.7a–c). As an experimental model, fresh anatomical specimens of the extended dynamical segment of human lumbar spine, which includes the second, third and fourth lumbar vertebra with the according intervertebral discs and ligaments, were used. These segments were in clear harmonious relations and the results obtained on each segment are comparable with other segments [32]. The cadaveric material was obtained from adults of both sexes, age 20–55 who did not have lumbar pain during life. After the approval from the Ethical Committee, the specimens were extracted and prepared at the Sestre milosrdnice University Hospital Center (formerly named "Dr. Mladen Stojanović Clinical Hospital"), Pathology Department "Prof. Ljudevit Jurak" in Zagreb. Thirty five fresh cadaveric specimens were used for this experiment. To simulate the conditions in vivo as accurately as possible, all the measurements were performed at the same day of the autopsy. All the experiments were conducted at the School of Medicine, University of Zagreb, at the Anatomy Department. Two 4.5 mm thick steel plates were connected to the upper and lower end of the examined specimens. They were fixed with Palacos material with the appropriate recesses formed to withstand centric and eccentric loads. Each of the prepared specimens has been exposed to axial forces between 0 and 1000 N. Under centric and eccentric loads, the tangential shifts were tested (Fig. 4.8a–c). The effects of these forces measured as angular displacements of lumbar vertebrae and expressed in degrees were calculated indirectly by assessing the deflection of laser beams on the screen. This method is based on the so-called Poggendorff light beam reflection method, where the rotation of the reflection mirror of 1 α results in the deflection of the reflected beam for the angle of 2 α. The use of a laser and an arbitrarily set screen enables an enormous increase in sensitivity while detecting vertebral displacements. A ruby laser of 0.5 mW power was used as a light source. A silver-nitrate covered 10×10 mm plate was inserted using a cortical screw in the middle of each vertebral body. This plate reflected the laser beam on the screen. The pressure machine was made out of a metal frame and three crosspieces. In the upper field there was a ring shaped dynamometer. In the lower field there was a lumbar spine specimen on which a constant axial centric and eccentric loads of 0 to 1000 N were applied. (The apparatus based on the Poggendorff method was a part of instrumentation setup selected for this research by professors Muftić, Keros and Nikolić, due to its high resolution.) After these experiments were done, the following neurosurgical operations were performed on these models: flavectomy, microdisectomy, unilateral and bilateral interlaminectomy, hemilaminectomy and laminectomy. Postoperatively, the same centric and eccentric loads were applied to follow the deformation line [33–35].

a

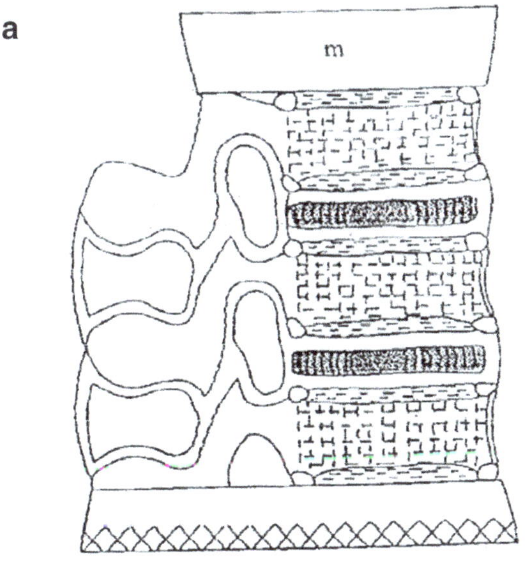

b

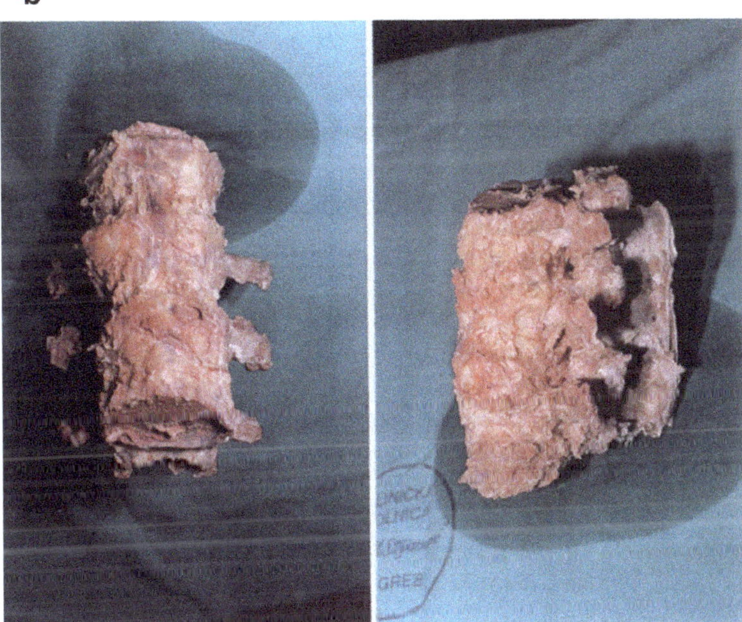

Fig. 4.7 **a** A model of an extended dynamical segment of human lumbar spine. **b** Anatomical specimen; view from the ventral (left) and lateral (right) side

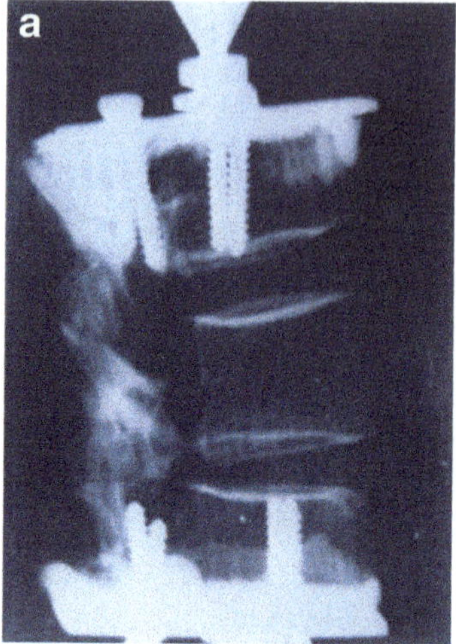

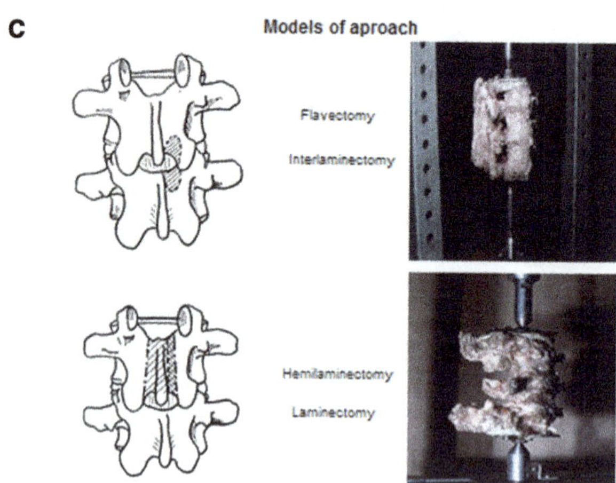

Fig. 4.8 **a** An X-ray image of the three-vertebra dynamical segment of human lumbar spine **b**. Apparatus for testing centric and eccentric loads on intact preparation of the lumbar spine **c**. Models of neurosurgical approach

4.2.3 Results and Discussion

The angles of rotation/deflection of laser beam on the screen were shown graphically as a line of load or deformation line of the lumbar spine exposed to centric and eccentric loads. The ordinate corresponds to the values of the applied load expressed in Newtons (N) and he abscissa corresponds to the angles of flexion, that is physiological deformity of lumbar spine, which is expressed in degrees. All the calculations were obtained by a computer (PC IBM compatible XT). The used program language was Quatro. The distance between the tested model of lumbar spine and the screen was constant. During centric loads on the lumbar spine ranging from 0 to 1000 N, physiological movements were 0 to 1° with the standard deviation of ±0.1 (Fig. 4.9a–c), and during eccentric loads forward and backward, the physiological movements were from 0.55° to exceptionally 1°. After minimally invasive neurosurgical operations (flavectomy, microdisectomy, unilateral and bilateral interlaminectomy) during centric loads, the physiological movements were up to 1.35°, and during eccentric loads forward and backward, the physiological movements were from 0.55° to exceptionally 1° (Fig. 4.10a–c). After hemilaminectomy during

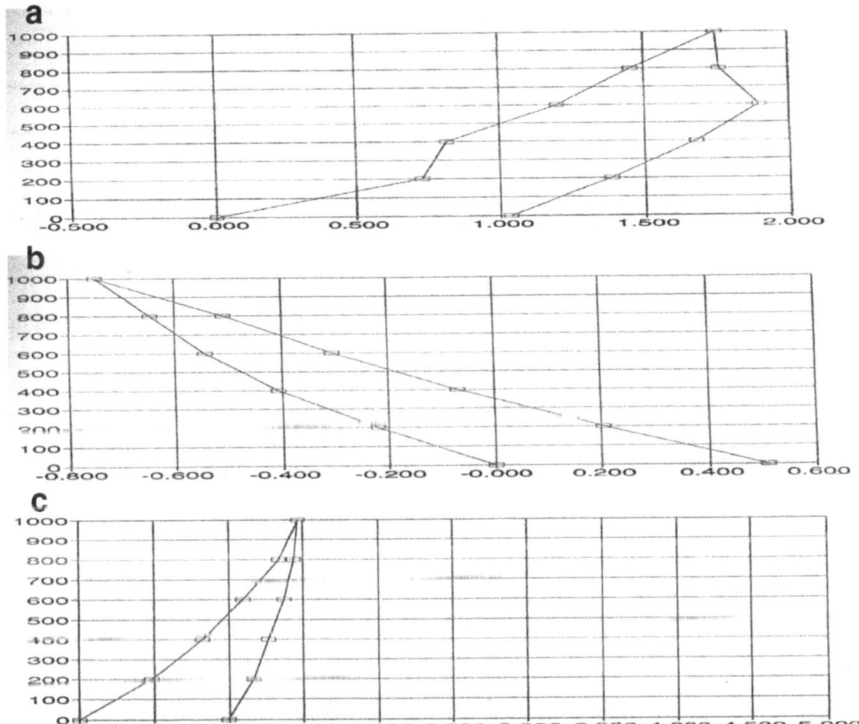

Fig. 4.9 Angles of rotation of the intact preparation. Y-coordinate-load (N); X-coordinate-angles (degrees) **a** Centric load; **b** Eccentric load direction forward; **c** Eccentric load direction backward

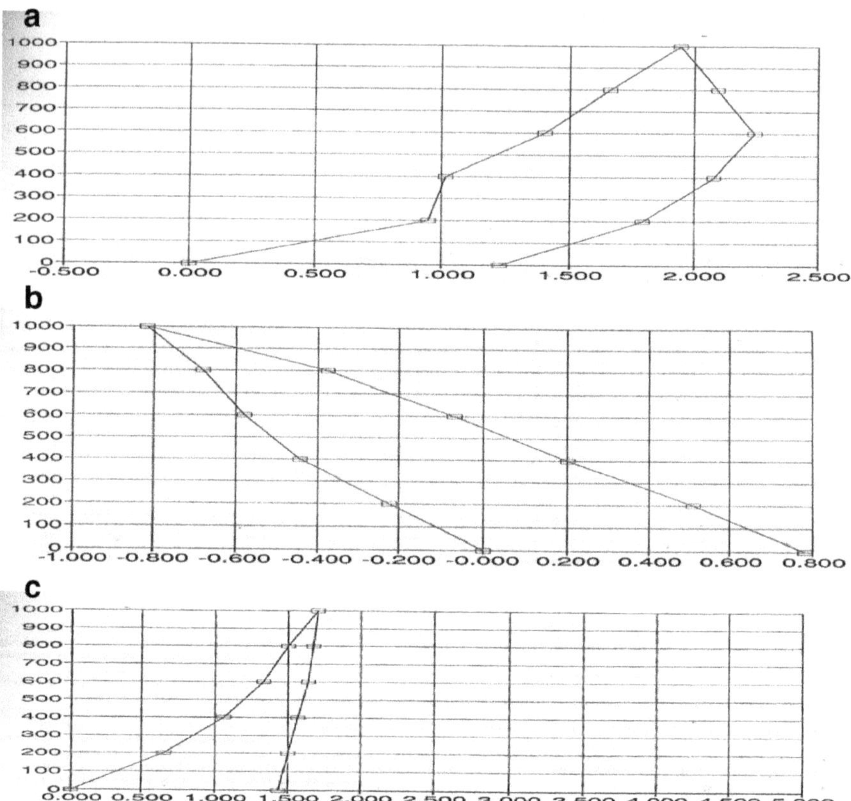

Fig. 4.10 Angles of rotation after interlaminectomy. Y-coordinate-load (N); X-coordinate-angles (degrees) **a** Centric load; **b** Eccentric load direction forward; **c** Eccentric load direction backward

centric and eccentric loads, the movements of the lumbar spine were up to 2.35° and during eccentric loads forward and backward from 2.75° to 4.2° (Fig. 4.11a–c). After laminectomy during centric loads, the movements were up to 4.1° and during eccentric loads forward and backward from 4.9 to 6.6° (Fig. 4.12a–c). Based on these results it can be concluded that flavectomy, microdisectomy and interlaminectomy are the most sparing operations in the lumbar spine surgery with the minimal effect on the changes of physiological movements of the lumbar spine ($p < 0.1$), while hemilaminectomy and laminectomy are more extensive surgical procedures with a significant effect on physiological stability of the lumbar spine ($p < 0.04$ and $p < 0.006$), and which therefore require stabilization of the treated region. It was also noticed that after hemilaminectomy and laminectomy, while greater loads were applied (800–1000 N), an anterior herniated disc regularly occurred, which was not the case (under the same conditions) after an interlaminectomy.

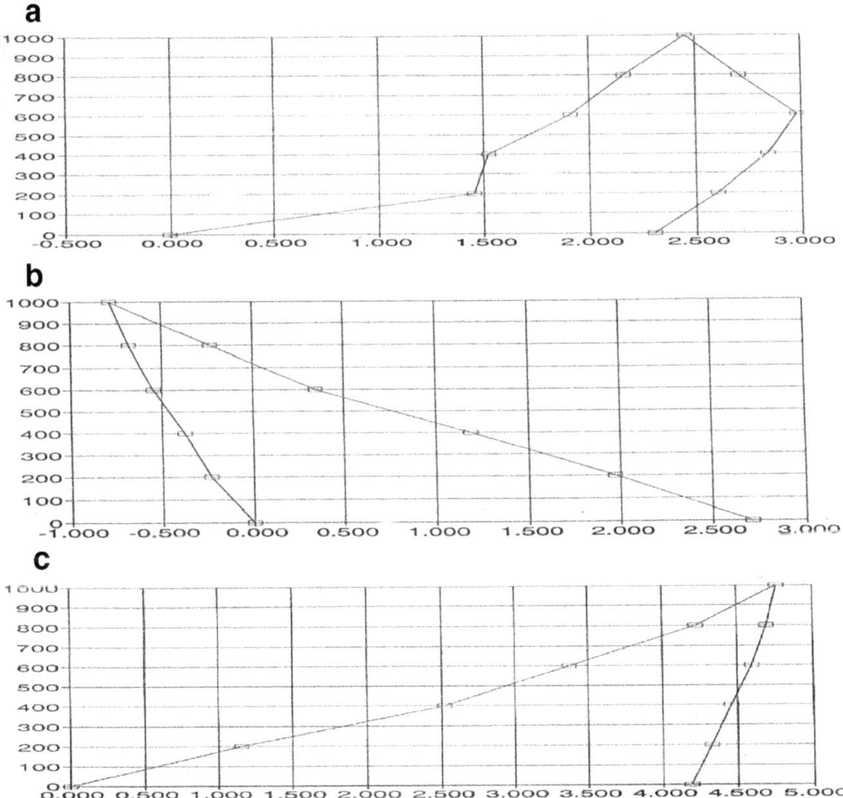

Fig. 4.11 Angles of rotation after hemilaminectomy. Y-coordinate-load (N); X-coordinate-angles (degrees) **a** Centric load; **b** Eccentric load direction forward; **c** Eccentric load direction backward

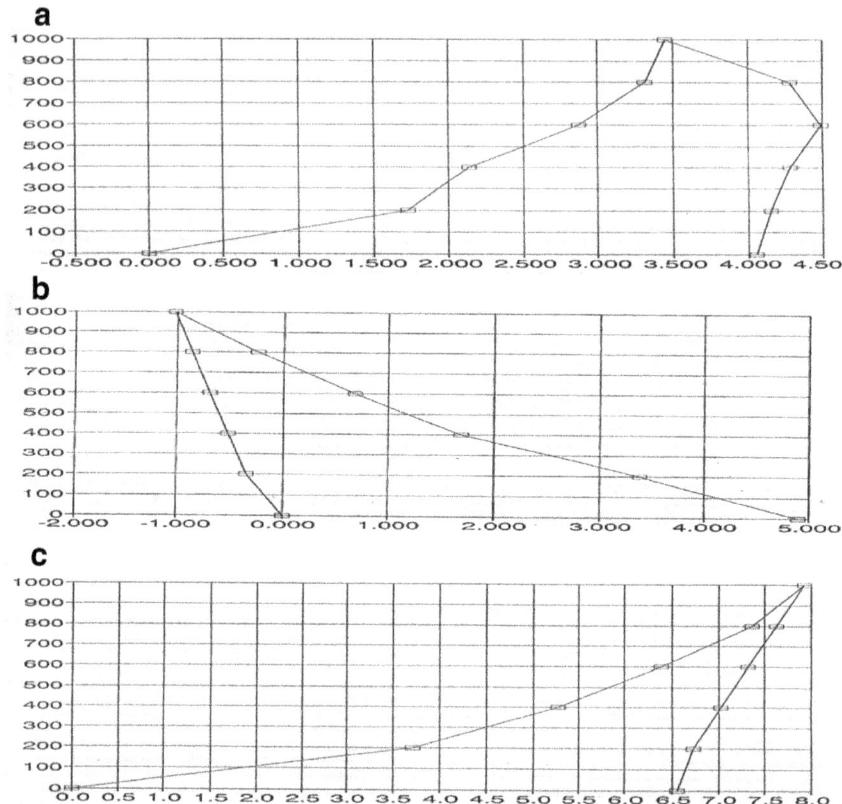

Fig. 4.12 Angles of rotation after laminectomy. Y-coordinate-load (N); X-coordinate-angles (degrees) **a** Centric load; **b** Eccentric load direction forward; **c** Eccentric load direction backward

4.2.4 Conclusion

We applied static centric and eccentric loads on the fresh anatomical lumbar spine specimens [1, 16]. The effects of the loads expressed as angular vertebral displacements were studied on the intact model and after different neurosurgical operations. With these experiments we have proven the following:

1. There are oblique and relative vertebral displacements, representing so-called physiological instability of the lumbar spine [19].
2. The values of angular displacements on the intact speciment range from 0° to 1.0° for centric loads, from 0.1 to 0.55 for forward eccentric loads, and from 0° to 1° for backward eccentric loads [20].
3. After interlaminectomy, there was no statistically relevant difference ($p < 0.1$) in the values of the angular displacements compared to the intact model during centric and also eccentric loads [21].

4. After hemilaminectomy, and especially after laminectomy, there were statistically relevant differences ($p < 0.004$ and $p < 0.006$) in the values of angular displacements compared to the intact speciment during centric and also eccentric loads [22].
5. During an extreme load from 800 to 1000 N on the models on which hemilaminectomies and laminectomies were performed, anterior herniated-extruded discs regularly occurred, while the same phenomenon did not occur under the same conditions on the intact speciment as well as on the ones on which interlaminectomy was performed [23, 24].
6. Due to the observed statistically significant differences and significantly increased instability of the spine after extensive operations (laminectomy), a spinal stabilization is indicated, mainly dynamical stabilization, to preserve the physiological dynamics of the spinal movements.

Acknowledgements Marija Dumančić is to be thanked for the artistic drawing interpretation of an anatomy model in Figs. 4.3–4.8.

References

1. Nachemson A (1981) The role of spine fusion. Spine 6:306
2. Kazarian (1975) Dynamic response characteristics of the human column. Acta Orthop Scand Suppl 146:1–53
3. Keros P, Rudež V, Tivanovac P, Pećina M (1969) The clinical importance of the lumbosacral angle and the possibility of surgical treatments of diseases of the lumbosacral curvature. Liječnički vjesnik 91(2):155–161
4. Keros P (1970) Funkcionalna anatomija lumbosakralnog prijevoja. JAZU 1970(358):97
5. Keros P, Krmpotić-Nemanić J (1972) Novija istraživanja u funkcionalnoj anatomiji kralješnice. Fol Anat 1/1 (2):17
6. Keros P, Krmpotić-Nemanić J, Mandić V, Barac B (1973) Certain morpholigical changes in the spine due to biomechanical factors. JAZU 366:55–63
7. Nikolić V, Hudec M, Hančević B, Vukičević D, Jo A, Vukičević S (1975) Anatomical specimen: a reliable model for biomechanicals investigations of the locomolor system. Folia Anat Iug 4:75–86
8. Ruszkowski I (1986) Ortopedija. Jumena, Zagreb
9. Keros P (1962) Anatomska podloga uspješne provodne anestezije u plastičnoj kirurgiji glave i vrata. Disertacija. Medicinski fakultet Sveučilišta u Zagrebu, Zagreb
10. Muftić O (1970) Razvoj tehnike mjerenja deformacije i naprezanja pomoću elektrootpornih mjernih traka. Elektrotehničar 22(3–4):1–4
11. Muftić O (1972) O vezi među zakonitostima teorije graničnog konstruiranja i strukture biološkog materijala. Disertacija. Fakultet strojarstva i brodogradnje. Zagreb
12. Muftić O (1974) Uvod u biomehaniku. Zagreb: vlastito izdanje
13. Muftić O, Keros P (1977) O utjecaju +g ubrzanja na građu čovjeka. Glasnik Antropološkog društva Jug 4169–173
14. Muftić O, Keros P, Husnjak M, Iveković V, Ivančić-Košuta M, Vidović M (1976) Morphological and functional distribution of lumbar vertebra strength. Folia Anat Iugosl 5:165–170
15. Muftić O, Keros P, Krmpotić-Nemanić J (1977) Anatomski uvjeti oblikovanja prirodnih zakrivljenosti kralješnice, skolioze i kifoze. Zbornik simpozija o skoliozama i kifozama. Medicinska naklada. Zagreb, 61–68

16. Junghanns H (1992) Der Lumbosacralwinkel. Deutsch Chir 4–6
17. Mijović B (1988) Extended conception of the dynamic spine segment. In: Proceedings of 12th International Congress of Anthropological and Ethnological Sciences. Zagreb, 1988:32–34
18. Božić B (1995) Istraživanja ponašanja slabinske kralješnice u dinamičkim uvjetima. Disertacija. Medicinski fakultet. Zagreb
19. Božić B, Negovetić L, Ivančić-Košuta M (1997) Biomechanical analysis of the lumbar spine. Minimal Invasive Neurosurg 40: 23–26
20. Bozic B, Negovetic L, Kogler A, Nikolic V (1997) Biomechanical analysis of the lumbar spine. 11th International Congress of Neurological Surgery, Amsterdam, The Netherlands, July 6–11 (1997), pp 1363–1367
21. Bozic B, Kogler A, Negovetic L, Skarica R, Vukic M, Ivancic-Kosuta M (1999) Biomechanical analysis of the lumbar spine. 11th European Congress of Neurosurgery, (1999), pp 529–533
22. Bozic B, Negovetic L, Kovac D, Kogler A, Sajko T (2003) Biomechanical analysis of the intact and operated human cadaveric lumbar spine. World Spine II - The Second Interdisciplinary Congress on Spine Care (2003), p 168
23. Božić B, Sajko T, Kogler A (2003) Sequestered extrusion of lumbar disc experimental model, clinical pictures, diagnosis and treatment. World Spine II-The Second Interdisciplinary Congress on Spine Care (2003), p 176
24. Grubišić F, Božić B, Nemčić T (2009) Funkcionalna anatomija lumbalne kralješnice. In: Grazio S, Buljan D (ur) (2009) Križobolja. Jastrebarsko: Naklada Slap, pp 41–54
25. Ivančić-Košuta M (1990) Istraživanja dimenzijskih funkcija u slabinskoj kralješnici odrasle muške osobe. Disertacija. Medicinski fakultet. Zagreb
26. Kasović M, Božić B, Vlahović H (2014) Biomehanika kralješnice u sportskim aktivnostima s osvrtom na mehanizam ozljede kralješnice. Fizikalna i rehabilitacijska medicina 26(3–4):99–100
27. Kasović M, Božić B, Cigrovski V (2014) Važnost pravilnog treninga i sportske opreme u prevenciji sportskih ozljeda kralješnice. Fizikalna i rehabilitacijska medicina 26(3–4):101–102
28. Anderson CK, Chaffin DB, Berrin GD, Matthe WS (1985) A biomechanical model of the lumbosacral joint during lifting activities. J Biomech 18(8):571–584
29. Panjabi MM, Krag M, Summers D, Videman T (1985) Biomechanical time. Tolerance of fresh cadaveric human spine specimens. J Orthop 3:292–300
30. Yang SW, Langrana NA, Lee CK (1986) Biomechanics of lumbosacral spinal fusion in combined compression torsion loads. Spine 11:337–342
31. Gunnar BJ (1990) Occupational biomechanics. In: Weinstein WJ, Wiesel S (eds) The lumbar spine. Saunders Co., Philadelphia-London-Toronto, pp 212–224
32. Hakim NS, King AI (1979) A three dimensional finite element dynamic response analysis of a vertebra with experimental verification. J Biomech 12(1979):277–292
33. Watts C, Smith H (1919) Disc disease. In: Grossmann RG, Hamilton WJ (eds) Principles of neurosurgery. Raven Press, New York, pp 415–435
34. Panjabi MM, Brandt A, White A (1976) Mechanical properties of the human spine. J Bone Surg 58:642–651
35. Bush-Joseph C, Shiplein OD, Andersson G, Andriacchi TP (1988) Influence of dynamic factors on the lumbar spine moment in lifting. Ergonomics 31:212–216

Chapter 5
On Measuring Kinematics and Kinetics of Human Locomotion

Vladimir Medved

Abstract Selected technical solutions aimed at measuring kinematics and kinetics of human locomotion are succinctly pointed to and illustrated. Among kinematic measurement systems the marker-based ones—as the most common—are put forward, making mention of related signal and data processing. Wearable type sensor measurement systems are commented next. Further, kinetic measurement instruments; a force measuring platform (force plate) and a pedobarograph are addressed shortly. Few examples of measurement results of sportive and normal locomotions are shown and commented to illustrate the methods. These measurement instruments, and related measurement protocols, are to be regarded as an integral part of an inventory at disposal to modern student of human movement.

5.1 Introduction

Approaching gross body movement and locomotion as a physical phenomenon, as pursued in the field of biomechanics (*Multiple rigid body modeling paradigm* in Sect. 2.2, Fig. 2.14), we combine modeling with measurement. A sort of melting of these two approaches has happened around the end of the nineteenth century, when Braune and Fischer for the first time have calculated respective physical quantities, forces and moments of forces, realizing inverse dynamics approach. The body has been modeled as a multi-segment mechanical system. Kinematic information was acquired by the chrono-photography method, and stereometry (3D measurement) has been realized. Rapid development of electronics since has brought new sensor technologies, digital computers and other technical hardware and software commodities (Chap. 1, Fig. 1.3), and along new measurement instruments such as a force plate being designed, enabled realizing today essentially the same approach much more accurately and much faster. Today, a modern researcher of human movement has a lot of measurement tools available for this task at disposal. Here we focus on what is usually the first grasp in measuring certain movement pattern; i.e. measuring its

V. Medved (✉)
Faculty of Kinesiology, University of Zagreb, Horvaćanski zavoj 15, 10000 Zagreb, Croatia
e-mail: vladimir.medved@kif.hr

V. Medved (ed.), *Measurement and Analysis of Human Locomotion*,
Series in Biomedical Engineering, https://doi.org/10.1007/978-3-030-79685-3_5

kinematics and/or its kinetics (dynamics). Our insight into measurement possibilities will not be comprehensive, though, as today's engineering and computing technology offers a plethora of diverse technical solutions, but instead we are only taking a sort of a „snapshot " of modern devices most in use today.

5.2 Measuring Kinematics of Human Locomotion

Kinematics is a part of mechanics concerned with geometry of movement, therefore the goal of systems aimed at measuring kinematics of human body during movement is to "capture" instantaneous spatial (3D) body position and follow it in time. Among various technical (engineering) solutions to accomplish the task, the most popular are the marker-based ones, employing some sort of an optoelectronic principle to automatically detect and track the positions of body mounted markers [1, 2]. Besides, methods using wearable technologies (magneto-inertial sensors) are getting ever more prominence lately [3–5].

5.2.1 Optoelectronic Methods

Optoelectronics-based methods belong to the group of stereophotogrammetric methods, which in turn are one type of stereometric methods (classification according to [6]). Historically—as already mentioned—stereophotogrammetric methods were first realized using the high-speed photography technique (Braune and Fischer; see Sect. 2.1 in Chap. 2). In a long time span up to the second half of the twentieth century, until 1970s, classical high-speed photography has been successfully used as a method to record kinematics of human locomotion. However, the process of extracting relevant information from a photographic record was cumbersome and slow, requiring intensive manual work by human operator, a task also prone to human error. Efforts to automatize kinematic signal extraction procedure began sometime during late 1960-es. Among first realizations was a marker detector circuit, a part of the Stratchylde kinematic measurement system, which functioned by detecting a light threshold in a TV image (Fig. 5.1).

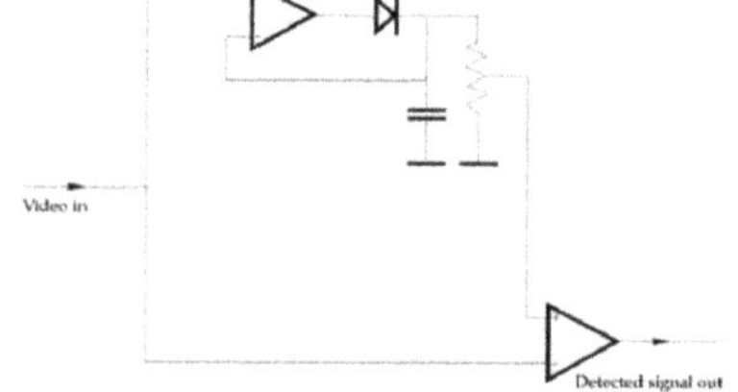

Fig. 5.1 Marker detector circuit in the Strachylde kinematic measurement system (from Jarett et al. [7], Reproduced with permission from the Institution of Engineering & Technology, 2021)

Fig. 5.2 An infra red (IR) camera equipped with a stroboscope comprised of light emitting diodes (LED)s, part of the Italian ELITE system (ELITE, commercial material) (With permission of BTS Bioengineering)

Subsequent technical solutions followed, based on various electronic technologies such as charged coupled device (CCD) or metal oxide semiconductor (MOS) matrix sensor based converters being used in the cameras. The systems were designed and developed as laboratory versions, while some have evolved to commercial products. The use of classical high-speed photography technique has since been overcome.

In [1], some technical details of different marker layouts and camera designs are described for a couple of systems at the time, covering both hardware and software aspects. A camera used in ELITE system is shown in Fig. 5.2.

A marker is detected using a principle of brightness of light recognition (light versus dark background), usually either using a light threshold principle or based on shape recognition. In this second variant markers are automatically recognized by means of dedicated computer algorithm using a pattern recognition technique for object identification in real-time and according to shape and size, not exclusively to light intensity, as in equipment in our Zagreb and Salerno situated laboratories: the ELITE System [8]. In this way, the system allows greater freedom with regard to lighting conditions and distribution of background illumination and is also applicable in sunlight.

A short description of the technical principle of marker detection in this system follows, while detailed elaboration of the technical solution realized may be found in (Ferrigno and Pedotti [8]), main steps also being reported in [1].

Video signal processing architecture includes two hierarchical levels (Fig. 5.3):

- The dedicated peripheral fast processor for shape recognition (FPSR), designed and accomplished by means of a fast digital VLSI chip
- The general purpose computer.

The FPSR is connected to a unit called ITE (Interface to the Environment) and calculates the cross-correlation of an input signal, previously digitized, and a stored mask (template). It thereby recognizes markers and calculates their coordinates, by which approximately 1000-fold data reduction is accomplished.

Chosen template which describes a marker is stored in the CPU. The shape detecting algorithm (SDA) is based on the 2D cross-correlation between digitized image and previously defined shape, dimensions 6×6 pixels. Only if and when the

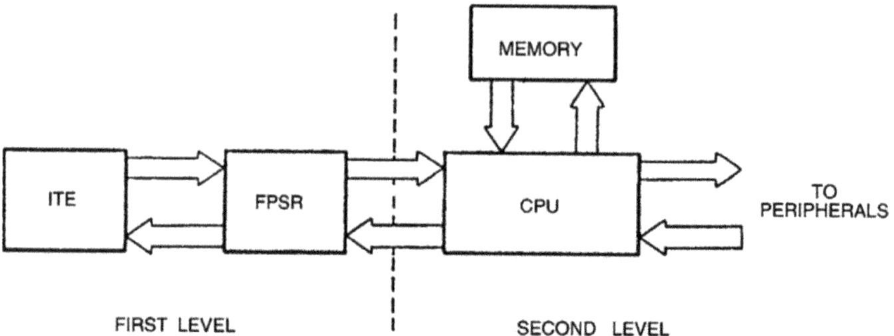

Fig. 5.3 Two hierarchical levels of video signal processing in the ELITE system. ITE goes for Interface to the Environment, FPSR goes for Fast Processor for Shape Recognition, and CPU goes for Central Processing Unit (from Ferrigno and Pedotti [8], with permission)

value of cross-correlation coefficient surpasses the previously determined threshold value are pixel coordinates digitized. If the threshold is surpassed and this does not correspond to marker image, e.g., as in the case of clutter, etc., it is rejected by this mechanism. The FPSR is implemented as a parallel hardware structure enabling real time processing. The output from the FPSR directly produces pairs of horizontal and vertical coordinates of the markers detected, which are sent to the CPU (level 2).

Because this principle is based on recognition of defined marker shape, and not their brightness, the reliability of the procedure is very high. Furthermore, because all levels of gray are used in an image, real resolution is significantly increased. In this system, however, the size of the monitored markers is critical and only limited marker movement along the line of vision of the camera can be tolerated. This would limit the device's use to approximately planar movements monitored in a sagittal plane. However, markers are small, even smaller than 1/256th of the viewing filed, which in practice is projected to a size smaller than the defined marker (subtemplate). The necessary number of pixels may be covered by intentional blooming and defocusing of such images. The authors consider this procedure to be a more reliable way of getting a subpixel resolution than by estimating geometrical centroids. The final accuracy, which includes distorsion corrections across the whole viewing field, is experimentally determined to amount to 1/2800 of the viewing field. In [8], a resolution test was described on the basis of minimal discernible linear marker displacement. Very good results are realized with characteristic small markers, which correspond to marker shape of 6×6 pixels.

Upon marker position detection in a 2D image in a camera (two or more cameras are to be used), a reconstruction of a spatial, 3D marker location is realized. Two cameras suffice for reconstruction of the location of one marker. To enable detection of multiple markers, a number of cameras have to be used. Each body marker must be seen by at least two cameras; adding more cameras further contributes to system's redundancy and to technical accuracy of the detection process—typically there are eight cameras properly positioned. In a laboratory measurement setting, principles of close range photogrammetry are to be implemented [9]. In Chap. 6 this technical issue is elaborated in detail, introducing quantitative relations describing relevant geometrical transformations.

A short comment on the whole technical procedure: markers are subsequently automatically followed based on a priori knowledge on their application, i.e. anthropomorphic characteristics of their locations. A special procedure of model definition requires all markers to be connected, while isolated points may also exist. A rather detailed description of hierarchical procedure of video signal processing, and of corresponding algorithms to realize this technique of marker recognition and follow-up is given in Sect. 4.2.1.3 in [1], with references to original literature. In practical terms, high spatial measurement accuracy on the order of mm is attained, with 100 Hz sampling frequency, quite appropriate for human locomotion measurements.

Laboratory facility ought to insure a dedicated, large enough space. The Zagreb Laboratory, for example, occupies a 15 × 15 m floor area equipped with multiple cameras, with force plate positioned in the middle of a walkway staged diagonally. Prior to measurement session, calibration of measurement space is to be performed. Retroreflective markers are to be properly mounted on subject's body, positioned at defined anatomical body landmark points, about 20 of them altogether—a procedure taking about 10 to 20 min of time of a skilled operator. General rule to be obeyed is that each body segment has to be equipped with at least three non-colinear markers, so that its spatial position is uniquely defined.

Figure 5.4 illustrates markers positioned on the body, together with sEMG electrodes glued on the right shoulder region and the right upper leg, for the purpose of measuring table tennis strokes.

In addition, video record may also be taken for later qualitative analysis purposes.

After the movement sequence has been performed and kinematic information (2D marker coordinate values) acquired by cameras, a 2D to 3D coordinate data transformation procedure for each marker has to be handled by the operator. Then, having determined 3D marker coordinate values, a geometrical transformation of marker coordinates' to (virtual) centers of joint coordinates is to be performed. So, in effect, gross body movement record is represented by a series of curves stored in a digitized form into computer memory. Customarily, 3D joint angles' changes are deduced (reconstructed) from this record and presented for analysis. The process by its nature introduces a spatio-temporal sampling of sorts (Sect. 5.5.1 in [1]). Modern systems achieve spatial accuracy of few mm, and sampling frequency of 100 to 200 Hz and more (100 Hz in the ELITE System as already mentioned).

In Chap. 6 a detailed quantitative elaboration of stereometry of human movement is given, illustrated by some applications. The examples of use of systems of this kind in gait analysis are presented in Chap. 10.

In the table tennis measurements mentioned the aim has been to compare movement kinematics of top spin strokes performed using a 38- and a 40-mm diameter ball [11].

This increase of ball diameter from 38 to 40 mm has been introduced in year 2000 in order to slow the rhythm of play. Greater ball diameter increases air resistance and decreases speed. It, however, also induces increased fatiguing of a shoulder region during playing of experienced athletes, the reason why we simultaneously have

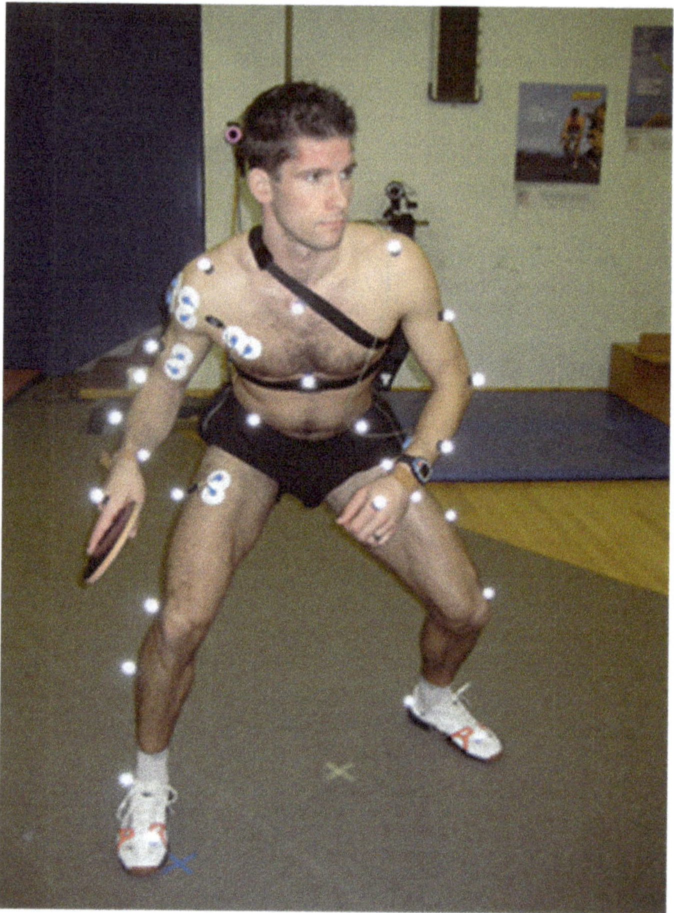

Fig. 5.4 Measuring kinematics and surface electromyography (sEMG) in table tennis. Top table tennis player. Biomechanics Laboratory, Faculty of Kinesiology, Zagreb (from [10], with permission)

measured sEMG with the aim to evaluate local muscle fatigue. The movement pattern was performed playing opposite to a table tennis robot machine. The comparison of selected kinematic parameters measured proved that differences in the amplitude of forehand stroke of the tested player increase due to increased ball size. As an illustration, one kinematic parameter is shown: elbow angle (in one plane) change in time when striking 38- and 40- mm diameter ball (Fig. 5.5). Larger amplitudes of movement are observed with increased ball size [11].

Another example of kinematics measurement comes from the sport of handball. Here, the aim has been to simulate a shot in laboratory conditions, however, therefore

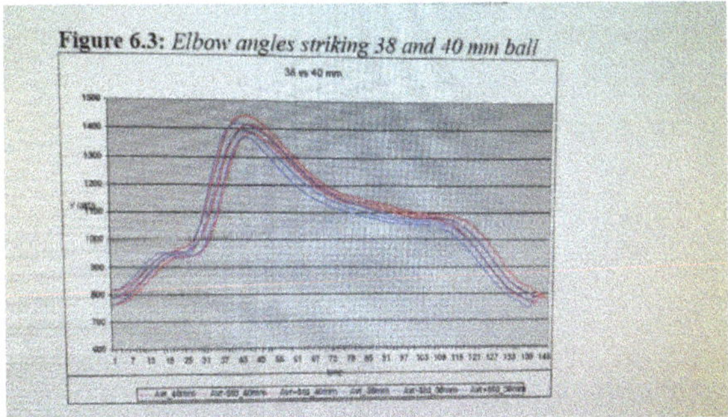

Fig. 5.5 Top spin stroke kinematics. Elbow angle (in one plane) as a function of time when striking with a 38-mm (blue) and a 40-mm (red) diameter ball. A subtle difference is present, i.e. an increase of movement amplitude (from Kondrič et al. [11], with permission)

mere attempted shots were performed, mimicking real movements in sports playground. A variant of Davis protocol for marker positioning suited for gross body movement measurement was applied prior to measurement ([12], Fig. 5.6).

As an illustration, a 3D kinogram of a simulated handball shot (without throwing a ball) is shown, recorded by ELITE System. The Biomechanics Laboratory, Faculty of Kinesiology, University of Zagreb (Fig. 5.7). Succinct kinesiological interpretation of this movement pattern analysis makes a part of figure caption.

It is clear that this kind of analysis of a movement sequence of interest requires adequate laboratory setting in the first place, and then also a thorough preparation of a subject including body marker positioning; a time consuming procedure by a skilled operator with good knowledge of anatomy. As such, this kind of procedure it is not a best candidate for a clinical type method of measurement/testing. On the other hand, it results with highly accurate 3D kinematic data, acquired with great spatial and temporal precision, of value in approaching kinesiological and motor learning aspects of skilled and complex movements.

Various movement patterns may be a subject of measurements: sports movements, normal human ambulation such as healthy gait, pathological gaits, etc. Further examples of measuring locomotion kinematics will be shown and commented in Chaps. 6, 10, and 12.

5.2.2 *On Kinematic Data Processing*

"Raw" kinematic data don't carry large information content, but are only a description of movement of sorts. They may be prone to direct interpretation by an expert in the field in the first place (sports coach, specialist physician). However, the true potential of kinematic measurement data is in their application possibilities, via an inverse dynamics approach, in the mathematical estimation of kinetic movement quantities.

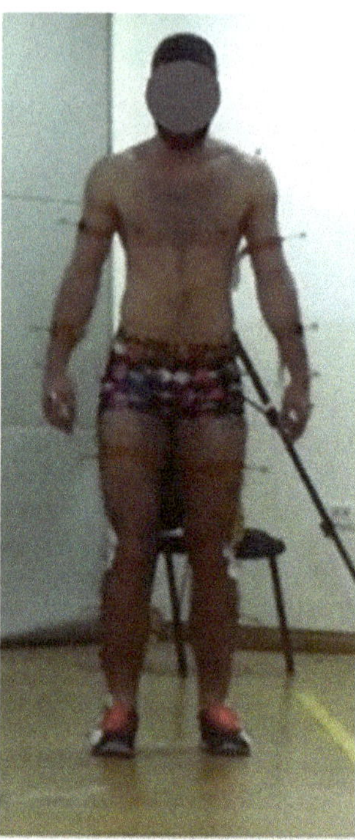

Fig. 5.6 Experienced younger senior handball player, subject specialized to be the central backcourt attacker and organizer, performed selection of most relevant specific throwing, jumping and moving handball techniques, amongst which "stance-shooting" technique was chosen (as a representative of most important basic throwing techniques) to be visualized (from Gruić and Medved [12], with permission)

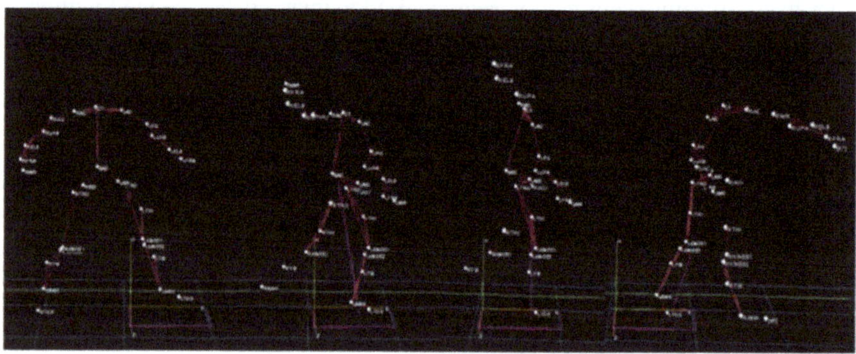

Fig. 5.7 3D kinogram of a simulated handball shot. Phases of basic stance-shot kinematics: (1) preparatory phase, (2) phase of initiation of throwing kinetic chain, (3) phase of crossing coronal (frontal plane) and (4) end of throwing phase are shown. Combining this kind of graphic visual representation with computer-stored numerical values of temporal and spatial movement variables enables quantitative analysis and interpretation by an expert, i.e. trainer. This particular recorded simulated shot is characteristic of a skilled handball player (from Pažin et al. [13], with permission)

The goal of processing kinematic signals is thus to calculate (compute) accelerations and to combine these data with estimated inertial body parameters.

A measurement process itself is not perfect, and is potentially prone to technical errors like missing data, superimposed noise, skin movement artifact caused by non-rigid character of body segments, etc. All these imperfections have to be taken care of. Classical treatment of kinematic signals includes, therefore, the following series of procedures [14]: reconstruction of locations of observed body markers in time (a part of a 3D kinematic measurement itself), signal smoothing/interpolation, application of human body model of the inertial segment type (by means of which anatomical signals are "produced" from measured ones), and differentiation of these signals.

The methodology of processing extracted kinematic signals has been standardized, both in terms of signal elaboration (smoothing, interpolation, etc.) and of signal processing for interpretation purposes and for inverse dynamics calculation ([14–18]; summarized in Sect. 4.4 in [1]. Due to problems caused by incompatibility of anatomical and geometrical concepts in the description of joints, as well as due to imperfections of the rigid body modeling assumption, however, not all technicalities are resolved.

Reader seaking for in-depth knowledge on stereofotogrammetry for human motion analysis is referred to the series of four papers by Professor Cappozzo's group which has critically elaborated practical technical aspects of stereofotogrametric kinematics measurement systems; their theoretical background [19], instrument errors [20], soft tissue artifact assessment and compenzation [21], and assessment of anatomical landmark misplacement and its effects on joint kinematics [22]. Specifically, soft tissue artifact problem related to measurement of gait has also been approached by Peters et al. [23], who gave a systematic review on the issue, and Camomilla et al. [24]. Recent contributions of the Cappozzo group [25] and others are documented in [26]. Researchers and clinicians alike have to be aware of limitations, imperfections and theoretical and practical applicative issues yet unresolved in this domain.

5.2.3 Wearable Technologies

An alternative means to measure and acquire body movement kinematics is to use wearable sensors. This type of sensors combine magneto-inertial sensing principle, yielding desired 3D kinematic information directly [3–5]. Mathematical basis of the method has been elaborated adequately in the literature [5, 27]. Systems of this sort do not require calibration of measurement space and are suitable for the outdoor use. Sensor units can be mounted on the body using velcro tracks or, alternatively, embedded into a body suit of a kind.

Figure 5.8 shows body suit equipped with inertial type kinematic sensors.

Physical basis of operation of sensors of this kind is commented shortly. Inertial sensors use the property of bodies to maintain constant translational and rotational velocity, unless disturbed by forces or torques, respectively. Inertial tracking is made

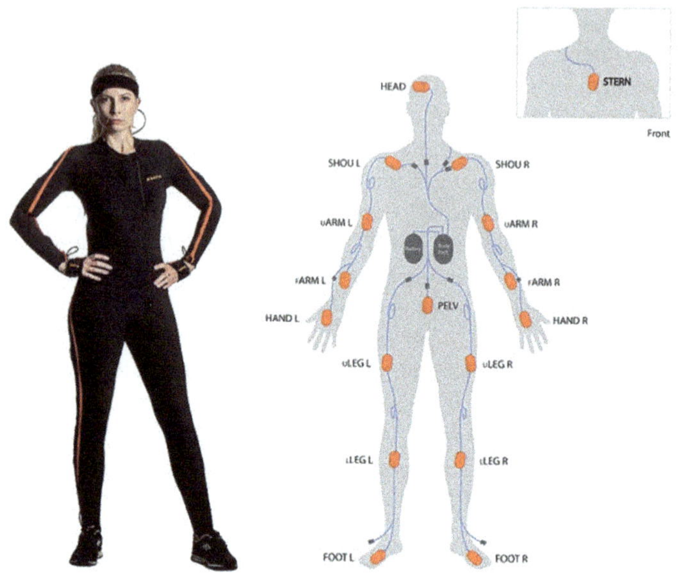

Fig. 5.8 Body suit equipped with inertial type kinematic sensors. Part of instrumentation accessory in the Salerno Movement Analysis Laboratory (Handicap Laboratory), Department of Human, Philosophical and Educational Sciences, University of Salerno (Xsens—commercial material. With permission)

possible in practice by advances in miniaturized and micromachined sensor technologies, particularly in silicon accelerometers and rate sensors. A rate gyroscope measures angular velocity, and if integrated over time provides the change in angle with respect to an initially known angle. An accelerometer measures accelerations, including gravitational acceleration g. If the angle of the sensor with respect to the vertical is known, the gravity component can be removed and by numerical integration velocity and position can be determined. By combining the signals from the inertial sensors with complementary sensors and using knowledge about their signal characteristics, drift and other errors can be minimized.

Wearable systems, providing 3D joint angle information, are finding multiple applications lately: in various sports, and in daily activities, often performed outdoors and in a harsh environment (alpine skiing). Specific medical applications also are being profiled, among other also there are gait analysis attempts, which will be commented in Chap. 10. However, lacking the possibility of inverse dynamics calculation, they can not compete with previously mentioned marker-based systems.

5.3 Measuring Kinetics of Human Locomotion

Kinetics of human locomotion comprises forces resulting either from body movement or being a cause of movement. As human body is regarded a multi-body system of rigid segments connected by joints (point contacts assumed in first iteration of modeling), following forces and moment of forces' components act (referring to Cartesian coordinate system representation): Fx, Fy, Fz, and Mx, My and Mz. Only those forces and moments that are manifested between the body and its environment are prone to measurement; in vast majority of situations this refers to the contact between the foot and the ground. The problem can be approached either considering it a point contact (Sect. 5.3.1) or—with refined modeling—a distributed pressure surface type contact (Sect. 5.3.2).

5.3.1 Force Measuring Platform

Following initial devices from „non-electrical" era and later, still mechanical (Chap. 2, Fig. 2.10), around second half of the twentieth century an instrument has been established in a form of a dynamometric platform to be positioned at the level of the floor in a laboratory: force platform or force plate. It literally takes advantage of the 3rd Newton's law, claiming that each acting force must have a counteracting one. During construction, force sensor units are adequately imbedded into a device and properly connected by Wheatstone bridges so as to secure uniform measuring of contact force, moment of force, and the instantaneous coordinates of mean point of force application. Human foot is considered here to exert a point contact with the floor surface, which certainly is an idealization, but consistent with the accepted body model (Chap. 2, Fig. 2.14) and suitable for application in inverse dynamics calculations. This instrument makes a standard piece of equipment in every motion analysis laboratory. The first professional grade six-component force plate that appeared on the market was a result of joint venture between Jürg Wartenweiler, of the ETH Zürich, and a private company [4].

Modern solutions most often employ strain gages or piezoelectric transducers as force sensors. Dimensions of contact surface are usually 60 × 40 cm. Platform is to be embedded into floor and may be covered with a mat so as to enable, preferably, as noninvasive as possible a measurement (subject not being aware of walking over a platform).

Details of platform construction may be found in [1], and here we present only essential design elements using the example of piezoelectric transducer-based type device. Usually, four identical force transducers are used, each positioned in one corner of the platform (Fig. 5.9). Moment values are deduced from the forces measured and from the relative positions of transducers in the platform. Twelve of the outputs from the four transducers are connected so that there are finally eight outputs.

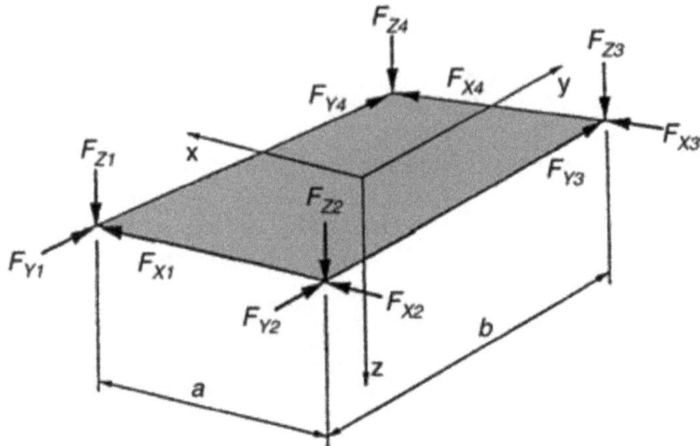

Fig. 5.9 Twelve individual force components measured by a piezoelectric force platform. The eight platform outputs are (Fx1 + Fx2), (Fx3 + Fx4), (Fy2 + Fy3), Fz1, Fz2, Fz3, and Fz4 (from Barnes and Berme [28], with permission)

There are four separate measurements of vertical force, two tangential components in the x direction and two tangential components in the y direction. In order to obtain six components of force and moment, data are further reduced to:

$$
\begin{aligned}
F_x &= (F_{x1} + F_{x2}) + (F_{x3} - F_{x4}) \\
F_y &= \left(F_{y1} + F_{y4}\right) + \left(F_{y2} + F_{y3}\right) \\
F_z &= (F_{z1} + F_{z2}) + (F_{z3} + F_{z4}) \\
M_x &= [(-F_{z1} - F_{z2}) + (F_{z3} + F_{z4})]\, b/2 \\
M_y &= [(-F_{z1} + F_{z2}) + F_{z3} - F_{z4})]\, a/2 \\
M_z &= [(F_{x1} + F_{x2}) - (F_{x3} + F_{x4})]\, b/2 \\
&\quad + \left[\left(F_{y1} + F_{y4}\right) - \left(F_{y2} + F_{y3}\right)\right] a/2
\end{aligned} \tag{5.1}
$$

When only compressive forces in the z-direction are applied to the platform, only the free moment in the x–y plane can be transferred to the platform. The point of application of the resultant force and the free moment can be calculated from the value of the force measured and the moment components expressed within a Cartesian coordinate system.

The piezoelectric platform by the firm "Kistler" provides measurement of the total vertical and horizontal components of the force applied with an accuracy better than 1%, nonlinearity and hysteresis less than 1%, and sensitivity up to 0.05 Pa, in a working range typically 200 kPa for the vertical and 50 kPa for the horizontal force components. A good repeatability and long-term stable properties make this instrument a standard piece of equipment in many a motion analysis laboratory.

There are specific comparative differences between measuring platforms based on the piezoelectric effect and those using strain gage-type transducers. Since frequency response of piezoelectric systems to mechanical excitation is very high, these transducers are indispensable for certain special applications. However, the piezoelectric system, as an active system, is suitable in the first place for dynamic measurements, but quartz as a piezoelectric material in combination with a charge amplifier nevertheless offers also the possibility of measuring approximately static phenomena that may last for a number of minutes or even hours. For biomechanical studies of human locomotion this is completely satisfactory. (Strain gage type transducer platforms are preferred for the purpose of static measurements).

Nilsson's research has addressed biomechanical and neurophysiological features (monitored by means of EMG) of the locomotor system during gait and running, specifically the transition between these two modes of locomotion [29]. In Fig. 5.10 an illustrative set of ground reaction force signals is shown (Fx, Fy and Fz force components) reflecting various speeds of walking and running, this later locomotion performed using two different techniques (rearfoot strike, forefoot strike).

Signal interpretation as reported in [1] follows.

While the first foot–ground contact in normal gait always occurs with the heal, in running this is not necessarily the case: one of the subjects chosen characteristically made first contact with the heel (rearfoot striker, the subject PS) while the other made first contact with the front part of the foot (forefoot striker, subject OS).

Ground reaction force in walking (above) and running (middle and below), at various speeds

F_x - lateral-medial
F_y - posterior-anterior
F_z - upward

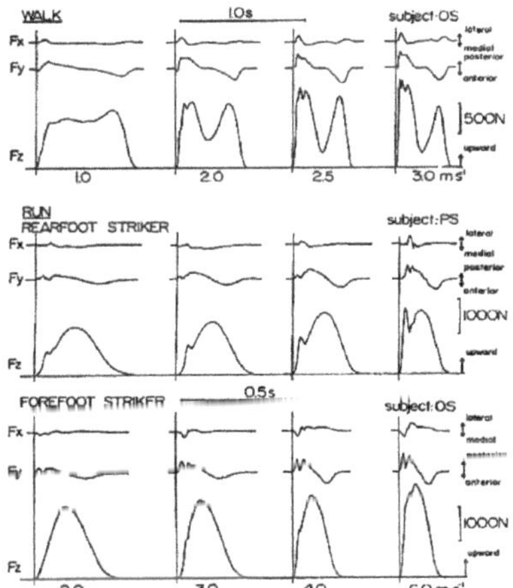

Fig. 5.10 Mediolateral (Fx), anteroposterior (Fy), and vertical (Fz) ground reaction force component waveforms exerted by the right foot in walking (1.0, 2.0, 2.5, 3.0 m/s) and running (2.0, 3.0, 4.0, 6.0 m/s), with rear- (PS) and forefoot (OS) strike, respectively. Healthy subjects wearing athletic footwear. The body weight of subject OS was 790 N and PS was 720 N. Kistler measuring platform. Details of signal interpretation in the text (from Nilsson [29], with permission of the author)

In walking (subject OS), the vertical ground reaction force component Fz displays a characteristic waveform with two maxima and a local minimum in between, occurring approximately halfway through the support phase. This instant marks the amortization of vertical body movement by knee joint activity (the third gait determinant described by The Berkeley Group (Chapter 2; discussed in more detail in Sect. 4.2.1 of 1).

Running, on the contrary, is characterized by a different ground reaction force signal waveform: the local minimum being absent and the signal being „monophasic ", with a more or less expressed impact impulse, more pronounced in rearfoot strikers (more rigid an impact) and in higher speeds.

Generally, vertical force peak assumes values between 1 and 1.5 times the body weight values in walking and between 2 and 3 times the body weight values in running. The vertical signal peak due to impact, characteristic of rearfoot strikers, increases with speed, from 1.3 to 2.6 times body weight value.

To enable intersubject comparisons, as well as intertrial comparisons in the same subject, it is possible to express the time scale in such representations in percentages of the values of the duration of one cycle, while force amplitude can be represented in percentage of body weight.

Inspecting the horizontal force signal components shown, it may be observed that, in walking, the anteroposterior component (Fy) displays a small initial peak in the anterior direction, which changes into a posteriorly directed braking force, which changes into a propulsive one in mid-support. Rising phases of the first vertical and braking horizontal force sometimes have small superimposed peaks, primarily at high speeds. The mediolateral ground reaction force component (Fx) allways shows a laterally directed peak at the instant of foot strike (i.e., the action force from the foot is directed medially), after which a medially directed reaction force generally follows, during a larger part of the support phase.

In running, the anteroposterior force signal component shows the same braking and propulsive pattern as in gait. In both subjects, typical representatives of their groups, braking anteroposterior force menifests double peak value. Medioateral force is a relatively complex waveform. Forefoot strikers allways manifest an initial, medially directed reaction force peak, while rearfoot strikers mostly manifest a typical laterally directed peak value. In this respect, the second phenomenon is similar to walking, in which foot-strike always occurs with the heel first. There is a striking difference in the timing of anteroposterior and mediolateral force peaks between these two individuals. The first and second anteroposterior and mediolateral force peaks coincide at all speeds among forefoot strikers, but are clearly separated in time among the rearfoot strikers. It is clear that, under the two conditions, the impact of the foot will differ significantly, but Nilsson did not establish corresponding kinematic correlates in his research.

In gait, as well as in running, ground reaction force signals reflect an increase in the movement speed through an ever-larger increase in peak values and a shortening of signal duration. The measurement signals of this kind in healthy subjects are important for the interpretation of signals obtained when measuring pathologies. In healthy locomotion, measurement data of normal running may be an important reference for analysis of running technique, the influence of athletic footwear etc. Significant values of transversal force components can therefore indicate inefficient propulsion, while areas below the courves measured can point to energy efficiency in walking.

Besides information gained from interpretation of force component signals, there are also information in the temporal curve of path of center of pressure, as well as in values of moment of force.

Finally, it is worth to emphasize again that it is understood that force signal waveforms such as these ought to be interpreted in combination with other (biomechanical) indices of movement, of kinematic and/or myoelectric (EMG) nature, if available.

5.3.2 *Pedobarography*

In addition to inverse dynamics approach in human locomotion biomechanics, and with refined biomechanical modeling, the attention is focused to the foot. With biomechanical model thus becoming more realistic, as the contact between the body and the ground happens via a contacting surface (of variable area during support), a measurement technique known as pedobarography has established itself. Using in the beginnings various physical transduction principles to detect and convert applied pressure into electrical signals, today a couple of transducer technologies have profiled themselves in applications producing a sensitive, accurate but at the same time robust enough measurement instrument, handling adequately a spectrum of input pressures encountered during human locomotor acts. These devices's layout may be in a form of insoles to be inserted in footwear, or in a form of a platform. The pedobarography method is elaborated in some detail from a methodological aspect, with special attention for evaluating an attained standardization level in use, in Chap. 7, and through clinical applications, showing a number of findings belonging to orthopedics and sports medicine domains, in Chap. 8.

5.4 Conclusion

To resume, kinematic and kinetic measurement instruments represent standard components to a set of equipment in biomechanics laboratories, sometimes called movement analysis laboratories or gait laboratories. Supplemented with EMG recording equipment, possibly also with a pedobarograph, a portable physiological instrumentation to asses pulmonary and cardiovascular systems' performance, and a video camera, they enable a comprehensive approach to human movement analysis. In a key note lecture at the IFMBE-MEDICON 2001 Symposium in Pula in 2001 (IFMBE goes for International Federation for Medical and Biological Engineering, and MEDICON goes for Mediterranean Conference on Medical and Biological Engineering and Computing), Håkan Lanshammar has given a short overview of the field of measurement and analysis of human motion which—in spite of a considerable time distance—still remains authoritative and timely [30]. It may be a sign of maturation of the field.

References

1. Medved V (2001) Measurement of human locomotion. CRC Press Inc., Boca Raton, FL
2. Selbie WS, Brown MJ (2018) 3D dynamic pose estimation from marker-based optical data. In: Muller B, Wolf SI (eds) Handbook of human motion. Springer, part of Springer Nature, pp 81–100
3. Cutti AG, Ferrari A, Garofalo P et al (2010) Outwalk": a protocol for clinical gait analysis based on inertial and magnetic sensors. Med Biol Eng Comput 48:17–25
4. Cappozzo A (2018) Observing and revealing the hidden structure of the human form in motion throught the centuries (Wartenweiler Memorial Lecture at ISB Symposium in Glasgow 2015). In: Muller B, Wolf SI (eds) Handbook of human motion. Springer, part of Springer Nature, pp 3–15
5. Cereatti A, Della Croce U, Sabatini AM (2018) Three-dimensional human kinematic estimation using magneto-inertial measurement units. In: Muller B, Wolf SI (eds) Handbook of human motion. Springer, part of Springer Nature, pp 221–244
6. Cappozzo A (1985) Experimental techniques, data acquisition and reduction. In: Berme N, Engin AE, Correia da Silva KM (eds) Biomechanics of normal and pathological human articulating joints. Martinuus Nijhoff, The Hague, pp 53–81
7. Jarett MO, Andrews BJ, Paul JP (1976) A television/computer system for the analysis of human locomotion. IERE Golden Jubilee Conference on the Applications of Electronics in Medicine, University of Southampton, England, April 6–8, 1976. IERE Conference Proceedings, vol 34, pp 357–370
8. Ferrigno G, Pedotti A (1985) ELITE: a digital dedicated hardware system for movement analysis via real time TV signal processing. IEEE Trans Biomed Eng 32:943–950
9. Abdel-Aziz YI, Karara HM (1971) Direct linear transformation from comparator coordinates into object-space coordinates in close-range photogrammetry. In Proceedings of the ASP/UI Symposium „Close-Range Photogrammetry“ Urbana, Il, American Society of Photogrammetry, Falls Church, VA, pp 1–18
10. Monografija - Sveučilište u Zagrebu Kineziološki fakultet 1959–2009 (2009) Zagreb: Sveučilište u Zagrebu Kineziološki fakultet. 435 str.
11. Kondrič M, Medved V, Baca A, Kasović M, Furjan-Mandić G, Slatinšek U (2009) Kinematic analysis of top spin stroke with balls of two different sizes. In: Kondrič M, Filipčić A (eds) Scientific approach in table tennis and tennis in Slovenia. Sport Books Publisher, Toronto, pp 53–60
12. Gruić I Medved V (2018) Computer-supported 3D kinematic patterns capture and analysis in handball. In: Andročec V (ed) Jubilee Annual 2017–2018 of the Croatian Academy of Engineering, Zagreb: Croatian Academy of Engineering, 2018, pp 217–222
13. Pažin K, Bolčević F, Gruić I (2016) Kinematička analiza tehnike u rukometu. U: Jukić I, Gregov C, Šalaj S, Milanović L, Wertheimer V, Knjaz D (ur) 14. međunarodna konferencija Kondicijska priprema sportaša, 26. i 27. veljače 2016. Kineziološki fakultet Sveučilišta u Zagrebu. Udruga kondicijskih trenera Hrvatske, pp 63–67
14. Woltring HJ (1987) Data acquisition and processing in functional movement analysis. Minerva Orthop Traumatol 38:703–716
15. Lanshammar H (1982) On practical evaluation of differentiation techniques for human gait analysis. J Biomech 15(2):99–105
16. Lanshammar H (1982) On precision limits for derivatives numerically calculated from noisy data. J Biomech 15(6):459–470

17. Wood GA (1982) Data smoothing and differentiation procedureds in biomechanics. Exer Sport Sci Rev 10:308–362
18. Woltring HJ (1985) On optimal smoothing and derivative estimation from noisy displacement data in biomechanics. Hum Mov Sci 4:229–245
19. Cappozzo A, Della Croce U, Leardini A (2005) Human movement analysis using stereophotogrammetry: part 1: theoretical background. Gait Posture 21(2):186–196
20. Chiari L, Della Croce U, Leardini A, Cappozzo A (2005) Human movement analysis using stereophotogrammetry: Ppart 2: instrumental errors. Gait Posture 21(2):197–211
21. Leardini A, Chiari L, Della Croce U, Cappozzo A (2005) Human movement analysis using stereophotogrammetry: Part 3. Soft tissue artifact assessment and compensation. Gait Posture 21(2):212–225
22. Della Croce U, Leardini A, Chiari L, Cappozzo A (2005) Human movement analysis using stereophotogrammetry: Part 4: assessment of anatomical landmark misplacement and its effects on joint kinematics. Gait Posture 21(2):226–237
23. Peters A, Galna B, Sangeux M, Morris M, Baker R (2010) Quantification of soft tissue artifact in lowerlimb human motion analysis: a systematic review. Gait Posture 31:1–8
24. Camomilla V, Dumas R, Cappozzo A (2017) Human movement analysis: the soft tissue artefact issue. J Biomech 62:1–172
25. Camomilla V, Cappozzo A, Verrozzi G (2018) Three-dimensional reconstruction of the human skeleton in motion. In: Muller B, Wolf SI (eds) Handbook of human motion. Springer, part of Springer Nature, pp 17–45
26. Muller B, Wolf SI (eds) (2018) Handbook of human motion. Springer, part of Springer Nature
27. Pennestri E, Valentini P (2009) Dual quaternions as a tool for rigid body motion analysis: A tutorial with an application to biomechanics. In: Arczewski K, Fraczek J, Wojtyra M (eds.) Multibody Dynamics 2009, ECCOMAS Thematic Conference (also published as: Pennestri E, Valentini P (2010) Dual quaternions as a tool for rigid body motion analysis: A tutorial with an application to biomechanics. Arch Mech Engin 57(2):187–205
28. Barnes SZ, Berme N (1995) Gait analysis: Theory and application. In: Craik RL, Oatis CA (eds) C.V. Mosby, St. Louis, MO, pp 239–251
29. Nilsson J (1985) On the adaptation to speed and mode of progression in human locomotion. Dissertation. Karolinska Institute, Stockholm, Sweden
30. Lanshammar H (2001) Measurement and analysis of human motion. An Uppsala perspective. In: Magjarević R, Tonković S, Bilas V, Lacković I (eds.) IFMBE Proceedings MEDICON 2001, 12–15 June, Pula, Croatia, pp 9–12

Chapter 6
The Principles of 3D Photogrammetry Systems Used in Human Motion Capture and Postural Assessment

Tomislav Pribanić

Abstract There are many different systems capable of objectively measuring and analyzing various motion parameters and signals. The focus of this chapter is on describing the measurements of kinematic parameters, e.g. spatial position, velocity and acceleration. To this end we introduce a photogrammetric approach where spatial three-dimensional (3D) measurements are derived from 2D camera images. The principle how a camera models 3D world into 2D image is described with the camera model parameters, and the computation of parameters through the camera calibration procedure is presented. Since 3D kinematic systems operate on a stereo principle, it requires imaging of the scene with minimum of two cameras and their spatial relationship is given in the form of epipolar geometry. We further turn our focus to the usage of 3D kinematic systems which is frequently defined by certain measurement protocol, tailored to analyzing a specific movement. Several popular possibilities where 3D kinematic system is jointly used with other types of devices are pointed out. Additionally, an inverse dynamics approach is briefly described and explained, allowing a derivation of internal kinetic data (joint forces and torques) combining an output of a 3D kinematic system with inertial body segment parameters. At the very end, several topics related to the research challenges and the anticipated development of 3D reconstruction systems are presented.

6.1 Introduction

A considerable interest in studying human motion has been present ever since times of Greek philosopher Aristotle, and later on, Italian polymath Leonardo da Vinci, whose many works confirm that. For example, Leonardo da Vinci was drawing flying birds and trying to understand a comprehensive mechanism of body motion and its structure [1]. Of course all such early attempts in studying motion were basically only of a qualitative nature, due to immature state of available technology. Nowadays in the modern

---

T. Pribanić (✉)
Faculty of Electrical Engineering and Computing, University of Zagreb, Unska 3, 10000 Zagreb, Croatia
e-mail: tomislav.pribanic@fer.hr

V. Medved (ed.), *Measurement and Analysis of Human Locomotion*,
Series in Biomedical Engineering, https://doi.org/10.1007/978-3-030-79685-3_6

era and already for several decades, there are many different systems capable of objectively measuring and analyzing various motion parameters and signals, for instance, kinematic, kinetic, myographic, cardiographic etc. [2]. Not surprisingly, a comprehensive human body postural assessment and/or motion capture may include the simultaneous measurement of all aforementioned parameters. Still, the focus of this chapter is on measuring kinematic parameters. Even considering kinematic parameters alone, there are many types of systems and principles available. Possible technical solutions use electromagnetic sensors [3], acoustic sensors [4], accelerometers [5] or process images acquired with the cameras [6, 7]. The latter approach is basically a photogrammetric approach where spatial three-dimensional (3D) measurements are derived from two-dimensional (2D) data, images (as pointed out in Chap. 5). A credit for adopting an original photogrammetric approach, specifically to designing 3D reconstruction system for biomechanical human motion analyses, is frequently given to pioneer work of Abdel-Aziz and Karara [8] from 1971.

One of the key advantages of 3D photogrammetry based reconstruction systems are: they are relatively easy to operate and they are essentially non-invasive w.r.t. to the subject of interest, therefore, allowing an undisrupted subject's performance during a motion capture. Consequently, these systems are not only used by physicians or physiotherapists, but also by kinesiologists, coaches, people from film industry, designers of sport equipment etc. A wide spread application resulted with such 3D reconstruction systems being in various communities often referred differently, using one or more attributes such as a 3D optoelectronic system, a 3D stereophotogrammetric system, a 3D kinematic system, a 3D motion capture system, and so forth, where each naming variant tries to emphasize one among many key operating features. In this text those names will be used interchangeably. The essential system hardware component is a camera which models and records 3D world in the form of 2D image and the next section will start explaining the principle of camera modeling.

The remainder of this chapter is structured as follows. After an explanation about camera modeling, the computational and procedural part of a camera calibration is explained, providing a concise overview on three different calibration possibilities in practice. Once 3D kinematic system is calibrated, the next section will explain the principles of 3D reconstruction using two or more cameras. A special emphasis will be given explaining the concept of epipolar geometry since it gives a fruitful insight into how 3D world is imaged on a pair of cameras' images and what the relationship between 3D world and cameras' images is. Afterwards it is presented how the actual 3D measurements are, in general, performed. The importance of passive body markers will be explained and the usual workflow when using a 3D optoelectronic system will be shortly commented. Nowadays 3D motion capture systems typically are upgraded with one or more systems measuring or computing various types of signals and data other than kinematic data. The most common upgrades will be mentioned in the Sect. 6.5.4. Inverse dynamics approach, being a central methodological paradigm, is described and explained next. This entire text, as a whole, touches several key issues related to working principles of 3D human motion capture, but it is by no means an exhaustive source of everything related to them. However, throughout the entire text an attempt is made directing a reader towards additional references about many key issues discussed. On the top of that, in the Sect. 6.6 a number of closely related topics and applications are mentioned as well. In the last section the main concluding remarks are expressed.

6.2 Camera Modeling

A camera models 3D world into 2D image by assigning every point from 3D space to its correspondent 2D image point. Algebraically speaking camera projects 3D projective space into 2D projective space using so-called projective matrix **P** (6.1):

$$\mathbf{P} = \begin{bmatrix} p_{11} & p_{12} & p_{13} & p_{14} \\ p_{21} & p_{22} & p_{23} & p_{24} \\ p_{31} & p_{32} & p_{33} & p_{34} \end{bmatrix} \tag{6.1}$$

A matrix **P** describes a general projective camera which transforms homogeneous point representation **X** from space into homogeneous point representation **x** in the image (6.2):

$$\begin{gathered} \mathbf{x} = \begin{bmatrix} x \cdot w & y \cdot w & w \end{bmatrix} \\ \mathbf{X} = \begin{bmatrix} X & Y & Z & W \end{bmatrix} \\ \mathbf{x} = \mathbf{P} \cdot \mathbf{X} \end{gathered} \tag{6.2}$$

where w is a scale factor different from zero, therefore, $x = [\mathrm{x}, \mathrm{y}]$ represents inhomogeneous point representation in the image. Similarly, $X = [\mathrm{X\ Y\ Z}]$ represents inhomogeneous point representation in 3D space. Since matrix P is defined only up to a scale, it has 11 degrees of freedom. There are many camera models which are special cases of a general camera model represented by (6.1). Many of those special cases allow decomposition of matrix elements from (6.1) into intuitive and physically meaningful representation. The most common assumption about the camera modeling points from 3D to 2D, follows the law of central projection under which points are projected along the straight lines and all lines intersect at the common point, i.e. at the center of projection. In general case, the center of projection may lie at infinity. Hence, one refers to such cameras as infinite cameras, as opposed to finite cameras having a finite center of projection. The simplest camera model within finite cameras is a pinhole camera model. It is so simple that it does not even require the presence of camera lenses. In more detail, a pinhole camera is a light proof box having just a tiny aperture (pinhole) through which light from a scene passes through and projects an inverted image on the opposite side of the box. Its image is also known under the name camera obscura image. A finite pinhole camera model serves as a basis for camera modeling in most applications and as such it will be described first.

Figure 6.1 shows the main components of a pinhole camera model. The entire camera lens system can be substituted with the center of projection **C**, known also as a camera (optical) center. In addition, a camera coordinate system (X_C, Y_C, Z_C) is defined where its two axes are parallel with the camera 2D image plane. The third axis represents a camera optical axis (also known as a principal axis) and it is perpendicular to 2D image plane into which all points from 3D space are projected. The intersection **P** of camera optical axis with the image plane is called a principal point. The distance between the camera center **C** and the image plane determines the camera (effective) focal length f. Given the image formation model above, the expression describing a projection of point **X** from a 3D space to **x** in 2D image plane can be derived using the law of similar triangles (6.3):

$$\begin{gathered}\mathbf{x} = \begin{bmatrix} x & y \end{bmatrix} \\ \mathbf{X} = \begin{bmatrix} X & Y & Z \end{bmatrix} \\ x = f \cdot \frac{X}{Z} \\ y = f \cdot \frac{Y}{Z}\end{gathered} \tag{6.3}$$

The expression (6.3) describes a projection from 3D Euclidean space to 2D Euclidean space. However, by taking the advantage of homogeneous point representations, the same process can be described in a compact form as a linear transformation from 3D projective space to 2D projective space (6.4):

$$\begin{gathered}\mathbf{x} = \begin{bmatrix} x \cdot w & y \cdot w & w \end{bmatrix}^T \quad \mathbf{X} = \begin{bmatrix} X & Y & Z & 1 \end{bmatrix}^T \\ \mathbf{x} = \begin{bmatrix} f & 0 & 0 & 0 \\ 0 & f & 0 & 0 \\ 0 & 0 & 1 & 0 \end{bmatrix} \qquad \mathbf{x} = \mathbf{P} \cdot \mathbf{X}\end{gathered} \tag{6.4}$$

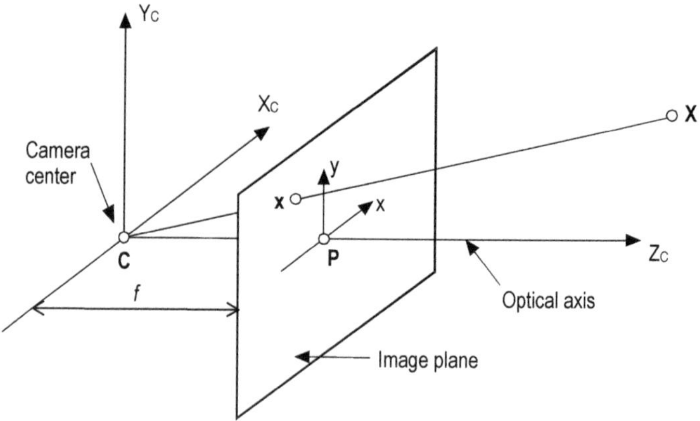

Fig. 6.1 Camera imaging under the central projection assumption

where a matrix $\mathbf{P}_{3\times4}$ is easily recognized as a projective matrix introduced already in (6.1), but this time with its elements given certain values and interpretation. The origin of image coordinate system is typically in one of the image corners, rather than at principal point **P**. Assigning the principal point image coordinates as (p_x, p_y), the expression (6.4) can be further augmented:

$$\mathbf{x} = \begin{bmatrix} f\,0 & p_x\,0 \\ 0\,f & p_y\,0 \\ 0\,0 & 1\,0 \end{bmatrix} \cdot \mathbf{X_{cam}} \qquad \mathbf{x} = \mathbf{P} \cdot \mathbf{X_{cam}}$$
$$\mathbf{P} = \mathbf{K} \cdot [\mathbf{I}|0] \qquad \mathbf{K} = \begin{bmatrix} f & 0 & p_x \\ 0 & f & p_x \\ 0 & 0 & 1 \end{bmatrix} \tag{6.5}$$

where a matrix **K** is referred to as a camera calibration matrix. Additionally, an expression (6.5) explicitly puts label $\mathbf{X_{cam}}$ since 3D spatial coordinates are given in the camera coordinate system (Fig. 6.1). However, in practice 3D spatial coordinates can be expressed in an arbitrary world coordinate system (X_W, Y_W, Z_W). Consequently, it is necessary first to transform a 3D point from some world coordinate system to the camera coordinate system. This transformation is achievable with the rotation matrix **R** and translation vector **t** (Fig. 6.2). The former describes three rotations of one system coordinate axes and the latter describes a translation of one coordinate system origin with the respect to another.

In the algebraic form the transformation of a point from one coordinate system to another can be expressed as (6.6):

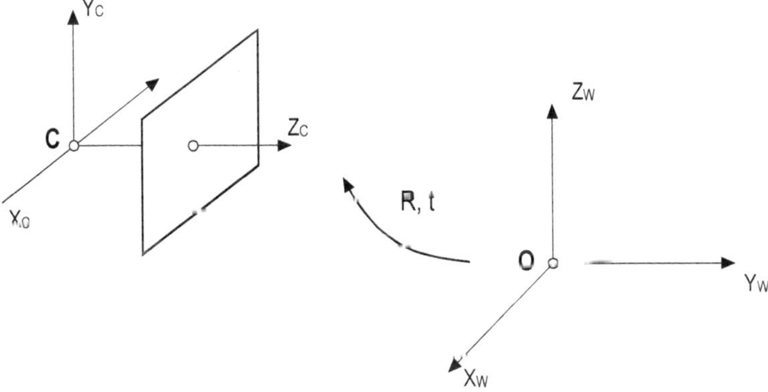

Fig. 6.2 The relationship between an arbitrary world coordinate system and a camera coordinate system using transformation parameters: a rotation matrix **R** and a translation vector **t**

$$\tilde{\mathbf{X}}_{\mathbf{cam}} = \mathbf{R} \cdot \left(\tilde{\mathbf{X}} - \check{\mathbf{C}}\right) = \mathbf{R} \cdot \tilde{\mathbf{X}} + \tilde{\mathbf{t}}$$
$$\mathbf{X}_{\mathbf{cam}} = \begin{bmatrix} \mathbf{R} & -\mathbf{R} \cdot \check{\mathbf{C}} \\ 0 & 1 \end{bmatrix} \cdot \begin{bmatrix} X \\ Y \\ Z \\ 1 \end{bmatrix} = \begin{bmatrix} \mathbf{R} & -\mathbf{R} \cdot \check{\mathbf{C}} \\ 0 & 1 \end{bmatrix} \cdot \mathbf{X} \tag{6.6}$$

where label ~ signifies the usage of inhomogeneous point representation and we recall that **R** is a rotation matrix, **t** is a translation vector and **C** is 3D position of camera optical center in the world coordinate system. Combining (6.5) and (6.6) provides the ultimate expression describing imaging of a point **X**, with a pinhole camera, from 3D space to its position **x** in 2D image plane (6.7):

$$\mathbf{x} = \mathbf{K} \cdot \mathbf{R} \cdot \left[\mathbf{I}| - \check{\mathbf{C}}\right] \cdot \mathbf{X} \; \mathbf{x} = \mathbf{K} \cdot [\mathbf{R}|\mathbf{t}] \cdot \mathbf{X}$$
$$\mathbf{P} = \mathbf{K} \cdot [\mathbf{R}|\mathbf{t}] \tag{6.7}$$

The entire projection is neatly encapsulated in the projective matrix **P**. Unlike in general case (6.1), where a projective matrix **P** could have 11 degrees of freedom, in the case of pinhole camera model a matrix **P** has 9 degrees of freedom. Three unknowns represent elements of a camera calibration matrix **K** (6.5): focal length f and principal point coordinates (p_x, p_y). The next three unknowns are three rotation angles of rotation matrix **R,** and finally, the last three unknowns are components of a translation vector **t**, i.e. a position of camera center **C** (Fig. 6.2). Elements of a camera calibration matrix **K** are frequently called camera *internal* parameters since they do not depend on the camera orientation and position in 3D space. On the other hand, three rotation angles of rotation matrix **R** and three components of a translation vector **t** do determine camera's orientation and position in 3D space with respect to some world coordinate system. Hence, they are called *external* camera parameters.

The above expressions assumed an infinite precision in terms of expressing the camera image coordinates in the image plane. Nevertheless, in practice most cameras have some type of an electronic sensor (usually CCD or CMOS, as mentioned in Chap. 5) which assumes a role of an image plane. An electronic sensor is represented by a finite number of pixel elements along horizontal and vertical axes. Thus, a resolution of an image plane is not infinite and image coordinates eventually are expressed as certain number of pixels. This introduces a slight change in the above expression for a camera calibration matrix **K**. Let factors m_x and m_y represent a number of pixel elements per unit measure of a sensor's width and height, respectively. Now camera calibration matrix **K** describes a conversion to a pixel finite coordinates as:

$$\mathbf{K} = \begin{bmatrix} \alpha_x & 0 & p_x \\ 0 & \alpha_y & p_x \\ 0 & 0 & 1 \end{bmatrix} \; \alpha_x = f \cdot m_x \; \alpha_y = f \cdot m_y \tag{6.8}$$

where a focal length f originally expressed in metric units is now replaced with parameters α_x and α_y. Parameters α_x and α_y express a focal length in the number of pixel elements (similarly, p_x and p_y are now expressed in pixels too), along the respective image axis. In the case of square pixels and an ideal noise free case the following should hold $\alpha_x = \alpha_y$. Another factor that is occasionally included in a camera calibration matrix **K**, is a skew parameter s (6.9):

$$\mathbf{K} = \begin{bmatrix} \alpha_x & s & p_x \\ 0 & \alpha_y & p_x \\ 0 & 0 & 1 \end{bmatrix} \tag{6.9}$$

A parameter s should compensate the effect of image axes not being mutually perfectly perpendicular. Still, nowadays most of newer cameras have a negligible skew parameter s and, consequently, it is not considered often.

This completes the most basic and typically used parameters during a camera modeling. The next chapter will address the problem of the actual computation of camera parameters through a process called camera calibration.

6.3 Camera Calibration

The task of camera calibration is computing the elements of projective matrix **P**. In turn, once we know for at least two cameras their projective matrices $\mathbf{P_1}$ and $\mathbf{P_2}$ along with their respective image coordinates $\mathbf{x_1}$ and $\mathbf{x_2}$ of some unknown point **X** in 3D space, it is possible to set a system of equations in order to retrieve 3D position of a point **X**.

6.3.1 Computation of a Camera Projective matrix—Minimizing the Algebraic Error

We recall that a 3D point transformation **X** from 3D space to 2D camera image **x** is neatly encapsulated using camera projective matrix $\mathbf{P_{3\times4}}$ by the following expression (6.10):

$$\mathbf{x} = \mathbf{P} \cdot \mathbf{X} \tag{6.10}$$

The simplest calibration procedure requires knowing positions of a sufficient number of points in 3D space (aka calibration points) and the identification of their image coordinates. It allows reformulation of expression (6.10) and writing the following homogeneous system of equations:

$$\mathbf{A} \cdot \mathbf{p} = 0 \tag{6.11}$$

where **p** is a vector containing the unknown elements of projective matrix **P** and evidently a vector **p** is a nontrivial nullspace vector of matrix **A**. Matrix **A** has dimension $2n \times 12$ where n is a number of calibration points utilized during a calibration. Equivalently, every calibration point provides two constraints (equations) about the elements of vector **p**. As it was already pointed out, the matrix **P** (6.1) in the general case has 11 degrees of freedom. Consequently the minimum of 6 calibration points are needed to compute its elements. In practice, in order to cope with the various sources of uncertainty, the number of calibration points is much larger than six. In fact due to the noise it is almost certain that no subset of 6 equations out of $2n$ total will perfectly satisfy expression (6.11). Hence, there is no unique solution for the elements of **p**. To remedy the problem a solution is searched that will minimize a so-called algebraic error:

$$\|\mathbf{A} \cdot \mathbf{p}\| \tag{6.12}$$

In order to avoid a trivial solution where all elements of vector **p** will be zero, during a computation of **p** usually an additional constraint is imposed stating that **p** vector norm should be 1. It can be shown that singular value decomposition (SVD) of a matrix **A** allows computing vector **p** satisfying above conditions. More complete explanation about SVD can be typically found in sources about linear algebra. Here we state in brief that SVD of **A** provides the decomposition in three matrices $\mathbf{A} = \mathbf{UDV}^{\mathrm{T}}$ where a solution for **p** is the column of matrix **V** corresponding to the smallest singular value on the diagonal of matrix **D**. The provided solution for **p** is a least squares solution.

On the more practical note and in terms of computation of elements of **p**, one of the natural questions is whether the particular choice of scale and origin of the image coordinate system, within which calibration points **x**'s are expressed, has any effect on the robustness of the final solution? It can be shown that the appropriate image point normalization cancels out the effect about the choice of a coordinate system origin and a scale [9]. Such normalization involves first computing the centroid of image calibration points and setting that centroid as the origin of image coordinate system. Next, scaling along image coordinate axis is carried out so that the average distance of a point from the image origin is the square root of two. A similar normalization of calibration point values **X**'s in 3D space can be carried out as well. 3D space point normalization seems to show some positive effects in the cases where a depth of calibration points in 3D is rather small. In practice, typically a normalization of calibration points in image coordinate system is performed. Nevertheless, for completeness, the expression below shows how to acquire camera projection matrix **P** from camera projective matrix **Pn. Pn** is computed using a normalization of calibration points in both image and 3D space using matrices **T** and **U**, respectively (6.13):

$$\begin{array}{ccc} \mathbf{x} = \mathbf{P} \cdot \mathbf{X} & \mathbf{xn} = \mathbf{T} \cdot \mathbf{x} & \mathbf{Xn} = \mathbf{U} \cdot \mathbf{X} \\ \mathbf{xn} = \mathbf{T} \cdot \mathbf{P} \cdot U^{-1} \cdot \mathbf{Xn} & \mathbf{Pn} = \mathbf{T} \cdot \mathbf{P} \cdot U^{-1} & \mathbf{P} = T^{-1} \cdot \mathbf{Pn} \cdot \mathbf{U} \end{array} \quad (6.13)$$

6.3.2 Computation of a Camera Projective Matrix—Minimizing the Geometric Error

Minimizing the algebraic error has an advantage of providing a straightforward linear solution for a calibration. On the other hand, a disadvantage is that we are not minimizing any actual quantity which has some physical (intuitive) interpretation. Therefore, an alternative approach is to minimize the geometric error of Euclidean distance between the detected calibration image coordinates and the one estimated by the camera model:

$$\sum_{i=1}^{N} d(\mathbf{x_i}, \mathbf{P} \cdot \mathbf{X_i}) \quad (6.14)$$

where N is a number of calibration points, d(**x, P·X**) is a measured Euclidean distance between detected image coordinate values **x** and ones provided by the camera model **P·X**. Minimizing an expression (6.14) becomes a non-linear minimization problem. Moreover, a projective matrix **P** is often presented using parameters which have some interpretation, such as in the form of internal and external camera parameters (6.7). Thus, the standard approach is first to initially compute the elements of matrix **P** (6.1) using some linear solution through the minimization of algebraic error (6.12). Then, the elements of matrix **P** can be decomposed [9] into camera model parameters such as (6.7), serving as an initial solution to be eventually refined by non-linear optimization of (6.14). Comparing to the minimization of algebraic error (6.12), minimizing the geometric error (6.14) should yield, in principle, a more accurate calibration solution and ultimately 3D reconstruction accuracy.

It should be noted that minimization primarily takes over camera model parameters, i.e. elements of projective matrix **P**. Still it is possible to let fluctuate (minimize) in (6.14) 3D positions of calibration points **X** too. Minimizing expressions of such form is frequently called Bundle adjustment [10] and it is routinely used in various other applications such as structure from motion problems [11]. However, 3D positions of calibration points are typically given with a very high degree of accuracy. Consequently, usually in practice the more accurate calibration results are acquired if 3D positions of calibration points **X** are kept fixed in (6.14). Given some initial parameter values and then finding a convergent set of solution through non-linear minimization of some objective function is not a trivial task. Out of many existing algorithms and for the purpose of camera calibration minimizing expression like (6.14), Levenberg–Marquardt algorithm [12] has proven to be a standard method of choice.

Finally, we note that camera modeling and calibration discussed so far did not consider the non-linear image distortions. These distortions are mainly due to the imperfection of camera lenses. They are compensated by introducing an additional polynomial expression relating ideal undistorted image coordinate positions with the actually (distorted) detected image coordinate positions. The mentioned polynomial expression usually describes an effect of two common types of non-linear distortions: radial and tangential [13]. Introducing additional parameters for non-linear distortions makes optimization of (6.14) even more complex. Fortunately at least for the initialization part of non-linear optimization, the actual values for non-linear distortions parameters are usually rather small and they can be safely initialized as zero values.

The next question to answer is how to come up with a certain number of calibration points in practice, i.e. how the calibration procedure is in fact performed.

6.3.3 Calibration Tools

In practice the source of calibration points is provided by some form of calibration object. It is an object having on itself clearly marked a certain number of calibration points whose 3D spatial positions $\mathbf{X}$ are known with a very high degree of accuracy. The calibration points should be clearly marked allowing relatively easy and automatic detection of calibration points in the image. Therefore, imaging calibration object and detection of calibration points' image coordinates $\mathbf{x}$, readily provides constraints to carry out a computational part of a camera calibration, as explained in the previous sections. The highest 3D reconstruction accuracy is expected within a volume occupied by calibration points. Thus, a general pre-requisite of all calibration methods is to cover with the calibration points the expected volume of interest. In addition, on rare occasions 3D scene itself may have a sufficient number of points with known locations usable as calibration points, but in most cases a particularly designed calibration structure has to be used. The size and shape of calibration object can quite vary, however, even more importantly the dimensionality of object itself. There are three common types: 3D calibration cage, 2D calibration plane and 1D calibration wand.

3D calibration cages have been used traditionally in the past. In that case a physical structure occupies all three spatial dimensions. There have been variants in form of cuboids or simply three orthogonal planes put together (Fig. 6.3). In either case, 3D calibration cage allowed imaging of large number spatially arranged calibration points, occupying the desired volume of calibration, which in turn typically did not require taking more than one image during a calibration. The disadvantage is a relatively large cost of fabrication, cumbersome storage and manipulation, and lack of flexibility to adapting for different sizes of calibration volume.

Alternatively, a calibration structure can be in the form of 2D calibration plane. Obviously a manipulation with something in the 2D form is simpler than a 3D structure. On the other hand 2D plane calibration requires usually taking more than

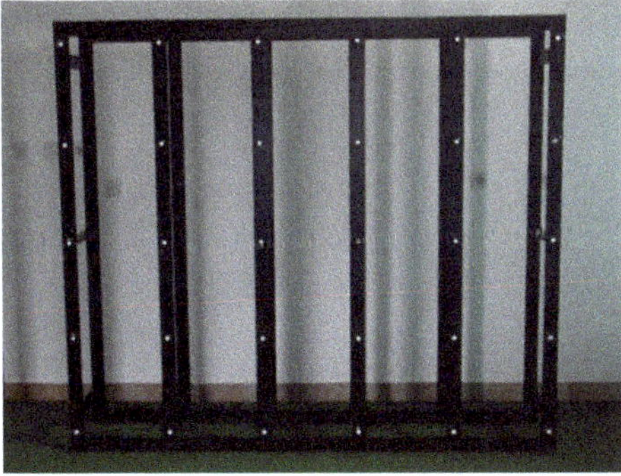

Fig. 6.3 3D calibration structures was originally manufactured as a cuboid [14, 15] which is later relaxed in the form of three orthogonal planes [16] (*source* Biomechanics Laboratory Polyclinic Peharec [17]. With permission)

one image, on various positions and orientations in space w.r.t. to camera, therefore, covering the desired size of calibration volume. On 2D calibration plane either a checkerboard or grid of circles is printed out. During processing of acquired images, centers of circles or square corners are detected. Combined with the known physical size and arrangement of square corners (circle centers), calibration data is provided to carry out the computational part of camera calibration (Fig. 6.4). Camera calibration using a 2D calibration plane is nowadays arguably the most popular approach, used in

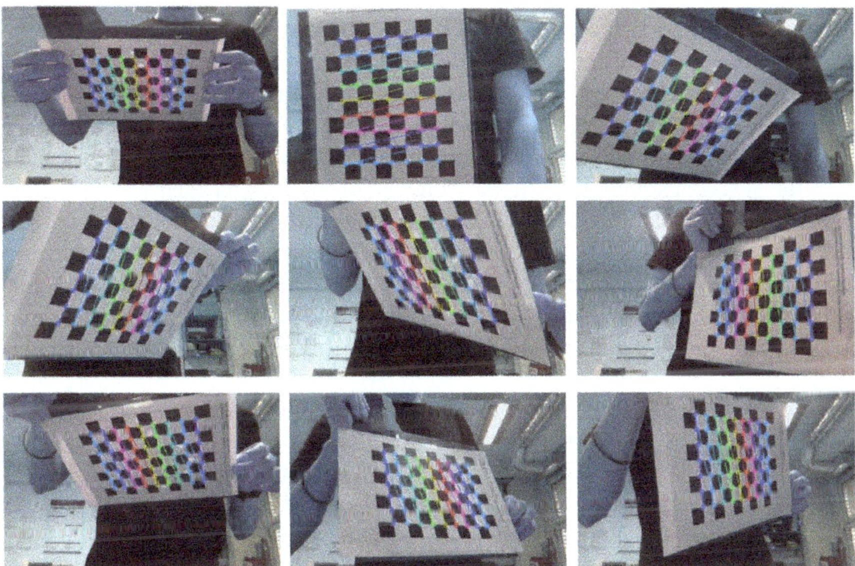

Fig. 6.4 Detected calibration points (corners) on the 2D calibration plane imaged on various positions and orientations w.r.t. camera

a) calibration wand

b) orthogonal triad of wands used for camera parameter initialization

Fig. 6.5 A calibration tools using wands (*source* Biomechanics Laboratory Polyclinic Peharec [17]. With permission)

numerous computer vision applications. Not surprisingly many open source libraries offer software packages for ready to use camera calibration procedure using 2D calibration plane [18]. Nevertheless, in the case of 3D kinematic systems specifically, it is even more popular to use simply 1D calibration structure, a calibration wand.

A calibration wand is a wand having just a few markers (calibration points) on itself (Fig. 6.5a). The known lengths between those markers are the actual ground true quantity that is used during a computational part of the calibration. The calibration procedure itself requires walking and waving with a wand in order to image a wand on as many locations as possible throughout the calibration volume. Walking and waving around with a wand is sometimes in jargon called a wand dance. For a calibration volume size of few cubic meters and 3D kinematic system with several cameras, a wand dance usually lasts 45–60 s. Furthermore, the complete procedure usually involves also an imaging of wand's triad, the purpose of which is to help initialization of certain camera parameters (Fig. 6.5b). The triad of wands is normally used to determine the origin and axes of 3D coordinate system that will be subsequently used for 3D reconstruction. Different numbers of markers on each wand of a triad can be utilized for automatic system determination of which wand defines which axis. In fact, the typical task of wand dance is refining the initially computed calibration parameters from an image of wand triads. Namely, due to various sources of errors and possible calibration volume much larger than triad of wands, it is hard to compute accurate calibration parameters only from a relatively few calibration points on the triad. On rare occasions the user is asked to repeat the calibration procedure which may be caused by one or more cameras not automatically recognizing calibration points of the wand triad (Fig. 6.5b) or poorly carried out wand dance part which did not provide a convergent set of solutions for camera parameters. In addition, commercial software often has a different measure of calibration quality allowing the user re-doing the calibration if not satisfied with the calibration quality evaluation score. As a rule, a wand calibration is generally regarded as the most user-friendly procedure for calibration of 3D kinematic systems. The storage and manipulation of 1D wand is obviously more convenient than using 2D calibration plane or even 3D calibration cage. At the same time 1D wand offers quick and easy system recalibration to various calibration volume sizes.

The aforementioned theory behind a camera modeling of points from 3D space to 2D image plane is basically invariant to the type of calibration structure used. However, it is useful to note that the previously shown expression (6.14) for camera calibration assumes 3D position of calibration points in space. Consequently, in the computational sense it is immediately applicable for calibration using 3D calibration cage which readily provides 3D positions of calibration points. On the other hand, 2D calibration plane provides calibration points only within a plane and a wand calibration essentially provide only length between calibration points along a wand. Nevertheless, both 2D plane calibration and 1D wand calibration can be eventually expressed using the same Eq. (6.14). The exact details and derivation include more involved concepts from a projective geometry and are out of the scope of this material. However, an interested reader can find out more technical and algorithmic details about 2D plane calibration and 1D wand calibration in [19] and [20, 21], respectively.

6.4 The 3D Reconstruction

The crux of the principle behind 3D photogrammetry based reconstruction systems is the triangulation. The triangulation is defined as the process of determining a point in 3D space given its projections onto two, or more, images. In order to solve this problem, it is necessary first to know the parameters of the camera projection function from 3D to 2D for the cameras involved. In the simplest case the camera projection matrices are acquired through the process of calibration, as explained before. The triangulation is essentially an implementation of stereo 3D reconstruction principle. Acknowledging that, it is interesting to note that human visual system perceives depth in a very much similar fashion. Two slightly different images are projected to the retinas of the eyes. The mostly horizontal differences of correspondence points between images are referred as horizontal disparities or, more generally, binocular disparities. Disparities are processed in the visual cortex of the brain to yield depth perception. Essentially, the same idea is recreated when designing a photogrammetric (computer vision) algorithm to process images acquired by two (or more) cameras, instead of human eyes. Actually, a depth from stereo is not the only cue to perceive depth. Both human visual system and certain computer vision system developed are capable of estimating depth using single image only. These are known as monocular cues and some of the most popular ones are shape from shading [22], shape from photometric stereo [23] and shape from texture [24]. However, the binocular cue, shape from stereo, is generally more robust and accurate than any of monocular cue principles.

3D kinematic reconstruction system usually consists of multiple cameras. But the principle behind triangulation is essentially defined with the spatial relationship between two (calibrated) cameras and their images. That relationship is defined by the so called epipolar geometry [25] which will be presented next.

6.4.1 Epipolar Geometry

Figure 6.6 visualizes the geometric relationship between two cameras, known as epipolar geometry. Consider a point **X** in space and its projections $\mathbf{x_1}$ and $\mathbf{x_2}$ in images of cameras $\mathbf{C_1}$ and $\mathbf{C_2}$, respectively. Based on the principle of a pinhole camera, points **X** and $\mathbf{x_1}$ with $\mathbf{C_1}$, and points **X** and $\mathbf{x_2}$ with $\mathbf{C_2}$, form the respective lines. Hence, all mentioned points are coplanar and they form a plane called epipolar plane. The line connecting camera centers $\mathbf{C_1}$ and $\mathbf{C_2}$ is called a baseline. The intersections of a baseline with the first camera image and second camera image are at points $\mathbf{e_1}$ and $\mathbf{e_2}$, respectively. $\mathbf{e_1}$ and $\mathbf{e_2}$ are known as epipoles. Considering single image only, for example camera $\mathbf{C_1}$ and projection of **X** in image point $\mathbf{x_1}$, it can be noticed that point **X** can be anywhere along the spatial line projecting into image point $\mathbf{x_1}$ (Fig. 6.6b). Obviously, we need an additional constraint such as one more image point $\mathbf{x_2}$ of another camera $\mathbf{C_2}$, to uniquely determine spatial position of point **X**. Determining that another correspondent image point $\mathbf{x_2}$ basically concludes the description and usage of triangulation principle, thus allowing 3D reconstruction of **X**.

From a computational perspective, having at least two calibrated cameras represented by their projective matrices $\mathbf{P_1}$ and $\mathbf{P_2}$, and acquiring their respective image coordinates $\mathbf{x_1}$ and $\mathbf{x_2}$, allows to reformulate the expression (6.10) and form the homogeneous system of equations where the unknowns are elements of spatial position **X**. It is useful to recall that a previous similar reformulation of the expression (6.10) allowed finding the initial values of camera parameters (6.11), computed through minimization of algebraic error (6.12). Actually the principle of triangulating 3D points from the calibrated cameras and their correspondent image points can be simply stated based only on the last few sentences mentioned above, i.e. without even mentioning the concept of epipolar geometry. However, it is still useful to study

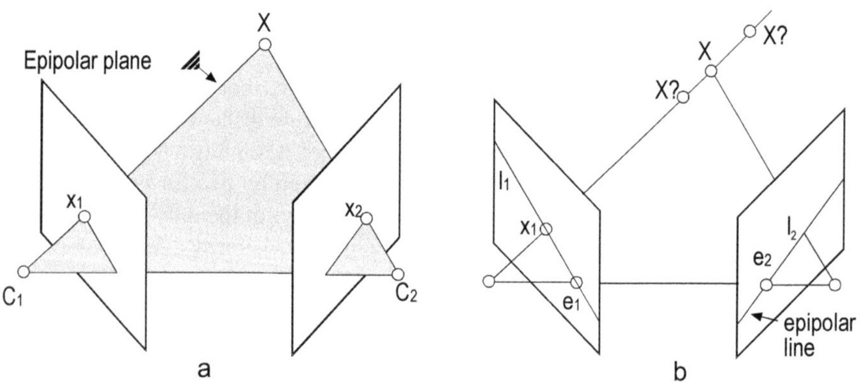

Fig. 6.6 Geometry of two cameras represented by their center of projections $\mathbf{C_1}$ and $\mathbf{C_2}$ and their images. **a** Cameras' centers and spatial point X all lie in the so-called epipolar plane **b** image point **x**, on a camera image $\mathbf{C_1}$, is backprojected in space as spatial line which is further projected as epipolar line l_2 in the image of camera $\mathbf{C_2}$

the principle of 3D reconstruction exactly thorough the concept of epipolar geometry since it takes analyses even one step further by analyzing where to look for a correspondent image point $\mathbf{x_2}$ in another image of camera $\mathbf{C_2}$.

It is clear from a construction of epipolar plane (Fig. 6.6) that a line connecting camera center $\mathbf{C_1}$, image point $\mathbf{x_1}$ and spatial point $\mathbf{X}$, is imaged in the second camera image as a line $\mathbf{l_2}$ connecting epipole $\mathbf{e_2}$ and the correspondent image point $\mathbf{x_2}$. A line $\mathbf{l_2}$ is called epipolar line. The common feature of all epipolar lines in some image is that they all pass through the epipole. The similar conclusion can be drawn about the existence of epipolar line $\mathbf{l_1}$ on which somewhere lies a point $\mathbf{x_1}$, a projection of $\mathbf{X}$ in the first camera image. Hence, the key insight of epipolar geometry is expressed through the epipolar constraint: given some image point $\mathbf{x}$ in one camera, the correspondent image point on other camera's image lies on the epipolar line $\mathbf{l}$. This means that given an image point $\mathbf{x_1}$ in the first camera image, a search for a correspondent $\mathbf{x_2}$ in the second camera image can be narrowed down to a single (epipolar) line only. Such search domain decrease, from 2D image to only 1D line, is a huge relaxation of a problem and it is heavily exploited in practical realizations of numerous computer vision algorithms, not just 3D reconstruction. To that end, the entire epipolar geometry relation is algebraically encapsulated in the form of fundamental matrix [24] which will be derived next.

Given some point $\mathbf{x_1}$ in the image of first camera, represented by the projective matrix $\mathbf{P_1}$ and the center of projection $\mathbf{C_1}$, we can write an equation of a line projecting a point $\mathbf{X}$ to its image $\mathbf{x_1}$ (6.15):

$$\mathbf{X}_1(\lambda) = \mathbf{P}_1^+ \cdot \mathbf{x}_1 + \lambda \cdot \mathbf{C}_1 \tag{6.15}$$

where a matrix $\mathbf{P_1}^+$ is a so called pseudo-inverse matrix of a projective matrix $\mathbf{P_1}$ defined as follows (6.16):

$$\begin{gathered}\mathbf{P}_1^+ = \mathbf{P}_1^\mathbf{T} \cdot \left(\mathbf{P}_1 \cdot \mathbf{P}_1^\mathbf{T}\right)^{-1} \\ \mathbf{P}_1 \cdot \mathbf{P}_1^+ = \mathbf{I} \\ \mathbf{x}_1 = \mathbf{P}_1 \cdot \mathbf{X} = \mathbf{P}_1 \cdot \mathbf{P}_1^+ \cdot \mathbf{x}_1 = \mathbf{P}_1 \cdot \mathbf{P}_1^\mathbf{T} \cdot \left(\mathbf{P}_1 \cdot \mathbf{P}_1^\mathbf{T}\right)^{-1} \cdot \mathbf{x}_1 = \mathbf{x}_1\end{gathered} \tag{6.16}$$

Points $\mathbf{P_1}^+\cdot\mathbf{x_1}$ and $\mathbf{C_1}$ will be projected in the image plane of a second camera, using its projective matrix $\mathbf{P_2}$, as points $\mathbf{P_2}\cdot(\mathbf{P_1}^+\cdot\mathbf{x_1})$ and $\mathbf{P_2}\cdot\mathbf{C_1}$, respectively. Moreover, both of those projections will lie on the epipolar line $\mathbf{l_2}$:

$$\begin{gathered}\mathbf{l}_2 = (\mathbf{P}_2 \cdot \mathbf{C}_1) \times \left(\mathbf{P}_2 \cdot \mathbf{P}_1^+ \cdot \mathbf{x}_1\right) \\ \mathbf{l}_2 = \mathbf{e}_2 \times \left(\mathbf{P}_2 \cdot \mathbf{P}_1^+ \cdot \mathbf{x}_1\right)\end{gathered} \tag{6.17}$$

where we emphasize again that a projection $\mathbf{P_2}\cdot\mathbf{C_1}$ is epipole $\mathbf{e_2}$. Next it is useful to recall that given some vector $\mathbf{a}$ it is possible to define antisymmetric matrix $[\mathbf{a}]_x$:

$$a = \begin{bmatrix} a_1 & a_2 & a_3 \end{bmatrix}$$
$$[\boldsymbol{a}]_x = \begin{bmatrix} 0 & -a_3 & a_2 \\ a_3 & 0 & -a_1 \\ -a_2 & a_1 & 0 \end{bmatrix} \tag{6.18}$$

where a vector **a** is null space vector of matrix $[\mathbf{a}]_x$. The practical significance of expression (6.18) is to replace vector product in (6.17) with a scalar product:

$$\mathbf{l}_2 = [\mathbf{e}_2]_x \cdot \mathbf{P}_2 \cdot \mathbf{P}_1^+ \cdot \mathbf{x}_1 = \mathbf{F} \cdot \mathbf{x}_1$$
$$\mathbf{F} = [\mathbf{e}_2]_x \cdot \mathbf{P}_2 \cdot \mathbf{P}_1^+ \tag{6.19}$$

where a matrix **F** is called *fundamental matrix*. It is convenient to write matrix **F** explicitly as a part of epipolar constraint where, given some spatial point **X**, and its correspondent image points $\mathbf{x_1}$ and $\mathbf{x_2}$, the following holds:

$$\begin{matrix} \mathbf{l}_2 = \mathbf{F} \cdot \mathbf{x}_1 \\ \boldsymbol{x}_2^T \cdot \mathbf{l}_2 = 0 \end{matrix} \Rightarrow \boldsymbol{x}_2^T \cdot \mathbf{F} \cdot \mathbf{x}_1 = 0 \tag{6.20}$$

Equation (6.20) is main result epipolar geometry describing relationship between two cameras. Assuming that spatial coordinate system coincides with positioned the first camera coordinate system it is possible to show how a fundamental matrix consists of cameras' model internal and external parameters:

$$\mathbf{P}_1 = \mathbf{K}_1 \cdot [\mathbf{I}|0] \; \mathbf{P}_2 = \mathbf{K}_2 \cdot [\mathbf{R}|\mathbf{t}] \;\; \mathbf{P}_1^+ = \begin{bmatrix} \mathbf{K}_1^{-1} \\ 0 \end{bmatrix} \; \mathbf{C}_1 = \begin{bmatrix} 0 \\ 1 \end{bmatrix}$$
$$\mathbf{F} = [\mathbf{e}_2]_x \cdot \mathbf{P}_2 \cdot \mathbf{P}_1^+ = [\mathbf{P}_2 \cdot \mathbf{C}_1]_x \cdot \mathbf{P}_2 \cdot \mathbf{P}_1^+ = [\mathbf{K}_2 \cdot \mathbf{t}]_x \cdot \mathbf{K}_2 \cdot \mathbf{R} \cdot \mathbf{K}_1^{-1} \tag{6.21}$$
$$\mathbf{F} = \mathbf{K}_2^{-\mathbf{T}} \cdot [\mathbf{t}]_x \cdot \mathbf{R} \cdot \mathbf{K}_1^{-1} = \mathbf{K}_2^{\mathbf{T}} \cdot \mathbf{R} \cdot \left[\mathbf{R}^{\mathbf{T}} \cdot \mathbf{t}\right]_x \cdot \mathbf{K}_1^{-1} = \mathbf{K}_2^{-\mathbf{T}} \cdot \mathbf{R} \cdot \mathbf{K}_1^{\mathbf{T}} \cdot \left[\mathbf{K}_1 \cdot \mathbf{R}^{\mathbf{T}} \cdot \mathbf{t}\right]_x$$

where $\mathbf{K_1}$ and $\mathbf{K_2}$ are camera calibration matrices containing the internal parameters of first and second camera, respectively. **R** is a rotation matrix explaining the relative orientation of the second camera with respect to the first one and **t** is a translation vector describing the displacement of a second camera center $\mathbf{C_2}$ with respect to the first camera center $\mathbf{C_1}$, i.e. in this case the origin of the spatial coordinate system. Occasionally, it is useful to express the fundamental matrix using cameras' epipoles $\mathbf{e_1}$ and $\mathbf{e_2}$ (6.22):

$$e_1 = \mathbf{P}_1 \cdot \begin{bmatrix} -\mathbf{R} \cdot \mathbf{t} \\ 1 \end{bmatrix} = \mathbf{K}_1 \cdot \mathbf{R}^{\mathbf{T}} \cdot t \;\; e_2 = \mathbf{P}_2 \cdot \begin{bmatrix} 0 \\ 1 \end{bmatrix} = \mathbf{K}_2 \cdot t$$
$$\mathbf{F} = [\mathbf{e}_2]_x \cdot \mathbf{K}_2 \cdot \mathbf{R} \cdot \mathbf{K}_1^{-1} = \mathbf{K}_2^{-\mathbf{T}} \cdot \mathbf{R} \cdot \mathbf{K}_1^{\mathbf{T}} \cdot [\mathbf{e}_1]_x \tag{6.22}$$

The main characteristics of a fundamental matrix can be summarized in several points:

- For every image point $\mathbf{x_1}$ in the image of the second camera and point $\mathbf{x_2}$ in the image of the second camera, fundamental matrices **F** satisfy the expression (6.20).
- if **F** is a fundamental matrix for a camera pair described by pair of projective matrices ($\mathbf{P_1}$, $\mathbf{P_2}$) then its transpose $\mathbf{F}^{\mathrm{T}}$ is a fundamental matrix for a camera pair described by pair of projective matrices ($\mathbf{P_2}$, $\mathbf{P_1}$) and the following condition holds (6.23):

$$x_1^T \cdot F^T \cdot \mathbf{x}_2 = 0 \tag{6.23}$$

- epipolar line $\mathbf{l_1}$ on which lies an image point $\mathbf{x_1}$, and epipolar line $\mathbf{l_2}$ on which lies a point $\mathbf{x_2}$ can be computed using fundamental matrix **F**, as shown in the expressions (6.24) and (6.25), respectively:

$$\mathbf{l}_1 = \mathbf{F}^{\mathrm{T}} \cdot \mathbf{x}_2 \tag{6.24}$$

$$\mathbf{l}_2 = \mathbf{F} \cdot \mathbf{x}_1 \tag{6.25}$$

- epipol $\mathbf{e_2}$ is a left null space vector of the fundamental matrix **F** (6.26) whereas epipole $\mathbf{e_1}$ is a right null space vector of the fundamental matrix **F** (6.27).

$$\mathbf{e}_2^{\mathrm{T}} \cdot \mathbf{l}_2 = 0\ \mathbf{l}_2 = \mathbf{F} \cdot \mathbf{x}_1\ \mathbf{e}_2^{\mathrm{T}} \cdot \mathbf{F} \cdot \mathbf{x}_1 = 0,\ \forall \mathbf{x}_1 \Rightarrow \mathbf{e}_2^{\mathrm{T}} \cdot \mathbf{F} = 0 \tag{6.26}$$

$$\mathbf{e}_1^{\mathrm{T}} \cdot \mathbf{l}_1 = 0\ \mathbf{l}_1 = \mathbf{F}^{\mathrm{T}} \cdot \mathbf{x}_2\ \mathbf{e}_1^{\mathrm{T}} \cdot \mathbf{F}^{\mathrm{T}} \cdot \mathbf{x}_2 = 0,\ \forall \mathbf{x}_2 \Rightarrow \mathbf{F} \cdot \mathbf{e}_1 = 0 \tag{6.27}$$

- fundamental matrix **F** has dimension 3×3 but it is a homogeneous and singular matrix. Hence, it has actually seven degrees of freedom.

fundamental matrix **F** can be computed either directly from the known camera projective matrices $\mathbf{P_1}$ and $\mathbf{P_2}$ (6.19) or using pairs of correspondent image points $\mathbf{x_1}$ and $\mathbf{x_2}$ and constraints based on the expression (6.20), i.e. (6.23). The latter case is common when there are only point correspondences between cameras available. A standard way to compute **F** using point correspondences is by so called *8 point algorithm* [26]. But in the context of 3D kinematic system and after the camera calibration, the camera projective matrices are typically known and **F** can be computed directly from it.

6.5 The Usage of 3D Kinematic System

The basic prerequisite before any actual usage is to have 3D kinematic system calibrated, using possibly any of camera calibration techniques explained above.

As succinctly presented in the preceding Chap. 5, a 3D kinematic system typically reconstructs on the human body a certain number of points. The actual number can vary from a few points up to several dozen of points. These points represent (in)directly some anatomical landmarks, positioned at body prominences. Traditionally, video-based cameras have been used [27], imaging a subject of interest in a standard video format. Since 3D triangulation requires identification of correspondent image points on at least two cameras, this kind of video based 3D systems typically require a manual digitization of each individual point on a frame by frame basis and throughout the imaged video. Assuming that video sequence lasts at least several seconds and with even a modest camera frame rate of only 25 Hz, evidently a manual digitization can be quite time consuming.

6.5.1 Introduction of Markers

In order to relax the problem of manual digitization of numerous points of interest to be potentially reconstructed, the passive markers have been introduced. The markers being attached to a human body are negligibly invasive. At the same time, automatic tracking of anatomical landmarks throughout the video frames have now been reduced to automatic tracking of distinctive markers which is significantly simpler from the image processing perspective. To make a marker tracking even simpler and more robust, standard video cameras are nowadays often replaced with infrared cameras (IR) (Fig. 6.7). As the name suggests, IR camera is particularly suited to be sensitive for an infrared part of the spectrum (e.g. using IR filter), resulting that surfaces which reflect these wavelengths will be clearly visible in its image, whereas

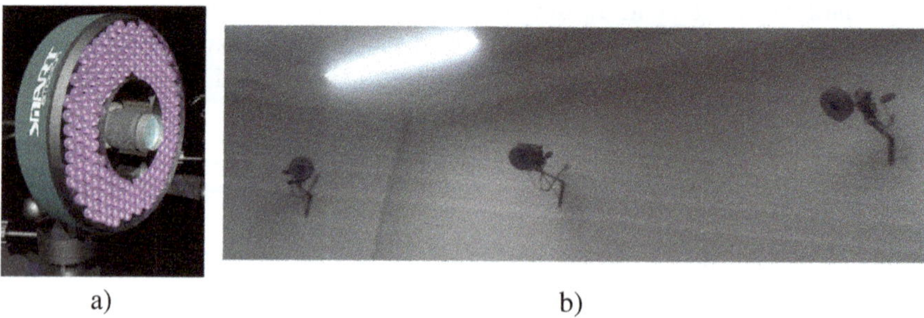

Fig. 6.7 **a** An example of infrared camera, frequently used in 3D motion capture systems. **b** The usual camera installment is on the walls of the biomechanics laboratory (*source* Biomechanics Laboratory Polyclinic Peharec [17]. With permission)

other wavelengths will be nearly to invisible in the image. Therefore in combination with such cameras, passive retro-reflective markers coated with the IR material are used. To further bust performance, a ring of IR led lights are often installed around camera's lens objective (Fig. 6.7). Systems equipped with IR cameras are particularly suited for indoors conditions. On the other hand, for outdoors the impact of sunlight may significantly impair their performance. A commercial 3D optoelectronic system usually consists of 8–12 cameras. Cameras' frame rate starts typically with 50 Hz and with the latest commercial variants can go easily up to several hundreds of Hz.

6.5.2 Measurement Protocols

Instead of just randomized markers positioning on a human body, most human motion analyses assume a certain protocol to be implemented. A protocol defines, among other things, number and placement of markers on human body. Occasionally marker

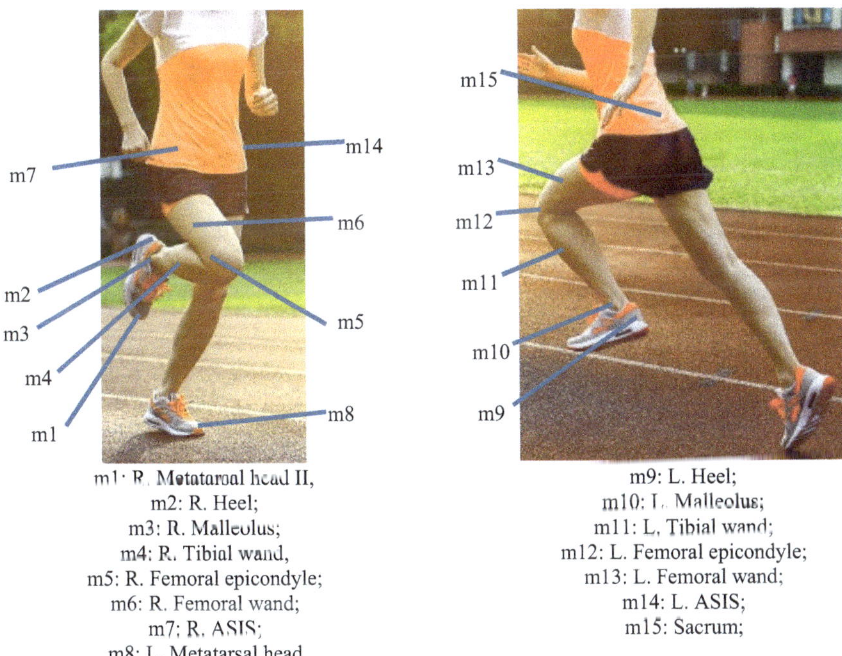

Fig. 6.8 Visualization of the modified Helen Hayes [28] hospital marker placement protocol, as used in [29]

are not directly put on the body but rather put on the wands which are then fastened on the body, in order to increase markers visibility w.r.t. the cameras. Perhaps one of the most popular protocols is Helen Hayes (HH) hospital protocol [28]. Variants of HH have been part of many a commercial software. Figure 6.8 shows one of the HH modifications designed for analyses, as proposed in [29].

The general idea is to minimize the necessary number of markers. It simplifies not only software processing but equally important it decreases time needed to prepare a subject (patient) of interest for imaging. For research purposes that may not sound extra critical. For instance and as a bit of extreme example, the work about parameter values of anthropometric segments in [30] required nearly 90 min to prepare a subject. On the other hand, on a system daily base usage in clinics, the time is of essence due to understandable reasons. In either case, protocols are designed often for some typical type of motion analyses. Next issue is placing the markers on the points on the body which are relatively easily palpated (especially in the case of obese patients) and unobscured during movement. To increase a visibility sometimes marker is not placed directly on the body, but on the small wand (e.g. Fig. 6.8, points for tibial and femoral wands). The solid marker placement is particularly important since markers are sometime considered merely as external anatomical landmarks. In that sense, the ultimate goal is frequently to compute internal landmarks such as joint centers, whose accurate estimate greatly depends on the reliable estimate of markers themselves. Next, during a longer and/or repetitive measurement of rapid movements (Fig. 6.9), a subject is likely to start sweeting. Special adhesive tapes can be used to diminish the chance of marker being displaced or falling off (Fig. 6.10).

Occasionally, the assessment is particularly concentrated around key frames of some movements, rather than around the entire sequence of frames. For example in postural analyses, besides analyzing a standard standing position, the measurement protocol may involve analyses of dynamics of movement (e.g. flexion, extension) (Fig. 6.10).

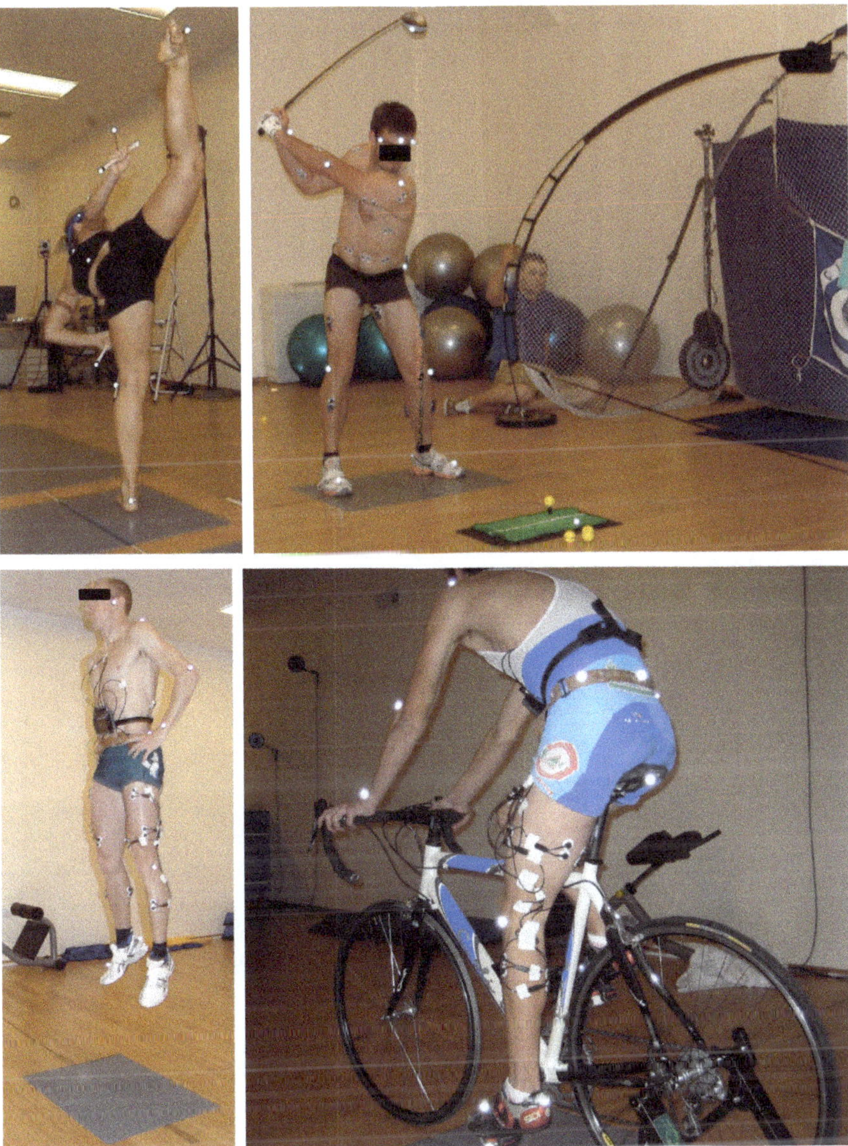

Fig. 6.9 Examples of motion recordings: Frame instances taken during a karate kick performance analysis, golf swing (top row), standing high jump and cycling (bottom row). On the top right and bottom images, EMG measurement is included too. Please refer to Sect. 6.5.4 Going beyond kinematic data for more details about EMG analyses enhancement (*source* Biomechanics Laboratory Polyclinic Peharec [17]. With permission)

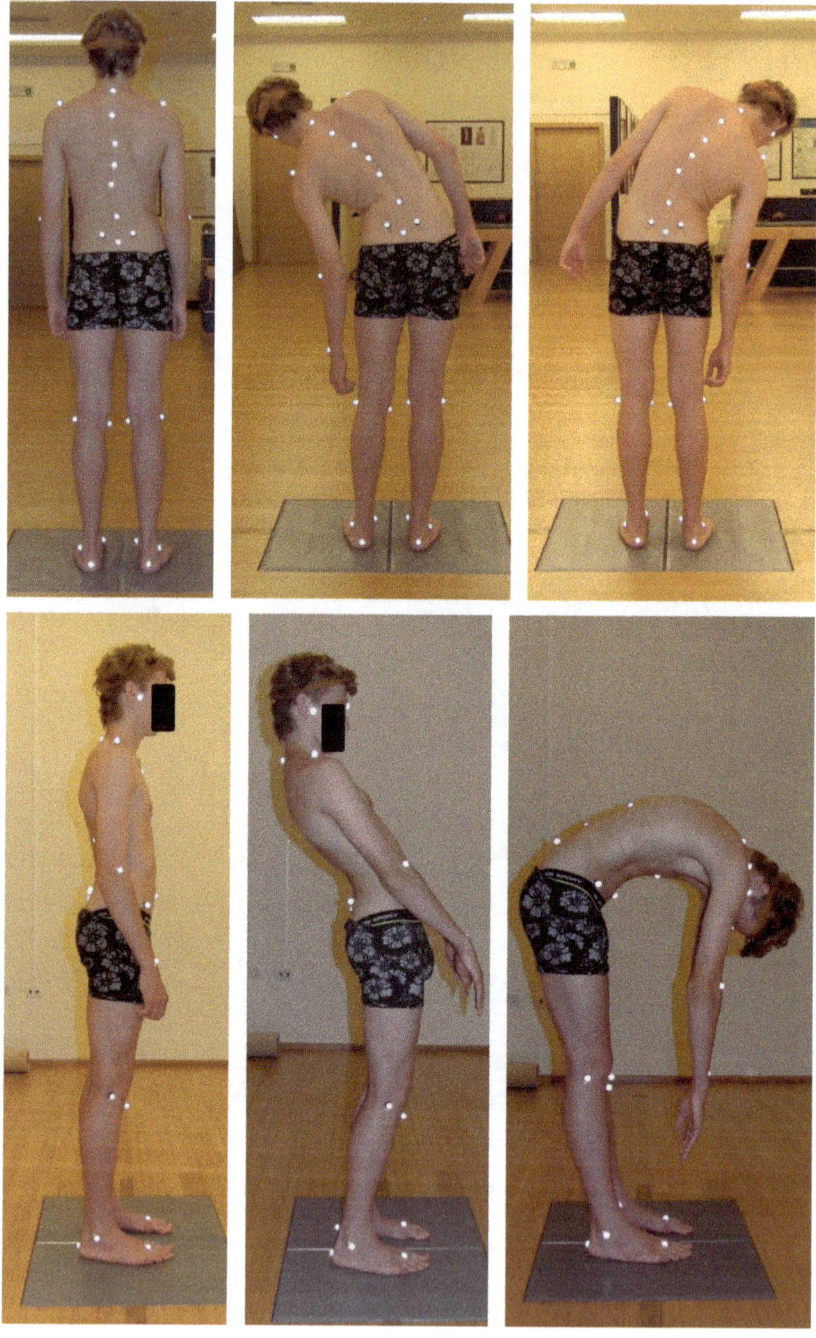

Fig. 6.10 Analyses of dynamics of movement during a postural assessment (*source* Biomechanics Laboratory Polyclinic Peharec [17]. With permission)

6.5.3 Example of Typical Workflow

Typical workflow of 3D kinematic system software consists of several steps. In the following text the usual workflow components will be presented based on Smart motion capture system that was installed and operational in biomechanics laboratory of Polyclinic Peharec in Pula, Croatia [17]. The initial step within a workflow is related to calibration where the appropriate graphical user interface (GUI) interface allows verification if an orthogonal triad of wands is properly placed within a calibration volume and visible in all cameras. As pointed out earlier, this serves to initialize cameras' calibration parameters. Immediately after follows a wand dance aimed at refining camera parameters (Fig. 6.11a). Once the system is calibrated and the markers are placed on the subject, imaging of the movement being studied can start. Next, the task of the tracker is to uniquely identify all markers throughout all imaged frames. The tracking phase is expected to be hopefully automatic, particularly in the case of lengthy image sequences. The constraints of epipolar geometry can be of great help in writing algorithms for automatic markers' tracking. Alternatively, if the system is unsure and/or the user notices an error in marker identification, the automatic procedure is interrupted, and the manual intervention is required. Upon finishing a successful tracking, 3D reconstruction of markers trajectories can be carried out (Fig. 6.11b). In majority of cases, marker placement is related to a measurement protocol characterized, among other things, by a certain body model. In the next step every marker will be assigned to a certain point on a model (Fig. 6.11c). On such a model a number of computations can be carried out and analyzed. This phase in a workflow presents perhaps the most informative and interesting part in terms motion analysis. The number of computations and analyses in this phase can be quite comprehensive. However, the concise findings and conclusion are frequently summarized in the final report.

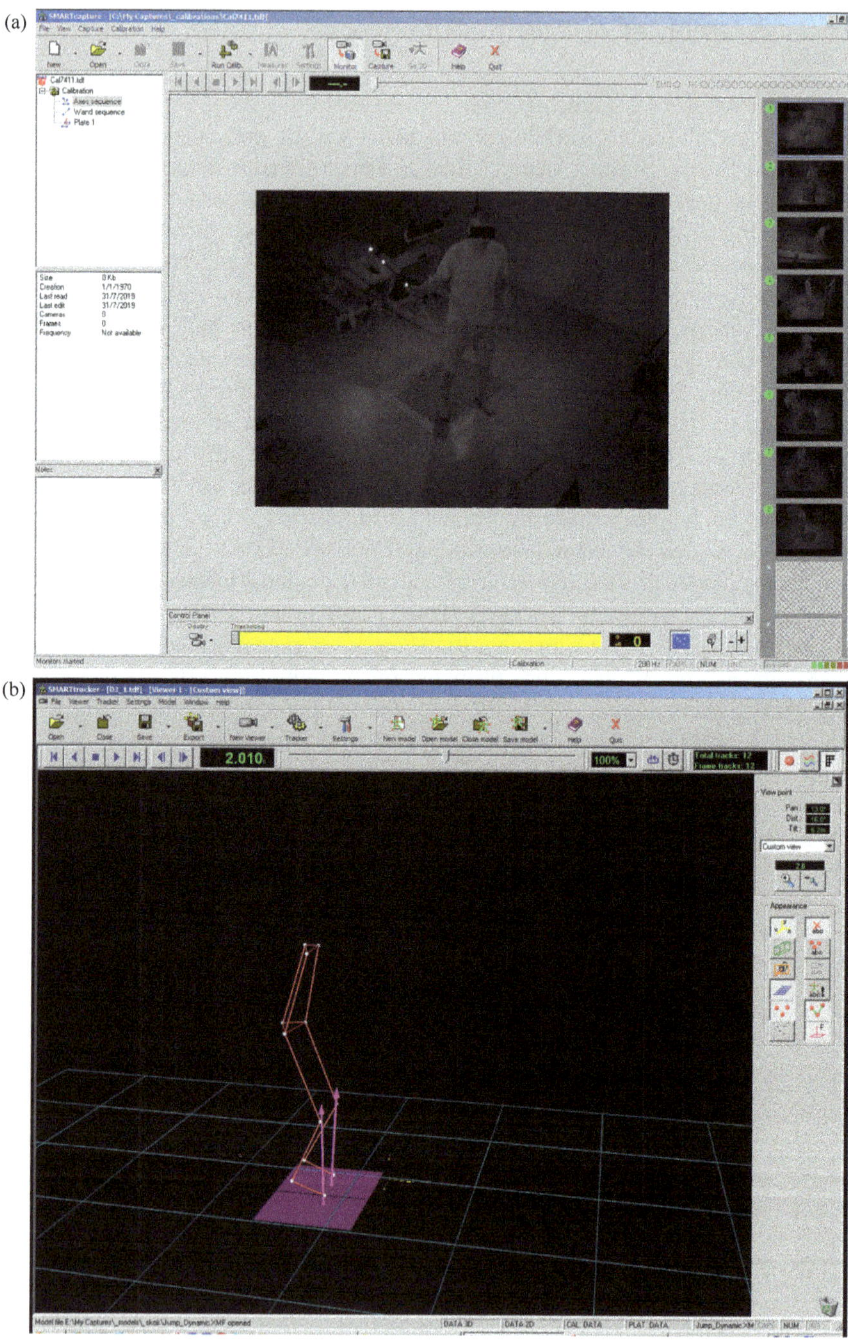

Fig. 6.11 Workflow steps when analyzing human motion. **a** Second part of calibration: camera parameter refinement by imaging of a wand dance **b** 3D reconstruction of the actual movement using markers **c** Assigning labels to markers based on some predefined body model (*source* Biomechanics Laboratory Polyclinic Peharec [17]. With permission)

(c)

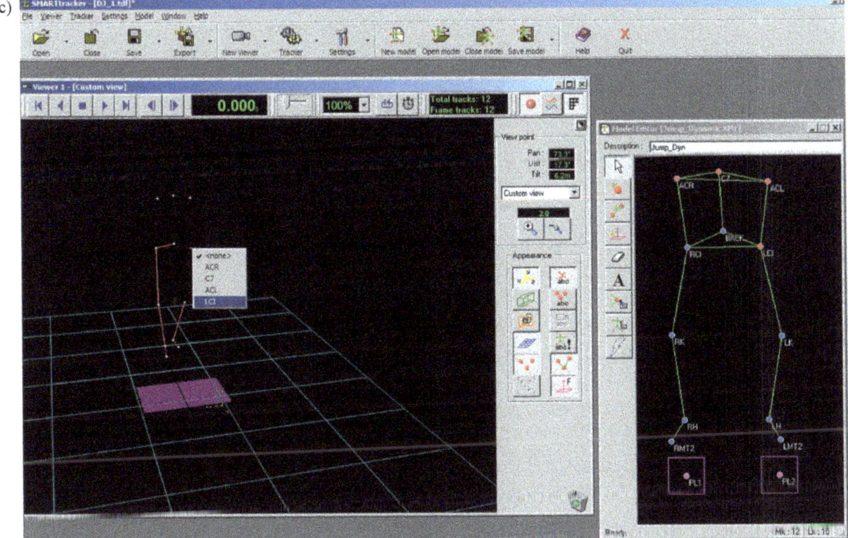

Fig. 6.11 (continued)

6.5.4 *Going Beyond Kinematic Data*

We recall that 3D optoelectronic systems readily output kinematic data. Additionally, modern motion analysis combines other measurement devices and signals to complement kinematic data. To that end, it is necessary to first synchronize in time 3D kinematic systems with other equipment. In the early days and/or using non-standard equipment, different software (manual) synchronization techniques have been used. Actually even the cameras themselves have been synchronized using the manual techniques. Namely, certain straightforward synchronization methods proposed simply imaging some light source on/off. Next, identifying the camera frames where the light source was fully lit, were taken as the correspondent frames between various cameras' video streams. That was essentially up to frame synchronization. Later, some more advanced software methods allowed estimating subframe difference as well [31, 32]. Nowadays, both 3D system camera synchronization and synchronization of 3D system with other measurement devices is done almost exclusively using hardware synchronization.

The most common add on measurement device is a force platform or a force plate. There are different variants, but one type commonly used explicitly uses triaxial force transducers at the corners of platform itself (Fig. 6.12). Hence, besides a vertical reaction force, the force is measured along horizontal and transverse direction. In addition, a center of pressure is computed and moments of force about the three platform axes too. A force platform is typically used to measure ground reaction generated in standing or moving across the platform. In the former case it may serve as an assessment technique used to quantify the central nervous system adaptive

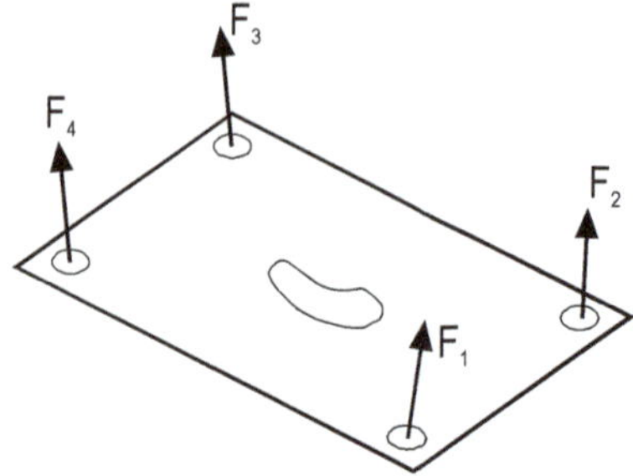

Fig. 6.12 A schematic principle of basic force platform measurements (please see more details in the text below and in Chap. 5)

mechanisms involved in the control of posture and balance. In the latter case, it may be used to study gait. Evidently, a force platform combined with data from 3D kinematic system allows more in-depth biomechanical analyses than either of measurement systems alone (Fig. 6.13). On a more technical side, it should be added

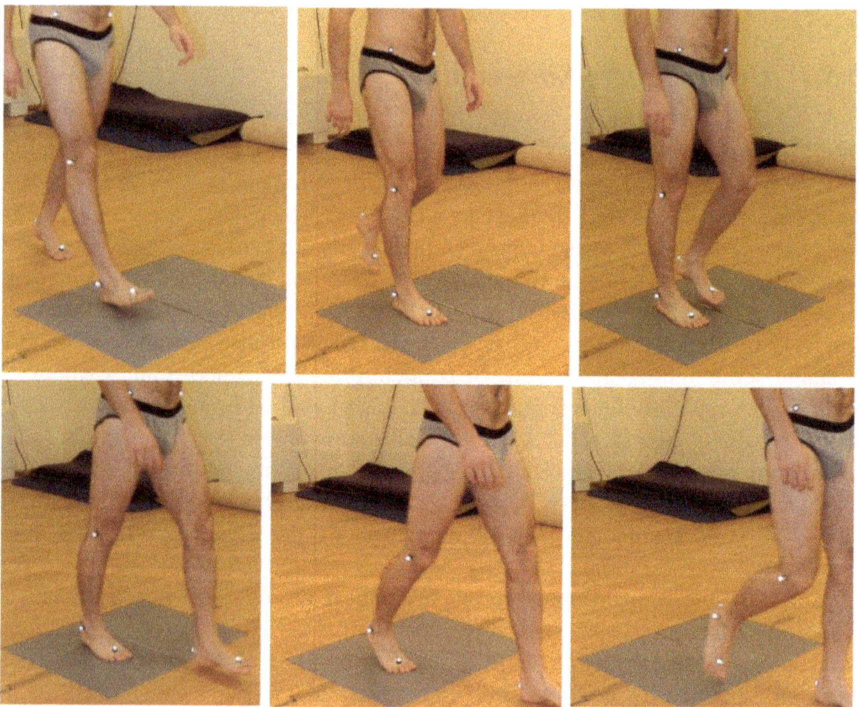

Fig. 6.13 Several frames of gait sequence, measuring 3D kinematics and ground reaction force, acquired by means of two force platforms (*source* Biomechanics Laboratory Polyclinic Peharec [17]. With permission)

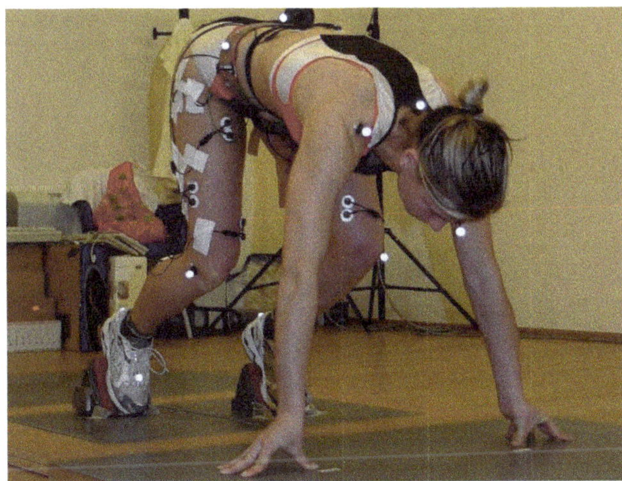
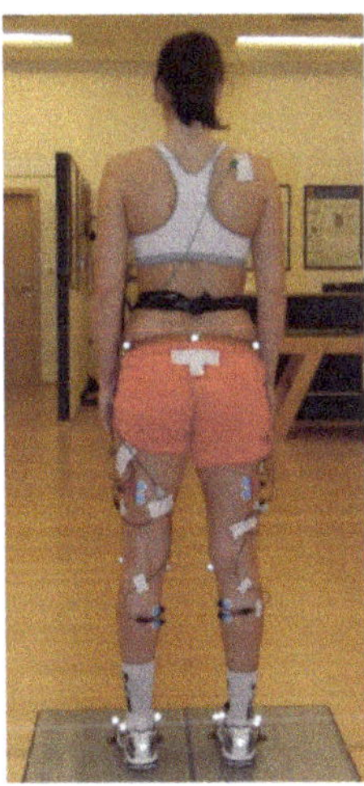

Fig. 6.14 Example of analyses of the sprint start (left image) and the vertical jump (right image) using combined measurement data originating from 3D kinematic system, force platforms and EMG device (*source* Biomechanics Laboratory Polyclinic Peharec [17]. With permission)

that it is necessary to calibrate (adjust) the force platform spatial coordinate system with the spatial coordinate system of 3D kinematic system.

3D kinematic data is frequently complemented with multichannel surface electromyographic (sEMG) recordings (Fig. 6.14). EMG is an electrodiagnostic medicine technique for evaluating and recording the electrical activity produced by muscles. Essentially it evaluates the health condition of muscles and the nerve cells that control them. To that end, an electromyography instruments are used to record an electromyogram. EMG surface signals are directly measured using skin surface electrodes where a correct placement of the electrodes on the skin is of crucial importance. Presently, multichannel (typically 4 or more channels) EMG instruments are used. During biomechanical analyses of specific exercises, it is highly desired to use wireless EMG. The wireless technology requires an attachment of only electrodes and small probes attached closely to them. Thus, it does not have any other wires, allowing the natural movement of analyzed subject (Fig. 6.15).

Perhaps the most natural extension of kinematic data is a computation of kinetic data from kinematic data [33, 34]. A computation of kinetic data from kinematic data is often referred as the inverse dynamics approach and can be briefly explained as follows. Human body is usually represented as a system of rigid body segments connected with joints [35]. Figure 6.16 shows a set of forces and torques in the form of segment's free body diagram.

$\mathbf{W}$ is a segment weight, $\mathbf{J_P}$ and $\mathbf{J_D}$ are joint reaction forces in proximal and distal joints, respectively. $\mathbf{M_P}$ and $\mathbf{M_D}$ are resultant muscle forces with the respect to proximal and distal segmented end, respectively. $\mathbf{M_P}$ and $\mathbf{M_D}$ can be approximated as the sum of all eccentric forces and which do not pass through the segment's joints whereas $\mathbf{J_P}$ and $\mathbf{J_D}$ are the sum of concentric forces. $\mathbf{m}$ and $\mathbf{j}$ are the vectors connecting

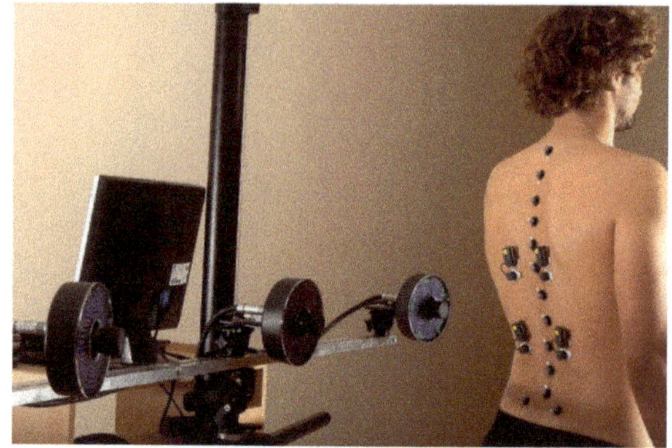

Fig. 6.15 The usage of wireless EMG electrodes, imposing a minimal intrusion on the movement (*source* Biomechanics Laboratory Polyclinic Peharec [17]. With permission)

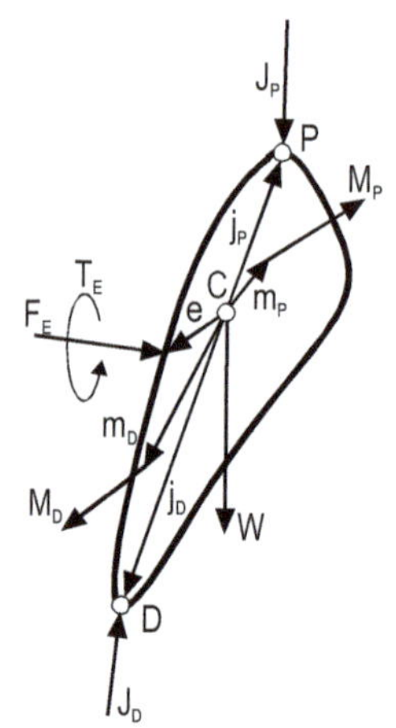

Fig. 6.16 A free body diagram of applied forces and torques for a body segment

the segment's center of mass C and the point of force application of $\mathbf{M_P}$ and $\mathbf{M_D}$, respectively. $\mathbf{F_E}$ is an external force and $\mathbf{T_E}$ is an external torque that a segment may be a subject to, as well. Rigid body movement can be decomposed into translation and rotational parts (Chasles' theorem [36]). For a translation part, the sum of forces (i.e. the net force) acting on the body is equal to the rate of change of linear (translational) momentum $\mathbf{p}$ (6.28). Likewise, for a rotational part the net torque around some fixed point is equal to the to the rate of change of angular (rotational) momentum $\mathbf{H}$ (6.29).

$$\sum_i \mathbf{F_i} = \mathbf{F_R} = \mathbf{J_P} + \mathbf{M_P} + \mathbf{J_D} + \mathbf{M_D} + \mathbf{W} + \mathbf{F_E} = \frac{dp}{dt} \tag{6.28}$$

$$\sum_i \mathbf{T_i} = \mathbf{T_R} = \mathbf{j_P} \times \mathbf{J_P} + \mathbf{m_P} \times \mathbf{M_P} + \mathbf{j_D} \times \mathbf{J_D} + \mathbf{m_D} \times \mathbf{M_D} + \mathbf{e} \times \mathbf{F_E} + \mathbf{T_E} = \frac{d\mathbf{H}}{dt} \tag{6.29}$$

Expressions (6.28) and (6.29) are basically Newton's second law of motion in algebraic forms. The rates of change of linear and angular momentum can be found from 3D kinematic system.

A simple algebraic manipulation of expressions (6.28) and (6.29) yields:

$$\mathbf{J_P} + \mathbf{M_P} = \frac{d\boldsymbol{p}}{dt} - \mathbf{J_D} - \mathbf{M_D} - \boldsymbol{W} - \boldsymbol{F}_\mathrm{E} \tag{6.30}$$

$$\mathbf{j_P} \times \mathbf{J_P} + \mathbf{m_P} \times \mathbf{M_P} = \frac{d\mathbf{H}}{dt} - \mathbf{j_D} \times \mathbf{J_D} - \mathbf{m_D} \times \mathbf{M_D} - \mathbf{T_E} - \mathbf{e} \times \mathbf{F_E} \tag{6.31}$$

Evidently there are only two constraints which are, in general, insufficient to solve for six unknowns. The next substitutions can provide a remedy for it:

$$\begin{array}{lll} \mathbf{F} \equiv \mathbf{J} + \mathbf{M} & \mathbf{F_P} = \mathbf{J_P} + \mathbf{M_P} & \mathbf{F_D} = \mathbf{J_D} + \mathbf{M_D} \\ \mathbf{T} \equiv (\mathbf{m} - \mathbf{j}) \times \mathbf{M} & \mathbf{T_P} = (\boldsymbol{m}_\mathrm{P} - \mathbf{j_P}) \times \mathbf{M_P} & \mathbf{T_D} = (\boldsymbol{m}_\mathrm{D} - \mathbf{j_D}) \times \mathbf{M_D} \end{array} \tag{6.32}$$

After introducing substitutions from (6.32) in (6.29) and (6.30) the following is obtained:

$$\mathbf{F_P} = \frac{d\boldsymbol{p}}{dt} - \boldsymbol{F}_\mathrm{D} - \boldsymbol{W} - \boldsymbol{F}_\mathrm{E} \tag{6.33}$$

$$\mathbf{j_P} \times \mathbf{F_P} + \mathbf{T_P} = \frac{d\mathbf{H}}{dt} - \mathbf{j_D} \times \mathbf{F_D} - \mathbf{T_D} - \mathbf{T_E} - \mathbf{e} \times \mathbf{F_E} \tag{6.34}$$

Assuming that $\mathbf{F_D}$ and $\mathbf{T_D}$ are known for a distal segment's joint, we can solve for the remaining two unknowns: a force $\mathbf{F_P}$ and torque $\mathbf{T_P}$ of a proximal segment's joint. The above substitutions neatly allowed a reduction of too many unknowns. In addition, it is useful to note that the substitutions $\mathbf{T_P}$ and $\mathbf{T_D}$ could be interpreted as torques produced by the muscle force about the proximal and distal segmented end, respectively. However, the interpretation for the introduced substitutions $\mathbf{F_P}$ and $\mathbf{F_D}$ is somewhat more complex. They are derived as net forces from joint reaction forces (**J**) and muscle forces (**M**) (6.32). The more comprehensive modeling and analyses are needed to identify each of them separately. In order to make an assumption in (6.33) and (6.34) about values $\mathbf{F_D}$ and $\mathbf{T_D}$ of a distal segment's joint, the usual practice is to start with the most distal body segment, i.e. with the segment which has no other segment attached to it in the distal direction. In that case values $\mathbf{F_D}$ and $\mathbf{T_D}$ can be set to zero and only perhaps $\mathbf{F_E}$ and $\mathbf{T_E}$ need to be known, usually from some measurement sensor. For example, a foot segment walking over force platform would have no distal segment attached to it. The external values $\mathbf{F_E}$ and $\mathbf{T_E}$ in (6.33) and (6.34) are in this case the ground reaction values and $\mathbf{F_E}$ and $\mathbf{T_E}$, readily provided by the force platform and shown as $\mathbf{F_{GR}}$ and $\mathbf{T_{GRF}}$ in Fig. 6.17.

This assures a straightforward computation of force and torque in the foot's proximal ankle joint (labeled as AK on in Fig. 6.17).

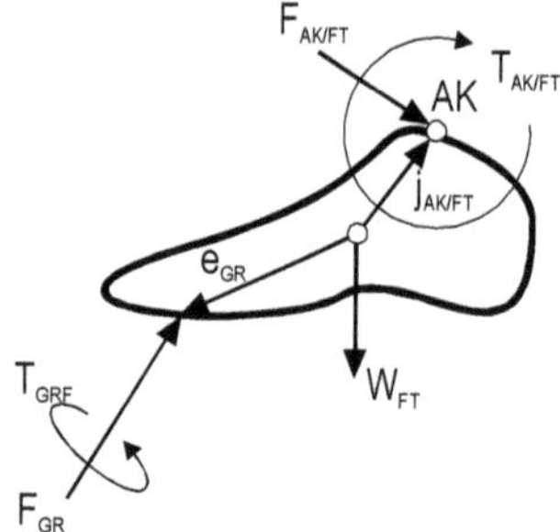

Fig. 6.17 A diagram of forces and torques for a foot segment

$$\mathbf{F_P} = \mathbf{F_{AK/FT}} = \frac{d\mathbf{p}_{FT}}{dt} - \mathbf{W_{FT}} - \mathbf{F_{GRF}} \tag{6.35}$$

$$\mathbf{T_P} = \mathbf{T_{AK/FT}} = \frac{d\mathbf{H}_{FT}}{dt} - \mathbf{T_{GRF}} - \mathbf{e_{GRF}} \times \mathbf{F_{GRF}} + \mathbf{j_{AK/FT}} \times \mathbf{F_{AK/FT}} \tag{6.36}$$

where $\mathbf{F_P} = \mathbf{F_{AK/FT}}$ is resultant force at proximal foot joint, $\mathbf{T_P} = \mathbf{T_{AK/FT}}$ is a foot torque around proximal joint, $\mathbf{j_{AK/FT}}$ is a vector from a foot center of mass to the ankle joint, AK/FT is a label introduced for proximal foot joint. Foot proximal joint is essentially a lower leg distant joint. Once the values $\mathbf{F_{AK/FT}}$ and $\mathbf{T_{AK/FT}}$ for the foot proximal joint are known, we can recall Newton's third law on action and reaction, allowing us to change only the sign of $\mathbf{F_{AK/FT}}$ and $\mathbf{T_{AK/FT}}$ and set them as lower leg values at distal joint. Therefore, another set of equations of similar form as (6.33) and (6.34) is possible to define for lower leg, yielding computation of proximal values of lower leg. It is easy to envision the rest of the procedure for the remaining body segments.

The crucial link allowing a computation of kinetic from kinematic data are body segments' parameters (BSP). BSP normally assume inertial body segment parameters such as segment's mass and moments of inertia [37], but in a broader sense BSP includes also segment's physical dimensions, volume, density, relative position of segment's center of mass wrt to segments end points etc. [38]. The ultimate accuracy of kinetic data computations directly depends on the reliability and accuracy of 3D kinematic system output and BSP estimates where the latter one is generally more prone to errors and consequently a weaker link. BSP measurements can be performed directly either on cadavers or on live subjects where in the former case evidently more invasive approaches are available, particularly it has been so in the early days of BSP research [39]. In terms of instrumentation, a part of BSP estimates such as segment volume or length of body segment, can be done relatively simply by taking advantage of Archimedes law or by direct measurements using calipers, respectively. Other more sophisticated instrumentation may include usage of CT scans to predict segments densities [40]. If such direct measurements are performed on the multiple subjects it is then convenient for the future estimations to derive a so called regression (prediction) equations [41]. Actually, for all practical purposes the usage of regression equations is preferred method since it takes as input some convenient and easily measurable quantity and output estimate for all kinds of BSP parameters that are necessary to carry out inverse dynamics calculations. For example in the simplest case, a total subject mass or height can be used to estimate specific segments masses and lengths, respectively. Of course, it is hard to obtain reliable BSP estimate from regression equations using only a single value as input. Therefore, regression equations are usually based on assumptions about certain body parts (segments) being modeled as particular geometric shapes which may take multiple inputs [42]. To that end, 3D kinematic system software typically uses several anthropometric measurement data of various body parts as input in BSP regression equations. These anthropometric measurements are computed either using 3D positions of specific markers' placement and/or entered in the software package from a direct measurement using calipers. In either case, 3D optoelectronic system by readily providing kinematic data, but which are further extended and processed in the context of inverse dynamics approach, is a great source to compute kinetic data as well. Moreover, it is even possible to compute muscle power exerted, energy consumption and energy transfer between segments etc. during a specific movement [43].

6.6 Further Reading

3D kinematic systems measurements considered so far were restricted to relatively small volumes where cameras can remain fixed. In cases where subject move in considerable larger volume, different solutions have been developed using pan and tilt cameras [44]. 3D motion capture systems have been used underwater as well [45]. In technological sense such systems have many peculiarities, starting with a camera underwater modeling. Therefore, a good start in analyzing the work of such a system is first to read about underwater camera calibration approaches [46]. Speaking of a camera calibration in general, it should be stressed that the ultimate goal would be to employ one of the self (auto) calibration techniques ([47, 48]). These techniques do not require any calibration object, but merely the accurate image point correspondences between camera views. Besides, providing 3D reconstruction up to unknown scale, which is less of a problem, selfcalibration approaches in addition tend to be very sensitive to noise which is likely the reason preventing them from being in practice widely implemented. Probably one of the most significant recent technological advances about 3D kinematic systems is the appearance of commercial markereless systems [49]. The image processing behind markerless motion capture systems [50] is, in general, more demanding than behind those using markers. A main challenge here is computing an accurate kinematic model without utilization of markers. To that end, many solutions take advantage of recent advances in machine (deep) learning (ML) area and, as ML algorithms keeps progressing, it is likely to see markerless systems becoming more and more popular. There is also a family of imaging techniques and associated 3D reconstruction systems which explicitly reconstruct the entire body surface, instead of just relatively few number of attached markers. There are frequently referred to as 3D scanners where 3D structured light (SL) based scanners are arguably the most accurate and reliable solutions. SL is an approach where in addition to a camera, a video projector is introduced with the intention of projecting one or more images (patterns) in the scene [51, 52]. By analyzing captured images of projected pattern(s) 3D depth can be very accurately extracted. For example, 3D SL scanners implemented on a smartphone can be used potentially in the fabrication of personalized shoe insoles [53]. Finally, 3D motion capture systems have not been exclusively used to study human gait, but animals as well [54]. Besides, an interested reader may look at the applications of 3D kinematic systems other than sport and medicine. The most popular alternative applications can be found in industry and entertainment [55].

6.7 Conclusion

3D optoelectronic system is a non-invasive 3D reconstruction system whose main component is a calibrated camera. Nowadays majority of 3D kinematic systems calibrate cameras using a calibration wand. The 3D reconstruction principle shares many similarities with the human vision where at least two camera images are needed and on which the correspondent image points need to be identified. To alleviate the correspondence problem search, the principles of epipolar geometry can be used and the passive markers are attached on anatomical points of a human body. 3D kinematic data are usually processed in accordance with some measurement protocol, which defines the number of markers, their placement on the human body and it aims at analyzing some specific movement. Although kinematic data by itself is very informative, 3D kinematic systems are frequently upgraded with hardware and software providing other types of data, e.g. EMG, reaction forces and kinetic data. 3D optoelectronic systems are constantly evolving and one of the research avenues is making an even more accurate and robust markerless 3D kinematic system. Unlike some other systems and devices in sport and medicine which are utilized for essentially injured people, 3D motion capture system can be used for healthy subject as well, e.g. improving the athletic performance. Moreover, 3D optoelectronic systems are widely applicable and therefore can be found in numerous other applications and areas, such as industry and entertainment.

References

1. Ferrigno G, Borghese NA, Pedotti A (1990) Pattern recognition in 3D automatic human motion analysis. ISPRS J Photogramm Rem S 45:227–246
2. Winter DA (1990) Biomechanics and motor control of human movement. Wiley, New York
3. TrakSTAR. https://est-kl.com/manufacturer/ascension/trakstar-drivebay.html. Accessed July 2019
4. Mazuryk T, Gervautz M (1996) Virtual reality-history, applications, technology and future

5. Daniel Dinu D, Fayolasb M, Jacquet M, Leguyb E, Jean Slavinski J, Houel N (2016) Accuracy of postural human-motion tracking using miniature inertial sensors. Proc Eng 147:655–658
6. Allen BD, Welch G (2005) A general method for comparing the expected performance of tracking and motion capture systems. In: VRST '05 Proceedings of the ACM symposium on Virtual reality software and technology, Monterey, CA, USA, pp 201–210
7. Kwon3D. http://www.kwon3d.com/. Accessed: July, 2019
8. Abdel-Aziz YI, Karara HM (1971) Direct linear transformation from comparator coordinates into object-space coordinates in close-range photogrammetry. In Proceedings of the ASP/UI Symposium „Close-Range Photogrammetry“ Urbana, Il, American Society of Photogrammetry, Falls Church, VA, pp 1–18
9. Hartley R, Zisserman A (2000) Multiple view geometry in computer vision. Cambridge University Press, Cambridge
10. Triggs B, McLauchlan P, Hartley R, Fitzgibbon A (1999) Bundle adjustment—a modern synthesis. In: Vision algorithms: theory and practice, pp 298–372
11. Bianco S, Gianluigi C, Marelli D (2018) Evaluating the performance of structure from motion pipelines. J Imag 4(8):98, 1–18
12. Press WH, Teukolsky SA, Vetterling WT, Flannery BP (1997) Numerical recipes in C. Cambridge University Press, New York
13. Weng J, Cohen P, Herniou M (1992) Camera calibration with distortion models and accuracy evaluation. IEEE Trans Pattern Anal Mach Intell 14:965–980
14. An example of 3D calibration cage. https://ceps.unh.edu/ocean-engineering/optimal-conditions-calibration-multi-camera-rigs. Accessed: July, 2019
15. An example of 3D calibration cage. http://www.kwon3d.com/. Accessed: July, 2019
16. An example of 3D calibration cage. https://web.eecs.umich.edu/~jjcorso/t/598F14/files/lecture_0917_calibration.pdf. Accessed: July, 2019
17. Biomechanics Laboratory Polyclinic Peharec, Pula, Croatia. https://www.peharec.com/. Accessed: July, 2019
18. OpenCV. https://opencv.org/. Accessed: July, 2019
19. Zhang Z (2000) A flexible new technique for camera calibration. IEEE Trans PAMI 22(11):1330–1334
20. Borghese NA, Cerveri P (2000) Calibrating a video camera pair with a rigid bar. Pattern Recogn 33:81–95
21. Pribanić T, Sturm P, Cifrek M (2007) Calibration of 3D kinematic systems using orthogonality constraints. Mach Vis Appl 18(6):367–381
22. Zhang R, Tsai PS, Cryer JE (1999) Shape from shading: a survey. IEEE Trans Pattern Anal Mach Intell 21(8):690–706
23. Basri R, Jacobs D, Kemelmacher I (2007) Photometric stereo with general unknown lighting. Int J Comput Vis 72(3):239–257
24. Kanatani K, Chou T-C (1989) Shape from texture: general principle. Artif Intell 38:1–48
25. Zhang Z (1996) Determining the epipolar geometry and its uncertainty: a review. RR-2927, INRIA. 1996. ⟨inria-00073771⟩
26. Hartley R (1997) In defense of the eight-point algorithm. PAMI 19(6):580–593
27. APAS system. http://www.arielnet.com/home/index. Accessed: July, 2019
28. Kadaba MP, Ramakrishnan HK, Wootten ME (1990) Measurement of lower extremity kinematics during level walking. J Orthop Res 8:383–392
29. Vaughan CL, Davis BL, O'Connor JC (1999) Dynamics of human gait. Kiboho Publishers, Cape Town
30. Hatze H (1980) A mathematical model for the computational determination of parameter values of anthropometric segments. J Biomech 13:833–843
31. Pourcelot P, Audigie F, Degueurce C, Geiger D, Denoix JM (2000) A method to synchronize cameras using the direct linear transformation technique. J Biomech 33:1751–1754
32. Pribanić T, Lelas M, Krois I (2015) Sequence-to-sequence alignment using a pendulum. IET Comput Vision 9(4):570–575
33. Zatsiorsky VM (1998) Kinematics of human motion. Human Kinetics, Champaign, Il

34. Zatsiorsky VM (2002) Kinetics of human motion. Human Kinetics, Champaign
35. Kwon Y-H, Theories and Practices of Motion Anal. http://www.kwon3d.com/theory/jntkin.html. Accessed: July, 2019
36. Baruh H (1999) Analytical dynamics. WCB/McGraw-Hill
37. Chandler RF, Clauser CE, McConville JT, Reynolds HM, Young JW (1975) Investigation of inertial properties of the human body. AMRL-TR-74–137, AD-A016–485. DOT-HS-801–430. Aerospace Medical Research Laboratories, Wright-Patterson Air Force Base, Ohio
38. Clauser CE, McConville JT, Young JW (1969) Weight, volume and center of mass of segments of the human body. AMRL-TR-69–70. Aerospace Medical Research Laboratory, Wright-Patterson Air Force Base, Ohio
39. Braune W, Fischer O (1889) The center of gravity of the human body as related to the equipment of the german infantryman. Treat. of the Math-Phys. Class of the Royal Acad. of Sc. of Saxony (ATI 138–452 Available from Defense Documentation Center), 26
40. Ackland TR, Henson PW, Bailey DA (1988) The uniform density assumption: its effect upon the estimation of body segment inertial parameters. Int J Sports Biomech 4:146–155
41. Zatsiorsky VM, Seluyanov VN (1985) Estimation of the mass and inertia characteristics of the human body by means of the best predictive regression equations. In: Winter DA, Norman RW, Wells RP, Hayes KC, Patla AE (eds) Biomechanics IX-B. Human Kinetics, Champaign, IL, pp 233–239
42. Rodrigue D, Gagnon M (1984) Validation of Weinbach's and Hanavan's models for computation of physical properties of the forearm. Res Q. Exercise Sport 55:272–277
43. Roberston DG, Winter D (1980) Mechanical energy generation, absorption and transfer amongst segments during walking. J Biomech 13:845–854
44. Kurihara K, Hoshino S, Yamane K, Nakamura Y (2002) Optical motion capture system with pan-tilt camera tracking and real time data processing. In: Proceedings 2002 IEEE international conference on robotics and automation. Washington DC
45. Qualisys. https://www.qualisys.com/hardware/underwater/
46. Shortis M (2019) Camera calibration techniques for accurate measurement underwater. In: McCarthy J, Benjamin J, Winton T, van Duivenvoorde W (eds) 3D recording and interpretation for maritime archaeology. Coastal Research Library, vol 31. Springer, Cham
47. Hartley R (1997) Kruppa's equations derived from the fundamental matrix. IEEE Trans Pattern Anal Mach Intell 19(2):133–135
48. Sturm P (2002) Critical motion sequences for the self calibration of cameras and stereo systems with variable focal length. Image Vis Comput 20(5–6):415–426
49. Examples of commercially available markerless motion capture systems. http://thecaptury.com/ http://darimotion.com/ https://www.zflomotion.com/. Accessed: July, 2019
50. Corazza S, Mündermann L, Chaudhari A, Demattio T, Cobelli C, Andriacchi T (2006) A markerless motion capture system to study musculoskeletal biomechanics: visual hull and simulated annealing approach. Ann Biomed Eng 34(6):1019–1029
51. Salvi J, Fernandez S, Pribanić T, Llado X (2010) A state of the art in structured light patterns for surface profilometry. Pattern Recogn 43(8):2666–2680
52. Petković T, Pribanić T, Đonlić M, D'Apuzzo N (2017) Multi-projector multi-camera structured light 3D body scanner. In: Proceedings of 8th international conference and exhibition on 3D body scanning and processing technologies (3DBODY.TECH), Montreal QC, Canada, 11–12 Oct. 2017
53. Pribanić T, Petković T, Đonlić M, Angladon V, Gasparini S (2016) 3D structured light scanner on the smartphone. Lecture notes in computer science (9730): image analysis and recognition. Póvoa de Varzim : Springer, ICIAR 2016, pp 443–450
54. Havemeyer Workshop. Motion Capture and 3D Analysis of Equine Locomotion. Tamarindo, Costa Rica, March 26–29, 2006
55. Sato H, Cohen M (2010) Using motion capture for real-time augmented reality scenes. In: Proceedings of the 13th international conference on humans and computers. Aizu-Wakamatsu, Japan, pp 58–62

Chapter 7
On Standardization of Pedobarographic Measurement Protocols

Vladimir Medved and Igor Gruić

Abstract The measurement method of pedobarography exists in modern form for a few decades already, but relevant actions towards its standardization were undertaken rather recently. The foot can be biomechanically modeled incorporating realistic geometric and material properties of both skeletal and soft tissue components. This may serve to generate simulated contact stress distributions, to be compared with measured pressure data obtained using pedobarography. The method may assume several technical design possibilities employing various types of pressure sensors, the common feature of these systems being that the process of measurement occurs via a contact of a firm (sensor) and a deformable (foot sole) surface. Critical issues regarding standardization of measurement equipment, measurement protocol, and signal processing and interpretation are commented in some detail. An attempted standardization consensus for devices' performance suited to biomedical instrumentation standards is documented. A short end view on the instrument device in Zagreb laboratory is put forward.

7.1 On Pedobarography: Relation to Foot Biomechanics, Types of Devices' Design, and Standardization Problems

First studies of dynamic contact between the body (the foot) and the ground were those by Marey, Carlet, and Vierort who developed methods based on pneumatic and similar principles ([1]; Chap. 2). The attempts have progressed over solutions such as the one by Elftman (1938; Chap. 2, Fig. 2.10). One line of development proceeded towards force platforms (Sect. 5.3.1, Fig. 5.9; Sect. 6.5.4, Fig. 6.12). Regarding, however, pressure distribution measurement, with plantar resolution required of an order of magnitude cm x cm, the real precursors appeared—coincidentally to the technological development of sensor materials—relatively late, only in the mid-1970s.

V. Medved (✉) · I. Gruić
Faculty of Kinesiology, University of Zagreb, Horvaćanski zavoj 15, 10000 Zagreb, Croatia
e-mail: vladimir.medved@kif.hr

V. Medved (ed.), *Measurement and Analysis of Human Locomotion*,
Series in Biomedical Engineering, https://doi.org/10.1007/978-3-030-79685-3_7

Since then, modern pedobarography method succeeds in supplying high quality experimental data of satisfactory planar and temporal resolution and obtainable in a simple and elegant manner.

Proper interpretation of experimental data of this sort requires clear analytical insight into the system studied (human foot), therefore to understand foot biomechanics becomes of crucial importance. Worth mentioning in this context is the contribution by Gefen, from 2000, assuming a bioengineering approach, whereby accurate modeling of foot biomechanics is attempted [2]. A 3D numerical model of the foot, incorporating realistic geometric and material properties of both skeletal and soft tissue components of the foot was developed. The task attempted has been to analyze biomechanically the structural behavior of the foot during gait. An experimental method, integrating the optical Contact Pressure Display (CPD) method for plantar pressure measurements and a digital Radiographic Fluoroscopy (DRF) instrument for acquisition of skeletal motion during gait, was also developed in this study and subsequently used to build the foot model and validate its predictions. The integrated CPD/DRF measurements provided detailed skeletal motion and joint behavior of the foot simultaneously with the evolution of the plantar pressure pattern during the stance phase of gait, as shown in Fig. 7.1.

In the study, following subphases of the stance phase of gait were defined: (1) Initial-contact, (2) Heel-strike, (3) Mid-stance, (4) Forefoot-contact, (5) Push-off, and (6) Toe-off. This presentation of the stance phase has integrated the common literature definitions with two new terms: Initial-contact and Forefoot-contact. Initial-contact describes the position of the foot during its very early interaction with the ground, when only a small part of the heel plantar area is in contact (e.g., see the first frame in Fig. 7.1), subsequently, Heel-strike denotes the position of the foot when the heel is in full impacting contact with the ground (e.g., the second frame in Fig. 7.1). It was decided to introduce an Initial-contact subphase, before the Heel-strike develops, in order to be able to record the whole evolution and time length of the stance, up to Toe-off. Forefoot-contact, an intermediate event between Mid-stance and Push-off, is the position in which both the forefoot and the anterior part of the midfoot (arch) are in contact with the ground. The model was adapted for each of the above-mentioned subphases by altering its skeletal/muscle loading characteristics and its geometric positioning, including both its gross inclination and the alignment of the phalanges in respect with the metatarsal bones.

Not pretending to present in detail this thorough, elaborate and meticulous biomechanical study of the foot during gait, combining faithfull biomechanical modeling of the foot anatomical structure and then developing numerical model capable of generating pressure parameters (elaborated in detail in [2]), we show, as an illustration, some results (Fig. 7.2).

The structure of the human foot is highly complex and, accordingly, to develop reliable experimental methods and realistic computational simulations describing its mechanical behavior during gait and locomotion is complicated. In this example, validation of the stress has been achieved by comparing model predictions of contact stress distribution with respective CPD measurements. The presented 3D numerical model was shown not only to react according to experimental evidence, but also to

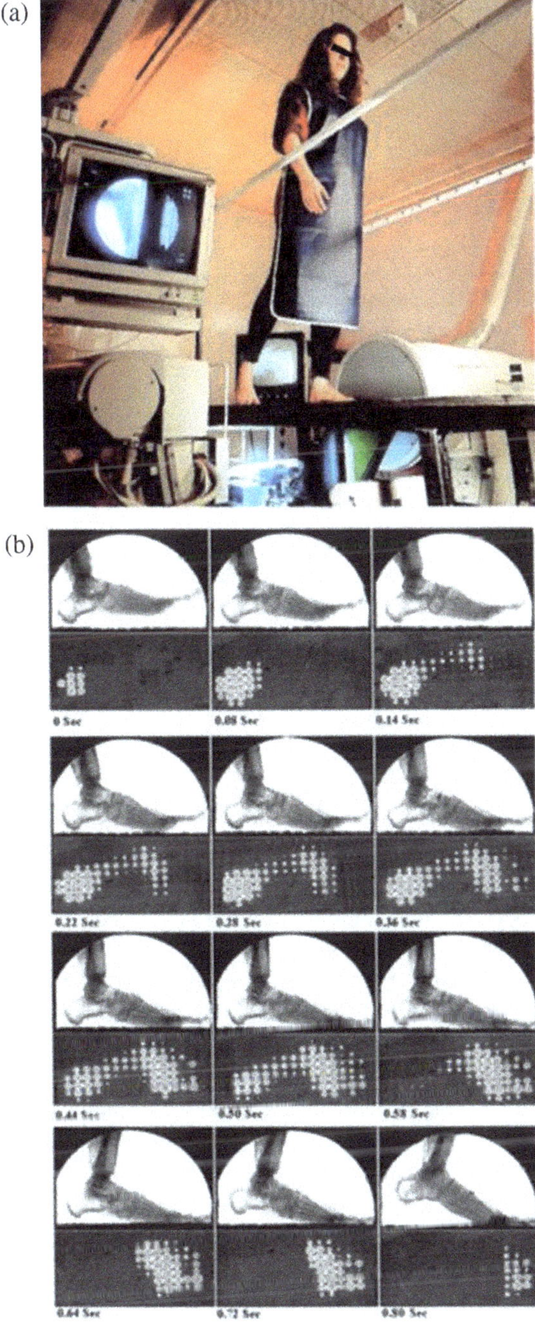

Fig. 7.1 The CPD/DRF integrated experimental analysis: "*a*… the experimental setup and" *b*… Representative CPD/DRF data of a normal foot structure during various discrete stages of the stance phase; gait velocity 0.5 m/s. Time intervals are different between frames to present the most characteristic/descriptive subphases of the stance (from Gefen et al. [2], with permission)

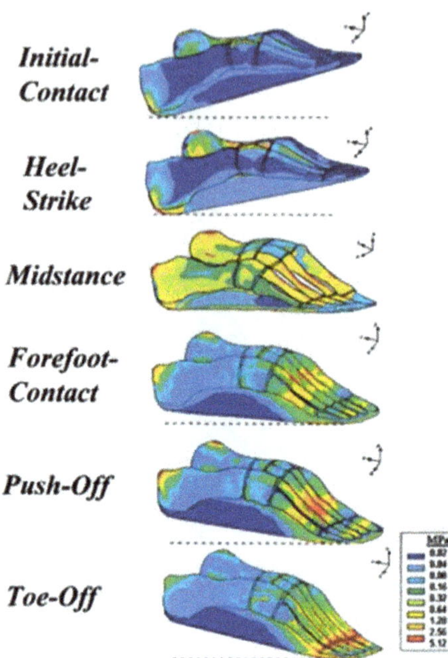

Fig. 7.2 The von Mises stress distribution in the foot model at the six characteristic subphases of stance. Von Mises equivalent stresses are explained in detail in 2 (In materials science and engineering the von Mises stress is used to predict yielding of materials under complex loading from the results of uniaxial tensile tests. https://en.wikipedia.org/wiki/Von_Mises_yield_criterion). (from Gefen et al. [2], with permission)

provide an integral approach to the study of structural behavior of the foot during gait. The ability to acquire data characterizing internal stress states and their evolution during gait makes the model a suitable clinical tool. Since the model can identify vulnerable skeletal and soft tissue components of the foot, it is potentially usable not only for understanding the development mechanism of some common disorders (like the ones associated with diabetes, arthritis, stress fractures, etc.) but also to serve as a tool for development of novel clinical decision making and foot treatment approaches.

Baker has critically commented the most used biomechanical foot models. When doing measurements for research or clinical purposes, several conventional models have been standardized; the two most popular being the Oxford Foot Model and IOR (Istituti Ortopedici Rizzoli) foot model [3, 4]. In last years, merging of classical inverse dynamics of the body as a whole with refined foot models, combined sometimes also with pedobarography and surface electromyography (sEMG) has taken place. A musculoskeletal foot model for clinical gait analysis has been proposed [5]. In the following Chap. 8 an anatomically based consideration of foot biomechanical function will be assumed, suitable for addressing medical clinical problems being diagnosed and treated.

An overview of the development of the measurement method of pedobarography and technical solutions attained at the time (2001) has been given in [1]. Beginning with a short historical outlook, a few instrument layouts were pinpointed including several types of platforms and of shoe insoles, all measuring vertical component of pressure only. Besides, two designs of systems sensitive also to tangential (shear) force component were also shortly described. For each presented solution technical data were documented. Technical possibilities to measure pressure distribution between the body and the ground may be classified into (Alexander et al. 1990, citation according to [1]):

- installing the sensor in the sole surface,
- installing the sensor in footware insoles or
- permanently embed the sensor in the ground (like it is typically done with force platforms as well (Sect. 5.3.1)). Solutions where the instrument is portable and is laid down on a floor before use may be considered equivalent.

All these variants of instrument solutions provide similar measurement information. The principal feature of these systems is that the process of measurement occurs via a contact of a firm (sensor) and a deformable (foot sole) surface.

Pedobarography, sometimes also named baropodometry, has increasingly been used during last several decades, in research and in clinics alike, and standardization of measurement protocols and reports has been attempted. In 1991, at the EMED meeting in Vienna, Ian Alexander suggested a standard form of reporting foot measurements. Authorities in the field still constate lack of full standardization however [6]. Relevant professional organizations and forums exist, such as Pedobarographic Group of the International Foot and Ankle Biomechanics Community (i-FAB-PG) attempting to form a Consensus Document on the topic. https://pubmed.ncbi.nlm.nih.gov/23007050/. The Italian National Institute of Health (ISS) hosts online moodle-based Forum dedicated to PMDs at http://vcms.oss.it/moodle19/. https://www.scielosp.org/article/aiss/2012.v48n3/259-271/en/.

In Giacomozzi et al. [7], an attempted standardization consensus at the time has been put forward. At least three fundamental steps have to be followed towards standardization in the use of pressure measurement devices (PMDs):

1. Definition and standardization of tools and protocols for the technical assessment of pressure measurement devices' hardware performance;
2. Definition and standardization of pressure acquisition protocols;
3. Definition and standardization of data processing and reporting.

The present document [7] only deals with the first step. It constitutes the final output of a dedicated Consensus Activity ("Agreement on pressure measurement devices' hardware performance"), which has been conducted by the above mentioned Pedobarographic Group (i-FEB-PG) of the International Foot and Ankle Biomechanics Community (moodle.i-fab.org). The outcomes of the described consensus activity have been collected in eight tables. Table 1 resumes the basic principles over which agreement has been reached, and which represent the basis for the PMD technical assessment approach; Table 2 describes the parameters which should be included in the minimum set to be assessed in a factory; Table 3 reports the technical requirements a proper test system should have to deliver the parameters listed in Table 2; Table 4 deals with a suggestion for implementing in factory testing protocols; Table 5 gives some recommendations in terms of minimum requirements for Medical PMDs assessed to barefoot gait analysis; Table 6 contains a proposal for the Manufacturer's technical report to be associated with the commercial PMD; Table 7 deals with on-site, periodical assessment of PMDs; Table 8 contains a sort of checklist for judging the consistency of data. (In each table, comments and explanations are reported when a full agreement had not been reached with respect to the specific point).

In continuation, the following paragraphs of Chap. 7 aim at giving the reader a brief description of the currently available pressure sensor measurement devices as well as comments on standardization of their use. Some comments on using the Zebris pedobarographic platform commercial system—the most recent addendum to the Zagreb Biomechanics Laboratory equipment accessory—follow.

According to Giacomozzi [6], the pedobarography measurement method, along to being relatively cheap compared to other standard biomechanical measurement instruments, simple to use, practical, and of satisfactory precision and accuracy, provides information that is meaningful and effective in the assessment and monitoring of any foot or ankle treatment. However, despite the great interest of the scientific community in this potentially valuable assessment tool, not enough evidence has been gathered of the effectiveness, appropriateness and generalizability of the plantar pressure measurement methodology. This fact is attributable to several reasons, the most important one being the lack of standardization (status in 2011).

7.2 On Pedobarographic Sensor Technology and Devices' Technical Assessment

In [1], several sensor principles used in design of pedobarographic systems at the time are described. Giacomozzi, further, has given a detailed elaboration of the subject matter assuming critical engineering design and evaluation standpoint [6]. Following types of sensors are used for pedobarography: optical devices (such as in [2]), pneumatic discrete sensors, discrete -or matrices of- resistive sensors and discrete -or matrices of- capacitive sensors. A textile sensor prototype has been developed, novel at the time. As already mentioned, only vertical pressure is being detected.

For proper detection of localized peak pressures under the foot, a linear dimension of 2–3 mm is considered reasonable for a sensor, basing this estimation on location and dimension of the metatarsal heads and on the expected local pressure curve [6].

Devices' technical assessment in a factory is to be provided, as stated in [7]. It becomes mandatory when such measurements are used for clinical purposes, since in this case the PMD is addressed as a medical device with measuring function, and must comply with specific market regulations. Recommended features of testing equipment for pressure platforms to be used for gait analysis in clinics (i.e. Medical PMDs) are, at minimum: pressure resolution below 10 kPa; force resolution below 1 N; spatial positioning re-positioning error lower than 2 mm, and sinusoidal loading–unloading cycles applied with a frequency in the range 0.5–1.0 Hz. PMD performance should be fully characterized in-factory before the device is delivered, or in a research laboratory in case of prototypes. Owing to ageing effects, it is not sufficient to guarantee its appropriate use over time. Thus, the periodic, simplified checking of technical features should also be implemented on-site, so as to early detect any deterioration of the device entailing the worsening of its measuring performance. In case of in-shoe systems, assessment is more complex since the devices should also be tested under flexed but repeatable conditions, and with respect to surfaces with adequate elasticity so as to reproduce as much as possible their most frequent working conditions.

7.3 Pressure Measurement Parameters and Protocols

Pedobarographic records are customarily interpreted by visual analysis, taking into account the range of pressures presented, supported by familiar color representation scale, as will be illustrated in Chap. 8 by numerous clinically relevant examples.

In an effort to better interpret data, to condense data on multiple walking trials, etc., some parameters obtained by signal processing were developed. The most often used parameters characterizing pedobarographic measurement information are listed below [6].

- *Peak pressure*: for each sensor, it is the highest pressure value experienced during the measurement. It is usually expressed in kPa, even though it is sometimes reported as PSI, N/cm^2, bar. A peak pressure (spatial) map may be conveniently presented as part of the report;
- *Mean pressure*: for each sensor, it is the pressure value averaged over the measurement period. Units are the same as for peak pressure. Averaging may be done over the entire measurement period or, alternatively, over the period when specific sensor has been loaded. Spatial map is typically a form of reporting;
- *Peak pressure curve* (usually known as PPC): shows instantaneous values of maximum pressure measured as a function of time;
- *Pressure–time integral or impulse* (usually indicated as PTI): this is the area under PPC;
- *Force–time integral or impulse* (usually indicated as FTI): obtained by calculating the area under the force curve, expressed in N * s;
- *Contact area*: the instantaneous value of loaded PMD area, expressed in cm^2, and
- *Center of pressure (COP) trajectory*: two dimensional array of values.

Besides to the above parameters, other interesting indicators were used in specific pressure-related studies, like specific pressure gradients or indicators related to CPO velocity or acceleration.

Giazomozzi, further, proposed some innovative parameters, at the time, naming them: a *Pressure-integral map*, an *Actual mean pressure map* and a *Loading time map*, each bringing its own potential advantages/disadvantages to interpretation of information contained in raw pedobarographic records [6].

Further considered are a so-called Geometry-based approach and an Anatomy-based approach.

It can be ascertained that pressure data processing remains a broad field where adequate mathematical/statistical methods are to be applied.

Finally, general thoughts on measurement and testing protocol follow, considering the conditions to be met in a laboratory environment in order to collect reliable and accurate measurement data in a non-invasive manner [7].

In the Zagreb Biomechanics Laboratory, the Zebris pedobarography system has been acquired and implemented [8–10]. Technical characteristics and sample of measurement signals are shown in Fig. 7.3.

We have tested metric characteristics of the method on normal population [8], measured and investigated gait in scoliosis [9] and in ankylosing spondylitis patients [10]. Zebris system has proven very handy, and, being portable, suitable for use in different environments. Focus of our research is in inclusion of pedobarographic evidence in biomechanical and kinesiological analyses of respective patient groups' gait. Researching ankylosing spondylitis, we collaborate closely with medical doctors specialized in physical medicine and rehabilitation.

Specifications

Measuring principle	capacitive	Hysteresis	< 3% (FS)
PC interface	USB	Interface	Video Modul Syncronisation
Measuring range	1 - 120 N/cm^2		Infrared transmitting
Accuracy	± 5 % (FS)		Sync. in / Sync. out

Type: FDM 1.5
Dimensions: 158.0 x 60.5 x 2.5 cm (L x W x H)
Sensor area: 149.0 x 54.2 cm (L x W)
Number of sensors: 11264
Sampling rate: 100 Hz, optional 200 Hz / 300 Hz

(a)

Fig. 7.3 Zebris platform. Technical characteristics (**a**) and sample of measurement signals (**b**) (The zebris FDM-System-Gait Analysis for Research and Clinical Applications, zebris Medical GmbH (Year of publication: 11/2008. With permission)

Fig. 7.3 (continued)

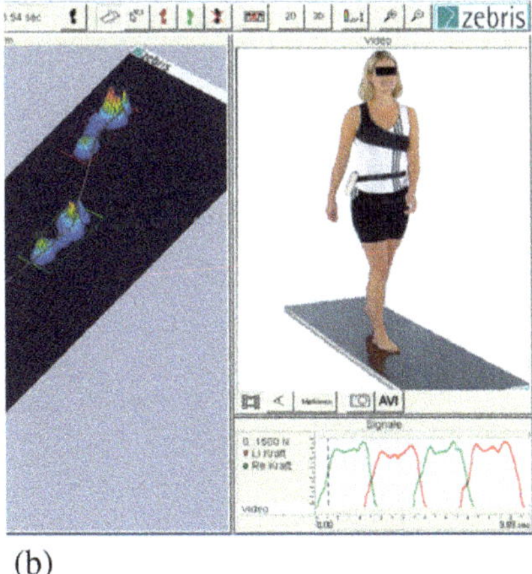

(b)

References

1. Medved V (2001) Measurement of kinetic variables. Medved V: Measurement of human locomotion. CRC Press, Boca Raton, FL, pp 154–168
2. Gefen A, Megido-Ravid M, Itzchak Y, Arcan M (2000) Biomechanical analysis of the three-dimensional foot structure during gait: a basic tool for clinical applications. Trans ASME 122:630–639
3. Baker R, Robb J (2006) Foot models for clinical gait analysis. Gait Posture 23:399–400
4. Baker R (2013) Measuring walking: a handbook of clinical gait analysis. Mac Keith Press, London
5. Andersen MS, MacWilliams BA (2010) A musculoskeletal foot model for clinical gait analysis. J Biomech 43(9): 1645–1652
6. Giacomozzi C (2011) Potentialities and criticalities of plantar pressure measurements in the study of foot biomechanics: devices, methodologies and applications. In: Klika V (ed) Biomechanics in applications. InTech, pp 249–274
7. Giacomozzi C, Keijsers N, Pataky T, Rosenbaum D (2012) International scientific consensus on medical plantar pressure measurement devices: technical requirements and performance. Ann Ist Super Sanita 48(3):259–271
8. Gruić I, Cebović K, Radaš J, Bolčević F, Medved V, Pedobarographic features of gait measured by FDM1.5 PMD. Regular Paper accepted as a Short Paper for a Poster Presentation. Paper Number: icSPORTS 2015 #82. Presented at: IcSports 2015 3rd International Congress on Sport Sciences Research and Technology Support 15–17 November 2015/Portugal, Lisbon, pp 66–71
9. Gruić I, Cebović K, Medved V (2016) Comparison of pedobarographic profile in young males with left and right scoliotic posture. In: Proceedings of the 4th International Congress on sport sciences research and technology support—vol 1: icSPORTS, ISBN 978-989-758-205-9, pp 89–95

10. Gruić I, Grubišić F, Grazio S, Medved V (2017) Pedobarographic profile of gait in patients with ankylosing spondylitis. In: Proceedings of the 5th International Congress on sport sciences research and technology support. Funchal, Madeira, Porugal: SCITEPRESS—Science and technology publications, pp 121–125. ISBN: 978-989-758-269-1

Chapter 8
Pedobarography Combined with Computerized Shoe Insole Design and Manufacture: Clinical Applications in Orthopedics and in Sports Medicine

Marko Pećina and Maja Mirković

Abstract Pedobarography can be used in clinical practice in many areas, as a diagnostic tool, in monitoring of the effect of therapy, both surgical and nonsurgical such as physical therapy and rehabilitation of injuries, painful conditions, degenerative and rheumatic diseases, habilitation of the children with developmental disorders. Pedobarography can also be used as a combination of diagnostics of gait disorders and computer-aided design (CAD) with computer-aided manufacturing (CAM), resulting in producing orthopedic insoles. There is a broad spectrum of possibilities using orthopedic insoles. A number of clinical findings in orthopedics and sports medicine (sports traumatology) are shown and interpreted. The use of orthopedic insoles in the most common diagnoses in orthopedics as there are Flat transverse arch and metatarsalgia, Flat feet and High-arched foot (pes cavus) are presented. The application of orthopedic insoles in different entities in sports medicine like there are Plantar fasciitis, Achilles tendinitis/tendinosis, Flexor hallucis longus muscle dysfunction, Posterior tibial muscle dysfunction, Peroneal muscles dysfunction, Anterior impingement syndrome of the ankle, Sesamoiditis, Stress fractures, Sever disease, Patellar tendinitis/tendinosis (jumper's knee), Osgood-Schlatter disease and Groin pain are described.

8.1 Introduction

From a medical point of view, foot pressure measurement has been an area of interest of many different professional groups including orthopedic surgeons, physical and rehabilitation specialists, prosthetists and orthotists, footwear manufacturers and those involved in biomedical and biomechanical research.

M. Pećina (✉)
School of Medicine, University of Zagreb, Šalata 2, 10000 Zagreb, Croatia
e-mail: marko.pecina@zg.t-com.hr

M. Mirković
Kinematika-Polyclinic for Orthopedy, Physical Medicine and Rehabilitation, Laginjina 16, 10000 Zagreb, Croatia

V. Medved (ed.), *Measurement and Analysis of Human Locomotion*,
Series in Biomedical Engineering, https://doi.org/10.1007/978-3-030-79685-3_8

Pedobarography can be used in clinical practice in many areas, as a diagnostic tool, in monitoring of the effect of therapy, both surgical and nonsurgical such as physical therapy and rehabilitation of injuries, painful conditions, degenerative and rheumatic diseases, habilitation of the children with developmental disorders. Pedobarography can also be used as a combination of diagnostics of gait disorders and computer-aided design (CAD) with computer-aided manufacturing (CAM), resulting in producing orthopedic insoles [1]. There is a broad spectrum of possibilities using orthopedic insoles.

A lot of clinical experience has been accumulated in Kinematika—Polyclinic for Orthopedy, Physical Medicine and Rehabilitation in Zagreb, and here we present some results.

8.2 Foot

Human foot is a complex structure. The base of the foot is formed by 26 bones connected in joints with ligaments and joint capsules. Muscles are maintaining foot arches. There are four foot arches formed by bones and muscles. Three among them are important in clinical practice (Fig. 8.1):

- Longitudinal arch between the calcaneal tuberosity and first metatarsal head,
- External longitudinal arch between the calcaneal tuberosity and fifth metatarsal head,
- Transverse arch between the head of first and fifth metatarsal bone.

Optimal distribution of plantar pressures is 3:2:1. Largest is the pressure beneath the heel. In forefoot, the pressure beneath first metatarsal head has to be twice as

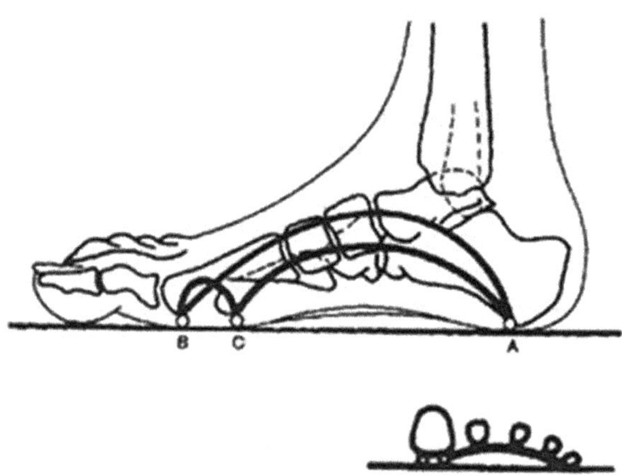

Fig. 8.1 Foot arches

large as beneath fifth metatarsal head (Figs. 8.2 and 8.3). At the beginning of stance phase there is a heel contact, then midstance when lateral part of foot is loaded, and propulsion through forefoot and big toe at the end of the stance phase [2–5] (Chap. 10).

Muscles' fatigue results in lowering of arches of the foot. That means that the foot is wider and longer when muscles are tired. When the foot and lower leg muscles are rested, foot is shortened and arches are formed.

The most important mechanisms in arch maintenance make *peroneus longus* muscle and *tibialis posterior* muscle (Fig. 8.4). They maintain longitudinal arches. Medial longitudinal arch is supported also by *flexor hallucis longus* tendon, *tibialis anterior* tendon, *flexor hallucis brevis* tendon and *abductor hallucis* tendon. Transverse arch is supported by transverse head of adductor hallucis, peroneus longus tendon and oblique head of *adductor hallucis* muscle tendon. Impaired function or strength of any of those muscles leads to the deformation of foot arches because of discrepancy of muscle strength and body weight [3–5].

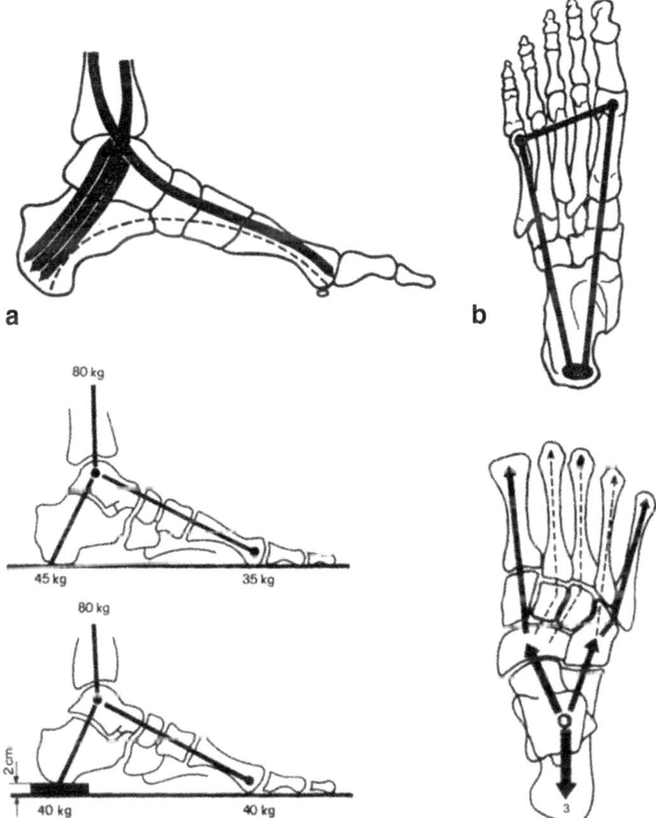

Fig. 8.2 Distribution of plantar pressure

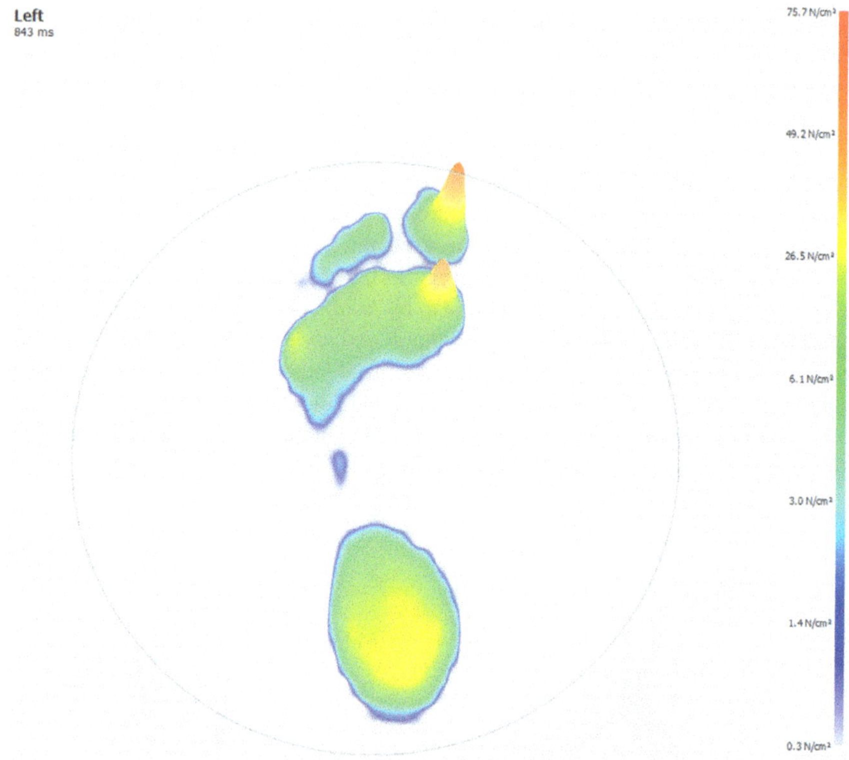

Fig. 8.3 Pedobarogram of normal distribution of foot pressure (pressure in N/cm^2), Footscan 9 (manufacturer: RS Scan)

Compared to everyday activities, athlete's feet are loaded much more, both in static and dynamic activities. More loaded feet leads to faster muscle fatigue. Long term results such as changes in bones, ligaments, muscles and tendons caused by fallen foot arches are mostly manifested as clinical symptoms or syndromes. The most frequent is faster muscle fatigue accompanied by pain and cramps. Those are acute difficulties. Long term foot overloading leads to the development of so-called overuse syndromes [6]. In those cases possibilities of injuries are increased.

Both prevention and treatment can be active and passive. Active prevention and treatment consists of strengthening and stretching exercises. Sometimes treatment also means rest, applying physical modalities or drugs. Passive prevention and treatment are supporting impaired function of the muscles by orthopedic insoles.

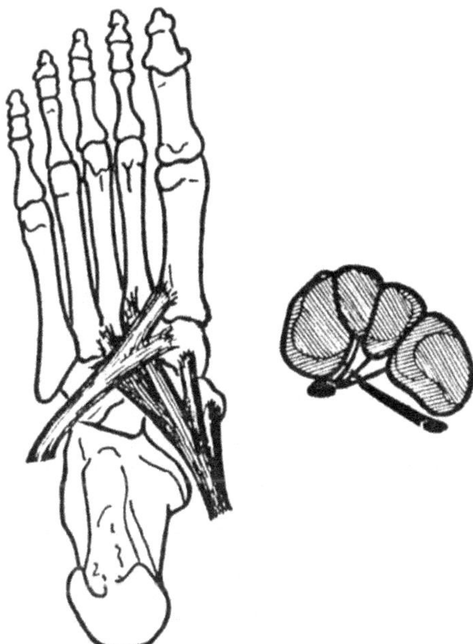

Fig. 8.4 Musculus *peroneus longus* and musculus *tibialis posterior* are most important in arch maintenance

8.3 Orthopedic Insoles

In childhood orthopedic insoles are used as corrective device. We often find planovalgus (flexible flat feet) which is physiological in first few years of the childhood and doesn't need correction with the orthopedic insoles if not symptomatic (Fig. 8.5). In spite of that, in many cases orthopedic insoles are unjustifiedly proscribed [2]. If the child has symptoms, like fatigue after walking or playing, if it has spasms, pain in the feet and lower legs, or wakes up during night with the pain in feet and lower legs asking to be massaged, then orthopedic insoles are to be applied as help for the muscles that are overloaded. Having flat feet as clinical or pedobarographic finding itself needs monitoring during the growth and intervention only if physiological improvement during the growth and development of the body fails.

In adulthood, orthopedic insoles can be used as a device for relieving the symptoms or as preventive ones, like in situations of practicing sports or if standing during working process (like it is the case with dentists, salesmen, hairdressers etc.). Symptoms may exist in feet and ankles, but also in other structures of the closed kinetic chain: knees, hips, lower back and all the muscles active during standing, walking, running or jumping.

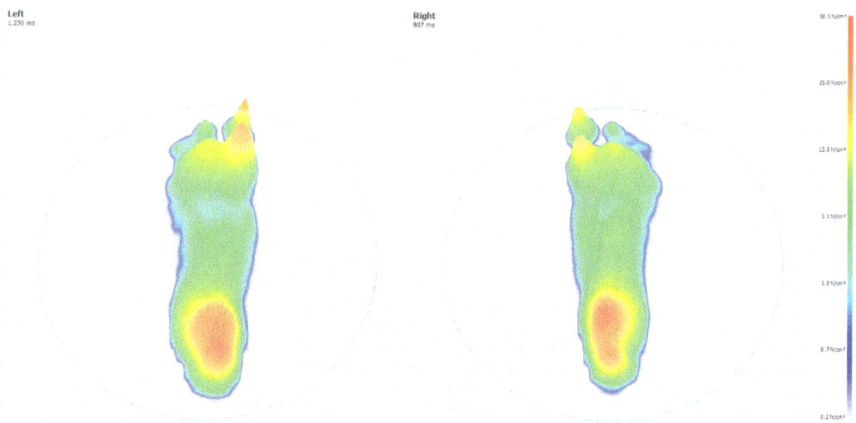

Fig. 8.5 Foot pressure images of child's feet

Orthopedic insoles for children are a corrective device. Those insoles are made of rigid materials with corrective function. They are usually covered with the layer of soft material to prevent discomfort.

In adults orthopedic insoles are supportive and provide relief. The materials used for adult orthopedic insoles are semirigid or soft, depending on the age, body weight and diagnosis. Insoles can be covered by leather and artificial materials (Fig. 8.6).

In athletes orthopedic insoles have a specific role. They are proscribed preventively at risk groups such as basketball, handball, volleyball players, tennis players, long distance racing athletes, high and long jumpers. The function of orthopedic insoles is to unload the specific muscle group known to be overloaded during the sport activity [7, 8]. The materials for sport insoles are softer with special shock absorbing function and usually covered with artificial material that prevents grinding. The materials for

Fig. 8.6 Orthopedic insoles

sport insoles should also be lighter than for everyday use, so that material wouldn't increase the footwear weight which is very important in athletics for instance.

8.4 Computer Aided Design (CAD) and Computer Aided Manufacturing (CAM) Insoles

Nowadays, using pedobarography makes possible to do much more accurate measurement than before [9–12]. Measurement can be both static and dynamic (Fig. 8.7). For the dynamic ones during walking or running at least five measurements should be taken (Fig. 8.8). Combination of the measurement and the clinical data give us the information about fallen foot arches, incorrect position for each foot; antalgic way of walking, the leg length discrepancy and moving center of gravity of the body to left or right side.

Those data have to be processed (Fig. 8.9). Digital foot pressure image is then converted to the extension applicable in orthopedic insole designing program (Figs. 8.10 and 8.11). Corrections can be made according to data and clinical finding. Designed orthopedic insoles are afterwards made by the Computer Numerical Control (CNC) milling machine.

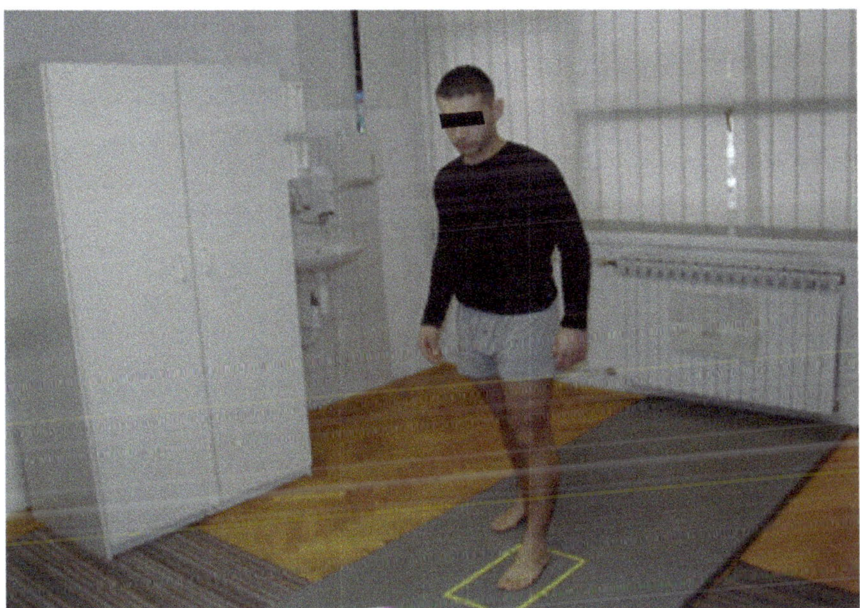

Fig. 8.7 Dynamic measurement of foot pressure on the pedobarographic platform. Kinematika—Polyclinic for Orthopedy, Physical Medicine and Rehabilitation in Zagreb (from Mirković et al. [13]. With permission)

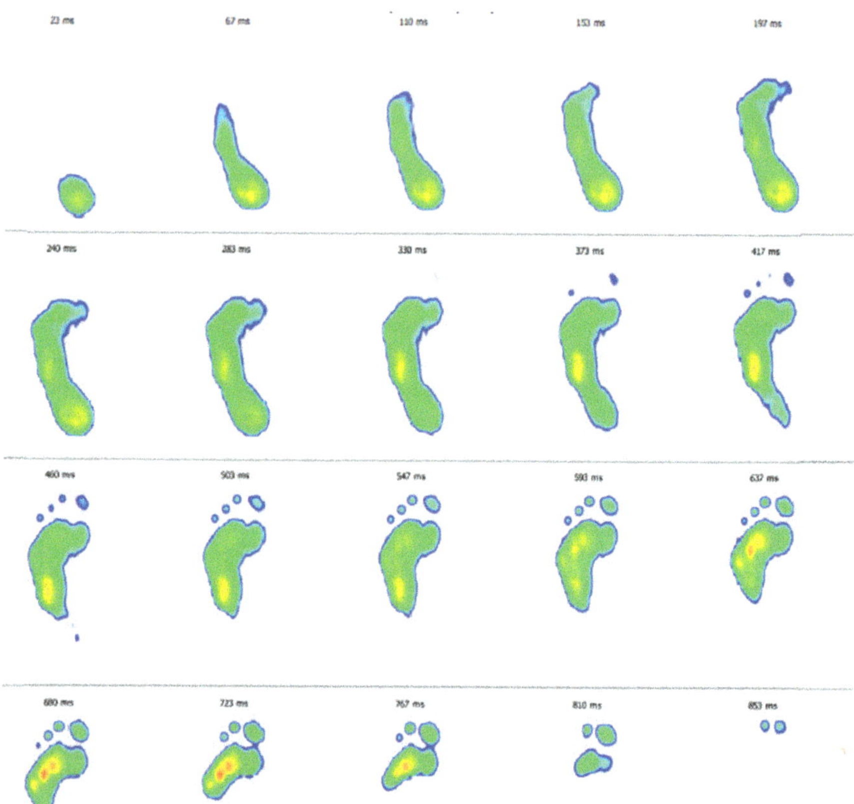

Fig. 8.8 Roll off printout of a step in 20 segments

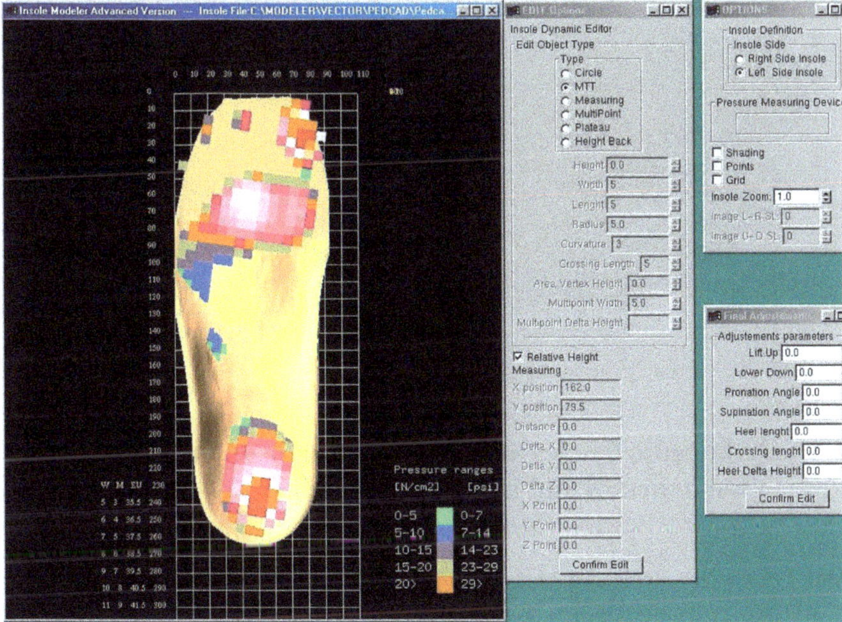

Fig. 8.9 CAD (Computer aided design) by modeler

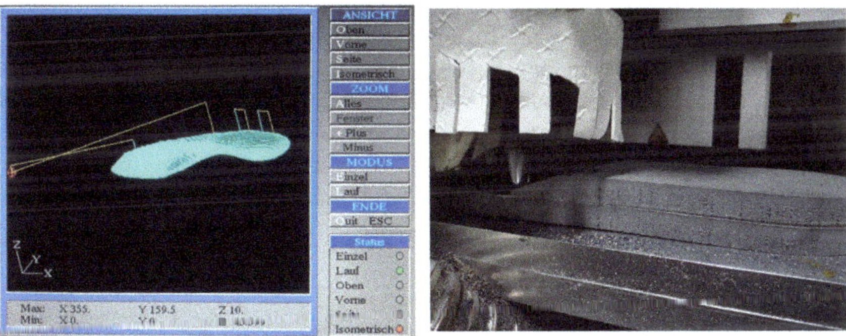

Fig. 8.10 CAM (Computer aided manufacturing) and CNC (Computer numerical control) milling machine (manufacturer: Isel)

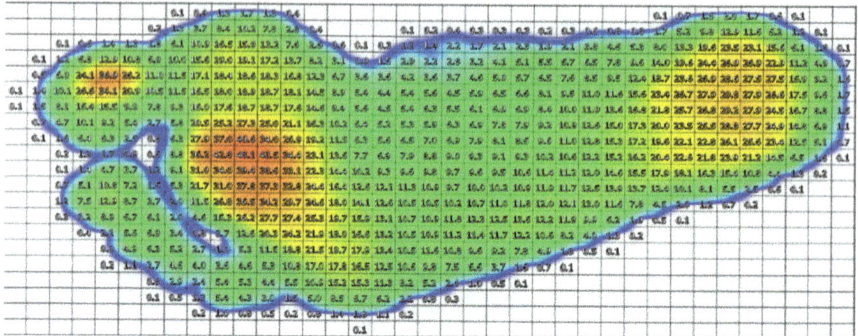

Fig. 8.11 Pressures in N/cm^2 at each sensor

8.5 Orthopedic Insoles and Most Common Diagnoses in Orthopedics and in Sports Medicine

Flat Transverse Arch and Metatarsalgia

The most common finding according to the pedobarographic measurement is flat transverse arch. That can be understandable if we consider some facts. The human being is not completely adapted to two-legged gait [4, Chap. 3]. We are surrounded mostly with flat surfaces for walking. We wear shoes that hinder some muscles to do their "job", especially in women when wearing high heel shoes.

Some people have symptoms according to the finding such as pain or burning sensation in forefoot area. Usually pain occurs beneath second and third metatarsal bone head because those metatarsal bones are longer then others and too much pressure takes place (Fig. 8.12). It is described as the forefoot pain (metatarsalgia). Blisters, hard skin, calluses or corns can occur. Deformation of big toe (bunion), little toe (bunionette) and other toes (flexed toes) can develop [14, 15].

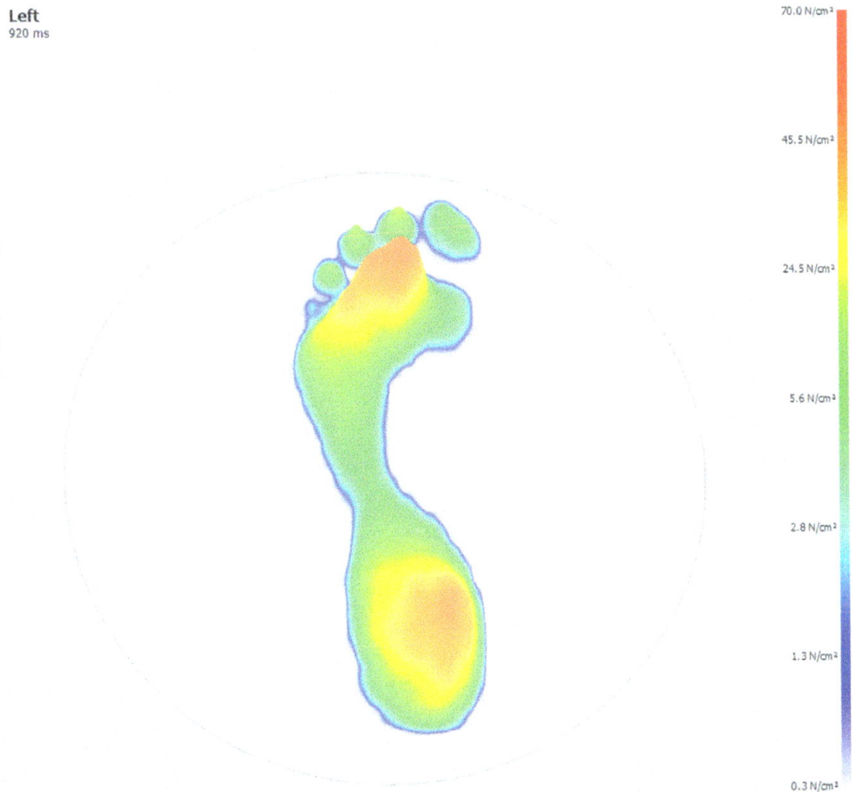

Fig. 8.12 Pedobarogram of fallen transverse arch: increased pressures beneath heads of second and third metatarsal bone

In those cases orthopedic insoles are applied. Metatarsal pad reduces the pressure beneath metatarsal bone heads and pain can be controlled by that. Thickening of the skin appears slower then without wearing insoles.

Flat Feet

As mentioned already, the flat feet (pes planus) in longitudinal arch during childhood is physiological in first few years of childhood and doesn't need correction with insoles if not symptomatic. It is usually accompanied with valgus position of the heel or foot: inner part of heel or foot is more loaded, so-called pes planovalgus (Fig. 8.13).

Flat feet, both in children or adults can produce symptoms like the pain in sole area, cramps, muscle fatigue. Adult patients with flat feet have to be suplied with orthopedic insoles which provide support of longitudinal arch and correction of valgus position if required (Fig. 8.14).

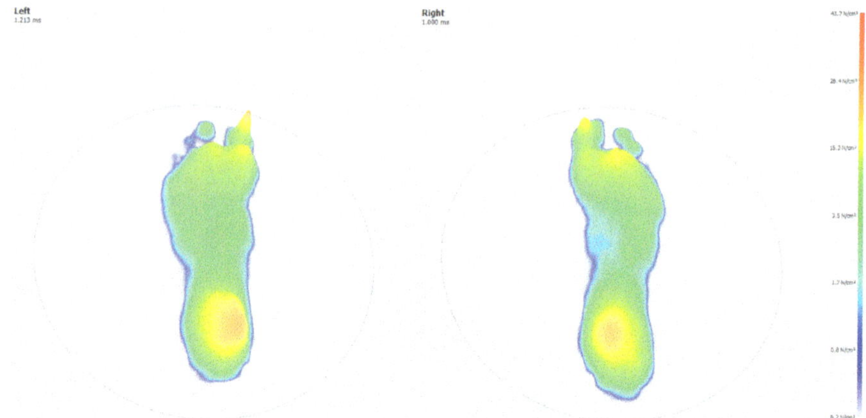

Fig. 8.13 Pedobarograms of flat feet in childhood: foot pressure image wider than normal in area of the longitudinal arch

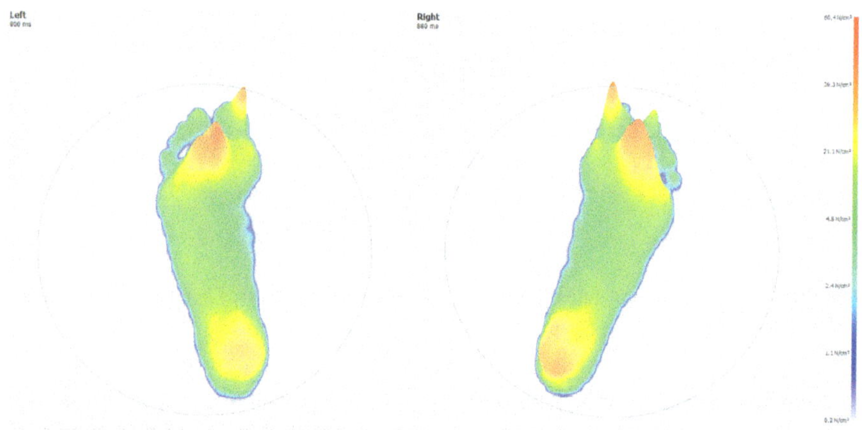

Fig. 8.14 Pedobarograms of flat feet in adulthood: foot pressure image wider than normal in area of the longitudinal arch

High-Arched Foot (Pes Cavus)

Opposite type to the flat foot is high-arched foot (pes cavus) with high longitudinal arch, higher than normally. It appears in 10% of people as morphological type of the foot. In these cases pes cavus can be the sign of neurological or neuromuscular disease. Because of high longitudinal arch, foot pressure images manifest a gap between forefoot and hindfoot (Fig. 8.15). Sometimes it is difficult for the muscles to maintain that height and longitudinal arch descends which causes tense sole, sometimes painful.

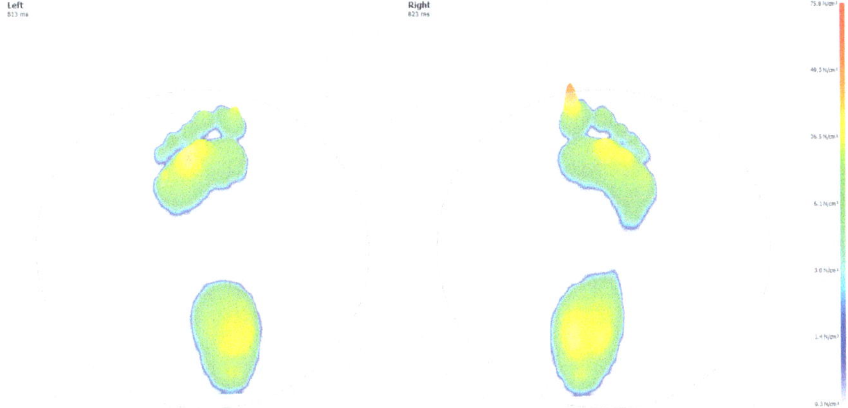

Fig. 8.15 Pedobarograms of feet with high longitudinal a rch (pes cavus): gap between forefoot and hindfoot

Orthopedic insoles for those patients must have support for longitudinal arch much higher than usual [16].

Plantar Fasciitis

One of the most common overuse syndromes of the foot is plantar fasciitis. Plantar fasciitis occurs when strong band of tissue (Fig. 8.16) that supports the arch of foot becomes irritated or inflamed. Plantar fascia elongates with increased loads to act as a shock absorber, but it's ability to elongate is limited (Fig. 8.17). Plantar fasciitis can develop both in flat feet or in those with high longitudinal arch (pes cavus). Valgus position of the heel or whole foot can contribute to the earlier manifestation of plantar fasciitis (Fig. 8.18). When inner part of the foot is overloaded medial bunch of plantar fascia is overloaded too [18]. Pain can occur in the morning when first few steps are made and then it disappears when muscles start to maintain longitudinal arch. But after some time during everyday or sport activities pain appears again. Muscles become tired and the arch descends. Tense painful sole or painful insertion of plantar fascia at calcaneus (heel bone) are symptoms characteristic for plantar fasciitis.

Conservative treatment includes rest, strengthening and stretching exercises and sometimes physical therapy procedures like extracorporeal shock wave. The use of plantar fascia night splint or orthosis has been effective in decreasing discomfort. Orthopedic insoles have to support both transverse and longitudinal arch and improve incorrect position of the foot. Usually they are made with higher heel support made of softer material for pain relief.

Surgical treatment is recommended when symptoms last longer than one year and do not recede in spite of conservative treatment. The open plantar fasciotomy is an effective surgical approach and recently techniques have been described whereby the plantar fasciotomy can be performed endoscopically.

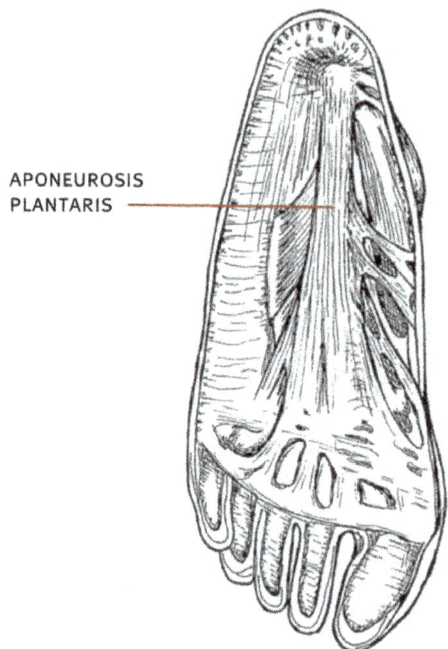

Fig. 8.16 Plantar fascia

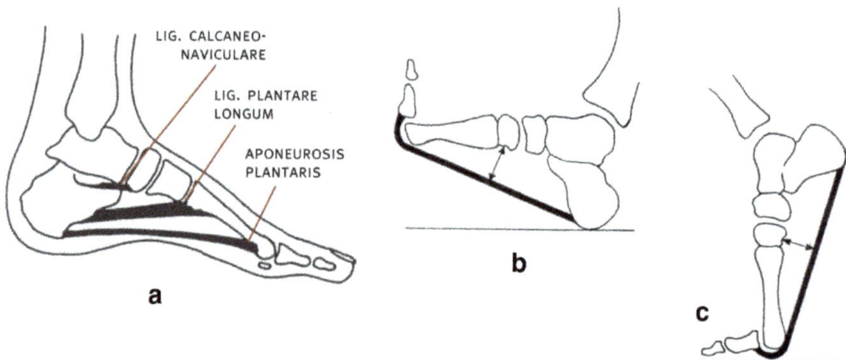

Fig. 8.17 Plantar fascia maintain foot arch together with long plantar ligament and calcaneonavicular ligament **a**, Action of the plantar fascia: active loading of the foot in the pre-stance phase as a shock-absorbing mechanism **b**, Action of plantar fascia: passive action in late stance phase producing so-called windlass effect **c** (**b** and **c** from Smerdelj and Mađarević [17]. With permission)

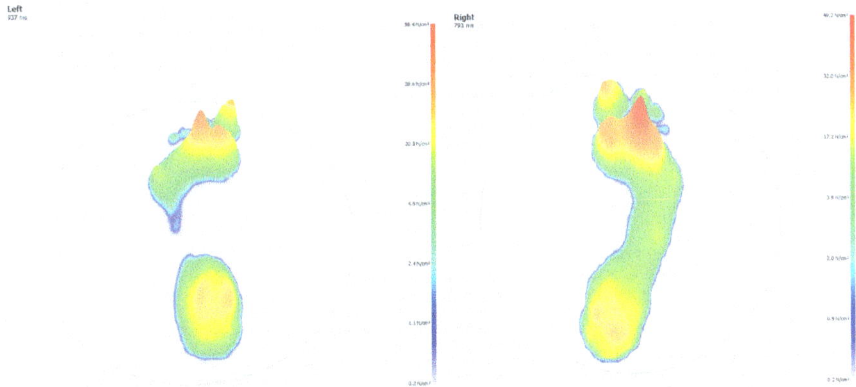

Fig. 8.18 Pedobarograms of two feet with plantar fasciitis, cavus foot and flat foot, both with valgus position of the foot: increased pressure of inner part of the foot

Achilles Tendinitis/Tendinosis

The Achilles tendon is a common site for the development of overuse injuries. Depending of localization the new terminology recommends the following terms and definition: paratenonitis—an inflammation of the paratenon; paratenonitis with tendinosis—paratenon inflammation associated with intratendinous degeneration; tendinosis—intratendinous degeneration due to atrophy (aging, microtrauma, vascular compromise, etc.); tendinitis—symptomatic degeneration of the tendon with vascular disruption and inflammatory repair response. The development of Achilles tendinitis/tendinosis is correlated with placement of excessive force on the Achilles tendon during walking and running (Fig. 8.19). It can be stretching force or compression force (the reactive force of surface) and torsion force (walking on uneven terrain). These forces can be increased by the predisposing factors like flat or pronated foot, highly arched foot, „tight" Achilles tendon, etc.

Achilles tendinitis/tendinosis appears in athletes practicing sports where repeated running or jumping causes inflammation/degeneration of tendon. It can also appear when intensive walking walking, especially on hard surfaces is present, like in soldiers, mountaineers or pilgrims. Sometimes it develops for years and the last stage of this overuse injury can be rupture of the tendon (Fig. 8.20). When we rest, muscles are recovering and next walk or sport activity can be performed. But at one moment, like when one drop pours a glass, symptoms appear. High longitudinal arch is a predisposing factor because usually in those people there is a too short gastrocnemius muscle (Fig. 8.21).

Treatment of Achilles tendinitis/tendinosis is directed toward lessening or eliminating pain, controlling the inflammatory reaction, facilitating the healing process and controlling the biomechanical parameters. Treatment is non-operative in the vast majority of cases; surgical treatment is indicated in exceptional cases.

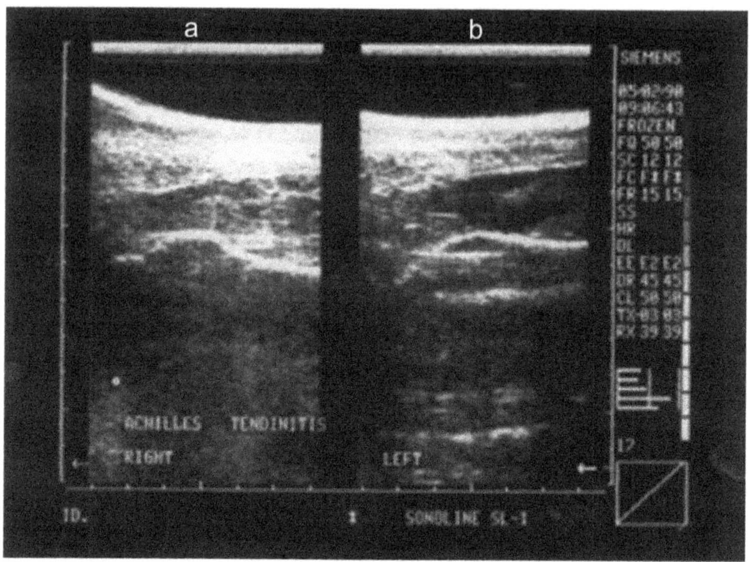

Fig. 8.19 Sonogram of the Achilles tendon. **a** Achilles tendinitis/tendinosis, **b** Normal tendon

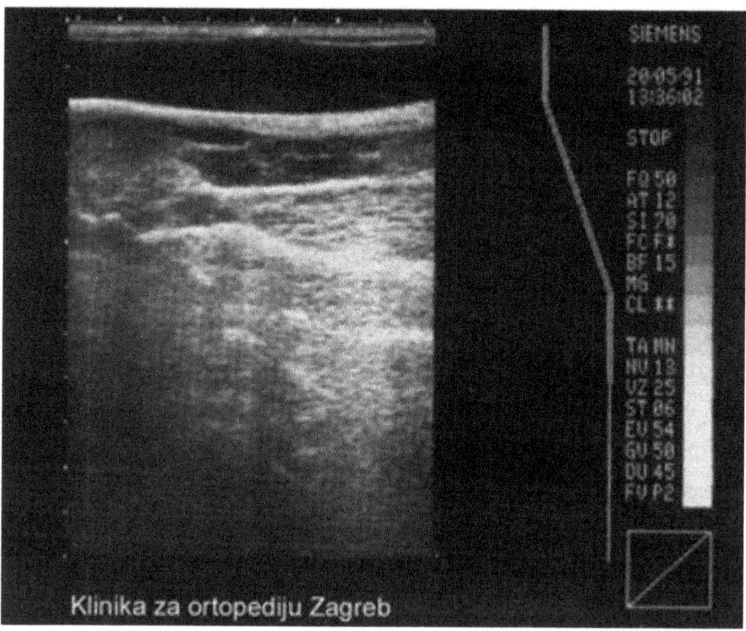

Fig. 8.20 Sonogram of the Achilles tendon. Acute rupture of the tendon

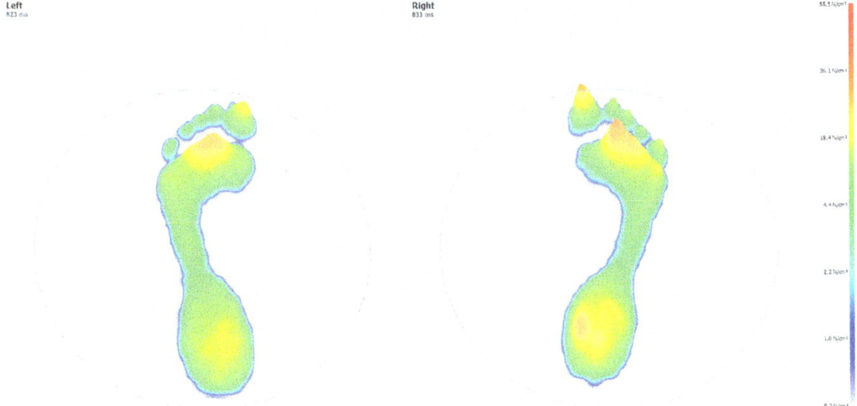

Fig. 8.21 Pedobarograms of patient suffering Achilles tendinosis of the left foot: less pressure beneath heel of affected leg

Orthopedic insoles with elevation of the heel (heel cushion) and softer heel support can reduce pulling force of gastrocnemius muscle at insertion on calcaneus (heel bone). Wearing orthopedic insoles should be accompanied by stretching and strengthening exercises (should be carried persistently).

Surgical treatment is based on the location and removal of the pathological changes in or on the tendon. Total ruptures of the tendon, which are the result of the final stage of Achilles tendinitis/tendinosis, are treated surgically using different operative techniques: open, percutaneous, percutaneous with mini-open, or endoscopy-assisted percutaneous repair of the tendon.

Flexor Hallucis Longus Muscle Dysfunction

When the *flexor hallucis longus* muscle is overloaded, that can cause pain in hindfoot and behind medial malleolus. *Flexor hallucis longus* provides support of longitudinal arch of the foot. Tendinitis/tendinosis of the *flexor hallucis longus* muscle is a common entity in ballet dancers; therefore, it is known as dancer's tendinitis.

Orthopedic insole has to support longitudinal arch and supinate the foot to provide correct position of the foot so that *flexor hallucis longus* muscle not be overloaded during activities as it is for example in long distance runners (Figs. 8.22 and 8.23). Treatment includes rest, strengthening and stretching exercises and sometimes physical therapy procedures and very rarely surgical treatment.

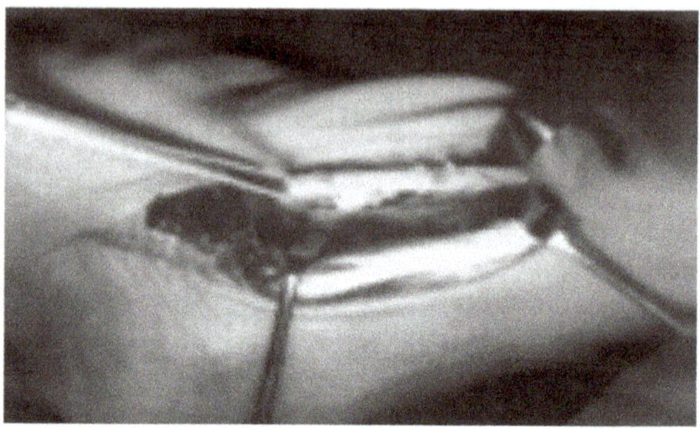

Fig. 8.22 Intraoperative finding of partial rupture of the *flexor halluces longus* muscle in a long-distance runner

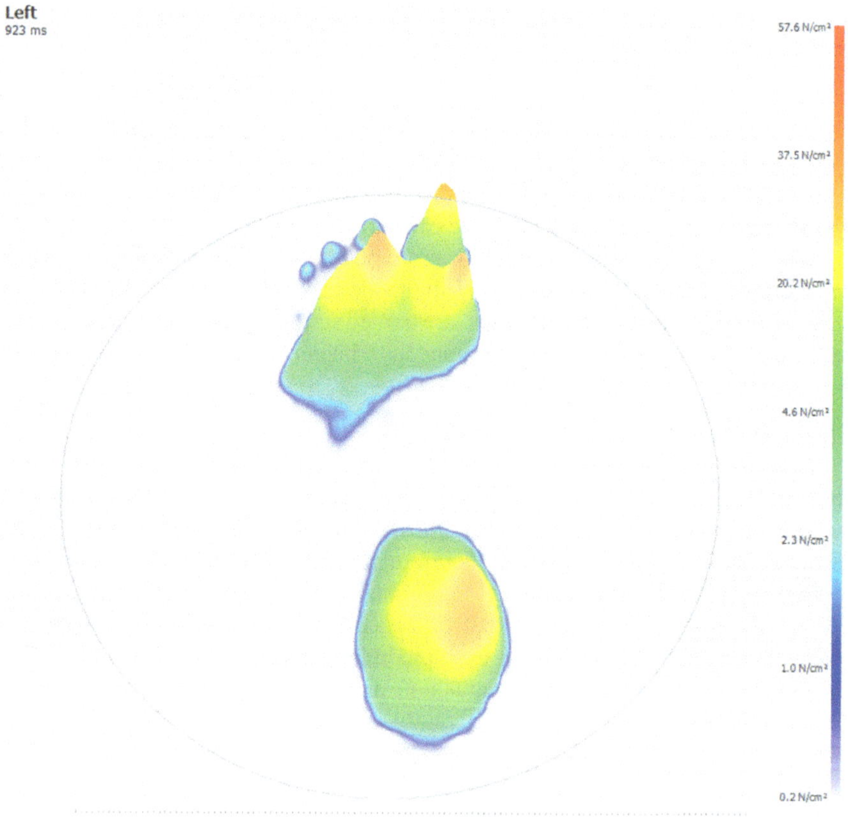

Fig. 8.23 Pedobarogram of foot with overloaded *flexor hallucis longus* muscle: high longitudinal arch and increased foot pressure of the inner part of foot

Posterior Tibial Muscle Dysfunction

The contribution of the *posterior tibial* muscle to forming the arch of the foot is of paramount importance. For this reason, the exertion of that muscle during athletic activities in individuals with flat-foot deformity is considerable (Fig. 8.24). Sometimes it is overloaded during walking especially when someone has flat feet or valgus position of the foot [18]. Dancers, long and high jumpers, runners and other athletes can develop *posterior tibial* dysfunction. Excessive exertion of this muscle can manifest either by the appearance of pain in the region of muscle origin on the posterior side of the tibia (runner's leg or shin splints) or on medial hindfoot and at insertion of *posterior tibial* muscle tendon at the navicular tuberosity.

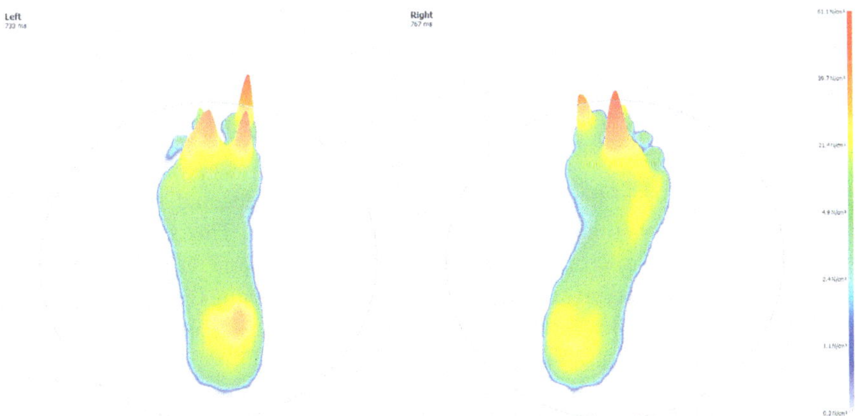

Fig. 8.24 Pedobarograms of feet with tibialis posterior muscle dysfunction of left foot: flat feet with overloaded medial part of heels and forefoot of the left foot

Like in *flexor hallucis longus* disorder orthopedic insole should provide longitudinal arch support and usually supinate the foot.

Peroneal Muscles Dysfunction

Peroneus brevis muscle is attached to the dorsolateral surface of the fifth metatarsal base. Overuse syndrome can cause minor painful swelling and skin redness of the area of insertion. When tendinopathy of *peroneus longus* and *peroneus brevis* muscles develops, pain and swelling of posterior lateral malleolus is present. The overloaded lateral part of the foot is the main reason of *peroneal* muscles dysfunction (Fig. 8.25). The entity of dislocation of the peroneal tendon and peroneal tenosynovitis occurs more frequently than is usually diagnosed especially in younger athletes.

Orthopedic insoles with applied pronation of the foot provide relief of overloaded exterior part of the foot.

Anterior Impingement Syndrome of the Ankle

Anterior impingement syndrome of the ankle develops when repeated maximal dorsal flexion of foot causes collision of anterior tibial edge and neck of the talus. It leads to formation of exostoses (osteophytes, bone spurs) on both bones. Exostoses

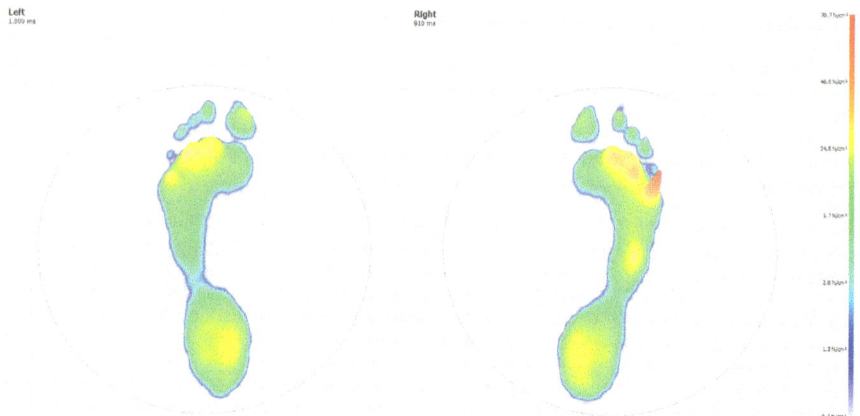

Fig. 8.25 Pedobarograms of feet with *peroneal muscles*' dysfunction of the right foot: overloaded lateral part of right foot

disable maximal dorsal flexion of the ankle and further repeated attempts bring to onset of pain and swelling of anterior part of ankle joint. This syndrome is most often encountered in soccer players and for this reason, some authors refer to this syndrome as soccer players' exostosis on the neck of the talus (Fig. 8.26).

Basis of treatment in anterior ankle impingement syndrome is avoiding maximal dorsal flexion of the foot. Orthopedic insoles with elevation of the heel (1 or even to 2 cm) have to be considered in treatment. Arthroscopic resection of the anterior tibial and talar spurs is successful treatment for the anterior impingement syndrome of the ankle.

Sesamoiditis

Sesamoid bones are inserted into tendons of certain muscles, usually where the tendon is passing over a joint, such as in the first metatarsophalangeal (MTP) joints.

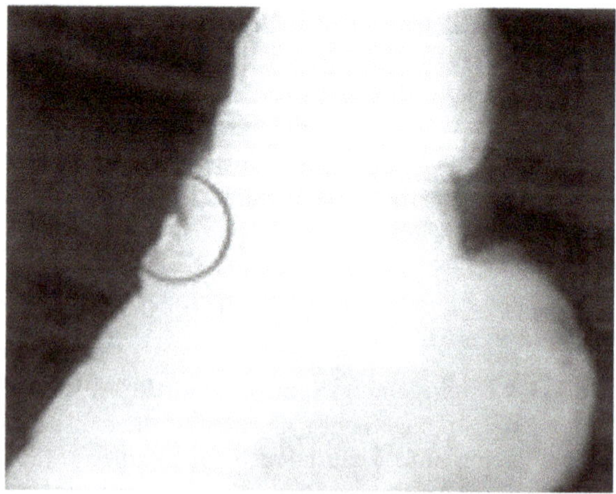

Fig. 8.26 Soccer players' exostosis on the neck of the talus

The knee cap (patella) is the largest one sesamoid bone. Sesamoid bones play an essential part in the first MTP joint biomechanics, together with other articular surfaces, joint capsule, plantar fascia, ligaments and tendons. In this complex we find two sesamoid bones, tibial (medial) sesamoid bone embedded into the medial head of the *flexor hallucis brevis* muscle (FHB) and fibular (lateral) sesamoid bone embedded into the lateral head of FHB. Both of these bones are located under the head of the first metatarsal bone distance. Their function is to distribute and ease the pressure produced during gait that affects the head of the first metatarsal bone, to lower the friction on the *flexor hallucis longus* muscle which passes between these two bones, as well as to act as a fulcrum for FHB. Sesamoids act like pulleys. They provide smooth surface over which tendons slide, thus increasing ability of the tendons to transmit muscle forces. Sesamoid bones of the great toe ossify at age of 8–9 in girls and at age of 11–13 in boys. During the ossification process the tibial sesamoid bone can be divided into two or more parts, thus forming a bipartite or multipartite bone.

As a part of the first MTP joint complex, sesamoid bones are prone to different acute and chronic injuries, such as acute fracture caused by trauma, stress fractures, chondromalacia, osseocartilaginous injury, osteonecrosis (Fig. 8.27), osteochondritis, bursitis, degenerative changes, inflammation, developmental anomalies, all of which is clinically diagnosed under a broad term called sesamoiditis [19, 20].

The mechanism of injury is most commonly associated with overuse of the anterior part of the sole of the foot accompanied by excessive dorsiflexion of the great toe. This usually occurs in sports that involve running and jumping, especially in the initial phase of running and during sprinting. Moreover, ballet dancers often use "relevé" position of the sole of the foot, thus putting a great amount of load on the head of the first metatarsal bone (Fig. 8.28). Furthermore, deformities such as pes cavus, conditions like a rigid sole of the foot and hard surfaces where different sports are played can facilitate development of sesamoiditis, often colloquially called turf toe in such cases.

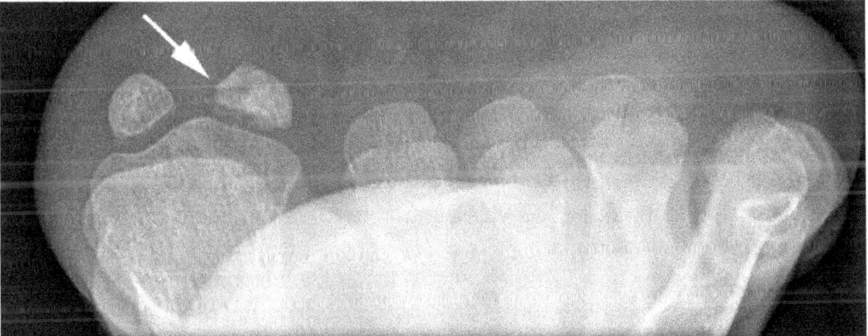

Fig. 8.27 Direct axial radiograph of sesamoid bones shows sclerotic and fragmented fibular sesamoid bone as a sign of osteonecrosis (arrow)

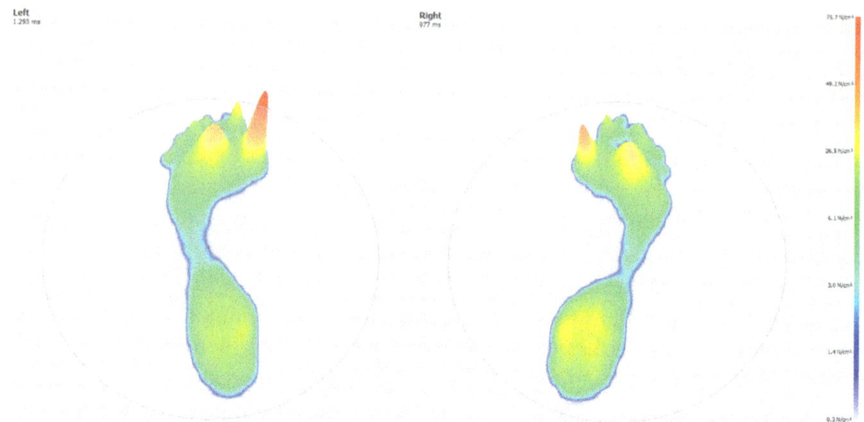

Fig. 8.28 Pedobarograms of ballet dancer with sesamoiditis of left foot: increased foot pressure beneath head of the first metatarsal bone

Typical symptom in sesamoiditis is pain beneath the head of first metatarsal bone which increases during walking on toes. Passive extension of the big toe increases pain too. Sometimes swelling and bruising occur.

Sesamoiditis is difficult to treat, primarily because it is hard to avoid placing pressure on the sesamoid bones when walking. Orthopedic insole with good support of longitudinal and transverse arch should be applied. Additionally soft pad in the area of head of the first metatarsal bone can be designed as dent in CAD and filled with softer material in CAM producing of insoles.

In long standing cases surgical excision of the sesamoid bone can be performed.

Stress Fractures

Stress fractures, which are classified among overuse injuries of bone, may be defined as partial or complete bone fracture that results from repeated application of stress of less strength than the stress required to fracture the bone in a single loading. Stress fractures have been described in nearly every bone of the human body but most commonly they occur in the lover extremity, that is in the foot (Fig. 8.29). They are common in runners and athletes who participate in running sports, such as soccer, basketball, etc. Stress fracture usually occurs when someone changes his activities, like trying new exercise or increasing the intensity of workout after period of rest (like increased activity during preparation of the athlete after period of rest [21]). Changing of the workout surface can contribute to development (jogging outside versus jogging on a treadmill).

The weight-bearing bones of the foot and lower leg are vulnerable to this type of fractures because of repetitive forces they must absorb during running and jumping. One of the frequent stress fractures is so-called Jones' fracture in soccer players (Fig. 8.30) Muscles get tired, they cannot maintain foot arches and stress forces occur.

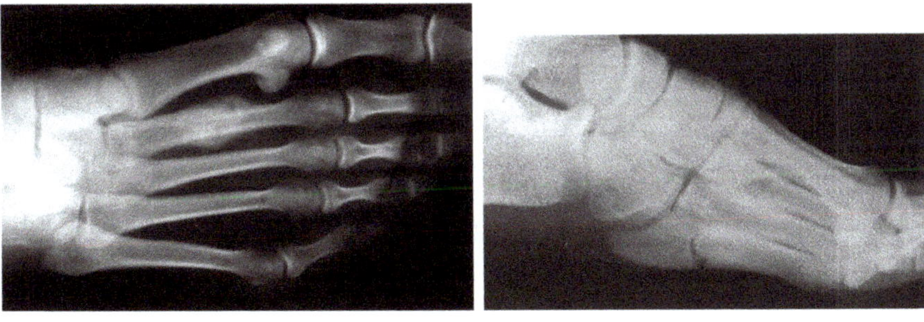

Fig. 8.29 Stress fracture of the second metatarsal bone and of the fifth metatarsal bone (so-called Jones' fracture) (Figure on right from Smerdelj and Mađarević [17]. With permission)

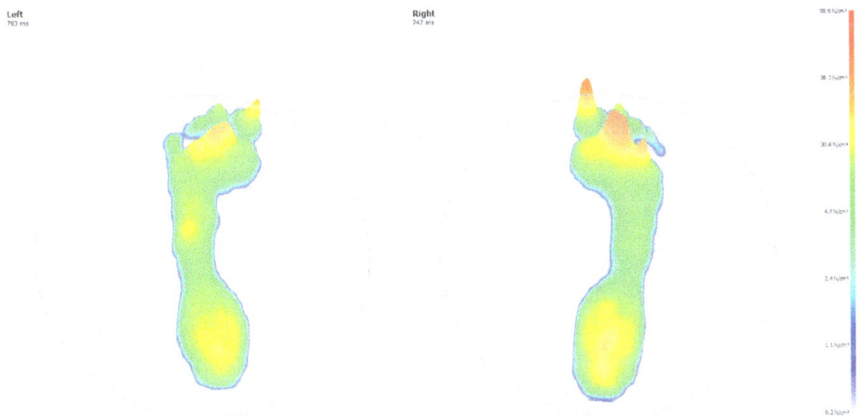

Fig. 8.30 Pedobarogram of soccer player with stress fracture of base of fifth metatarsal bone of the left foot: increased pressure of the outer part of the left foot

Refraining from high impact activities for an adequate period of time is key to recovering from a stress fracture. The majority of stress fractures are treated nonsurgically. It takes 6 to 8 weeks for a stress fracture to heal. During that time alternative activities like swimming and cycling can be performed.

Stress fracture of the foot can occur in a second [22] and third metatarsal bone, fifth metatarsal bone—Jones' fracture [23] and navicular bone. Not so common are stress fractures of talus and calcaneal bone.

Prevention and treatment of stress fractures is being implemented also by using orthopedic insoles which provide good force distribution, so that none area of the foot is overloaded [13, 24–27].

Sever Disease

Sever disease (calcaneal apophysitis) is one of the common causes of heel pain in children. It occurs during growth spurt. This group of disorders known as apophyseal injuries or traction apophysitis is characterized by irritation of apophyses due to excessive tensile forces at the point of attachment of a musculotendinous unit to the bone. Previously believed to be inflammatory in nature (thus the name apophysitis), they are now considered a series of microavulsion at the bone-cartilage junction, whereas inflammation is the body response to trauma. Running and jumping lead to repetitive stress on the growth plate as the foot strikes the ground. This disorder is frequent in athletes between the ages of 9 and 12 years. Clinical examination reveals a point of tenderness at the posterior aspect of the heel, and dorsiflexion of the ankle is limited. Pedobarography and radiography help in diagnosis [28]. Plain radiographs show some type of disorderly ossification of the calcaneal apophysis, either as fragmentation, incomplete appearance, or complete formation with increased sclerosis (Fig. 8.31). Pedobarogram of child with Sever disease reveals reduced heel contact of the foot (Fig. 8.32).

Besides the rest of the sport activity and stretching exercises, as additional treatment orthopedic insoles have to be used. Both elevation of the heel and controlling of usually valgus, or in some occasions varus position of foot by supination or pronation can result as pain relief, and eliminating incorrect position of the foot as the cause of this syndrome.

Patellar Tendinitis/Tendinosis (Jumper's Knee)

Next joint in a closed kinetic chain after foot and ankle joint is knee joint.

Patellar tendinitis/tendinosis (jumper's knee) is an overuse injury characterized by pathological changes in the distal parts of the extensor system of the knee joint: the *quadriceps* tendon and its insertion to the proximal pole of the patella, and patellar tendon (patellar ligament) and its proximal insertion to the apex of the patella or distal insertion to the tibial tubercle. Jumper's knee is a clinical entity most commonly found

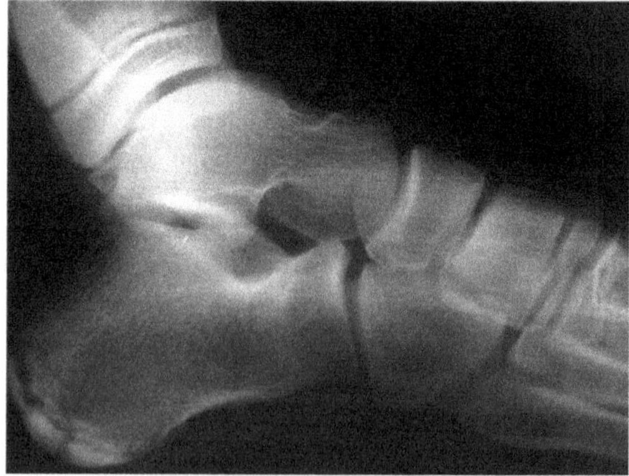

Fig. 8.31 X-ray of the foot with typical findings of Sever disease in 11-year soccer player

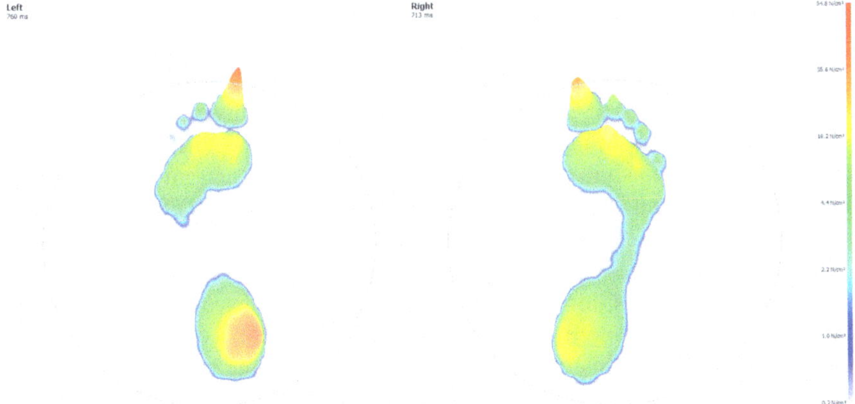

Fig. 8.32 Pedobarograms of child´s feet with Sever disease of right calcaneus: reduced heel contact of affected foot

in athletes who, during their athletic activities, habitually place excessive strain on the extensor system of their knees with numerous jump or long periods of running.

In growing children, the patellar tendon attaches to the growth plate of the kneecap and repetitive stress can irritate the growth plate. This condition is called Sinding-Larsen-Johansson disease.

The clinical picture of jumper's knee is characterized by the presence of pain as the basic symptom, and decreased functional ability of the afflicted lower extremity and patellar tendinitis/tendinosis is an important cause of the very well known syndrome of anterior knee pain [29]. According to incidence and progression of symptoms there are 5 clinical stages of patellar tendinitis/tendinosis and the last stage means complete rupture of patellar tendon. There is also ultrasound classification of the patellar tendinitis/tendinosis in 3 or 5 stages.

Also patellar tendinitis/tendinosis is relatively easy to diagnose, its treatment is more difficult. In acute stages of the disease, the physician generally recommends cessation of sporting activities that place strain on the afflicted extremity. When inflammation and pain have been reduced a rehabilitation program is introduced and it consists of stretching exercises and strengthening of the extensor muscle system of the knee and also increasing of the flexibility of the posterior area of the lower extremity. There is also another approach to the strengthening of the extensor system of the knee joint—concentric and eccentric exercises. Wearing a patellar brace knee strap is also highly recommended as it is also a recommended alternative training during the treatment of jumper's knee in athlete.

Orthopedic insoles with elevation of the heel is used, to unload pulling force of the extensor system of the knee where it attaches to kneecap or tibial tuberosity.

Surgical treatment is indicated only if a prolonged and well-supervised conservative treatment program fails. Today is preferred arthroscopically assisted apicotomy of the patella.

Iliotibial Band Friction Syndrome

The iliotibial band is thick band of fibrous tissue that is running from the hip down outside of the thigh, and attaches to the shinbone (tibia). Iliotibial band friction syndrome (ITBFS) results from activity comprising many repetitive flexion and extension movements of the knee, during which rubbing of the band against the lateral femoral epicondyle occurs; this produces irritation and inflammatory reaction with the iliotibial band or the underlying bursa (Fig. 8.33). ITBFS is one of the most common overuse injuries in professional and recreational joggers and in other athletes whose activities entail a lot of running.

Among other causes, inward rotation of leg can cause excessive friction during running and cycling (Fig. 8.34). Symptoms like dull aching or burning sensation on the outside of the knee during or after activity; pain in the hip known as referred pain. Like in most overuse syndromes the rest of activity is proper treatment. Exercise program for stretching and strengthening muscles is necessary. Surgery is recommended in resistant cases, a transverse cut is made in the 2 cm long posterior portion of the iliotibial band at the level of the lateral femoral epicondyle.

Orthopedic insoles for stabilizing the foot and knee is part of the treatment. In most of the cases varus position of foot is found and has to be corrected. Leg length discrepancy as one of the causes of excessive friction on one side has to be diagnosed and corrected if found.

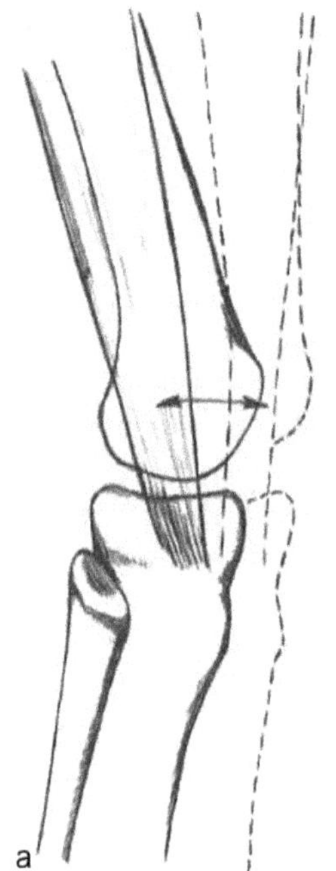

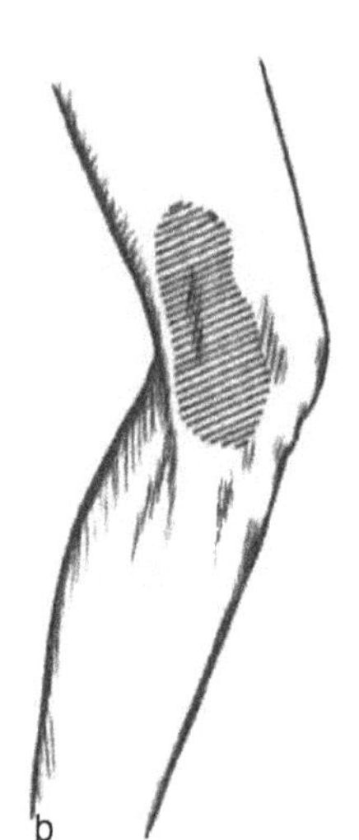

Fig. 8.33 Iliotibial band friction syndrome. **a** Mechanism of development; **b** Area of pain

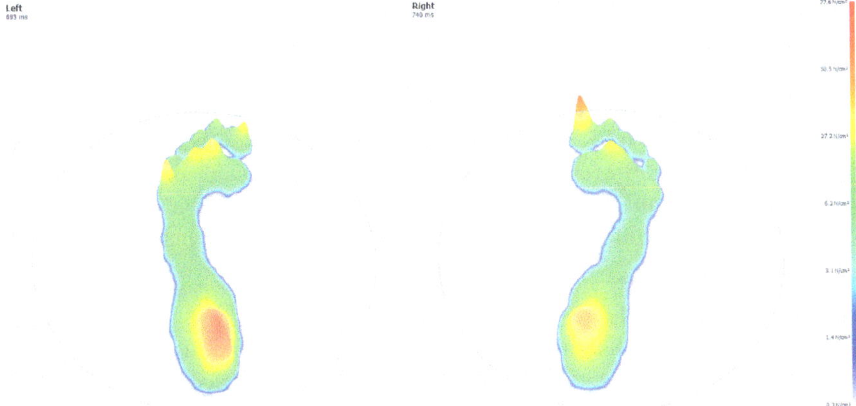

Fig. 8.34 Pedobarogram of runner with iliotibial band frictione syndrome of the left leg: overloaded left leg with increased pressure of the outer part of the foot

Osgood-Schlatter Disease

Osgood-Schlatter disease develops as apophyseal injury (traction apophysitis) of the growth plate at the proximal part of the shinbone (tibia). This growth plate, known as tibial tubercle is the place where the patellar tendon attaches to the tibia. During running and jumping, the quadriceps muscle of the thigh pull on the patellar tendon which, in turn, pulls on the tibial tubercle. The musculotendinous extensor apparatus of the knee (m.quadriceps) inflicts very strong tensile forces on the relatively small site of insertion of the patellar tendon to the tibial tubercle. These forces cause microavulsions of the tibial tubercle, and the prominence of tibial tubercle occurs as a consequence of the process of healing and ossification of these microavulsion (Fig. 8.35). In some children repetitive traction of the tubercle leads

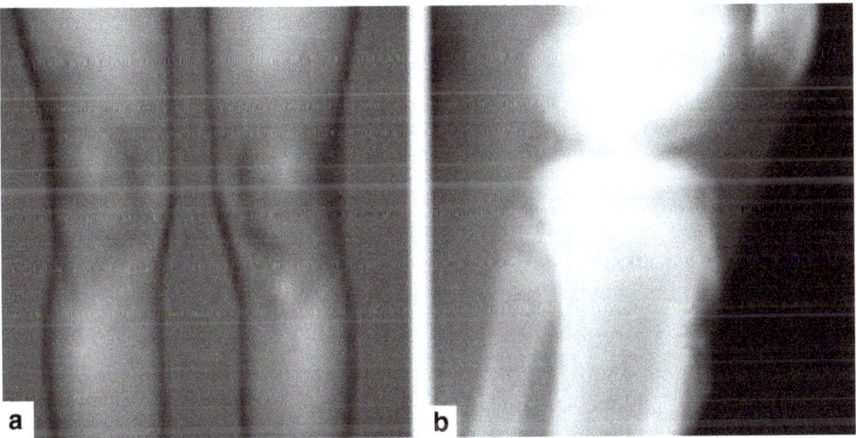

Fig. 8.35 Prominence of the left tibial tuberosity in young athlete with Osgood-Schlatter disease

to inflammation, swelling and tenderness causing pain. The prominence of the tibial tubercle may become very pronounced. Osgood-Schlatter disease is strongly associated with sporting activity in adolescent athletes between the age of 10 and 15, and is predominant in in disciplines that involve jumping, kicking and running.

Treatment of Osgood-Schlatter disease focuses on limiting exercises activity and including stretching. Braces, tape, or a slip-on knee support with an infrapatellar pad or strap may be useful if the athlete proposes to continue with the sporting activity. Orthopedic insoles with elevation of the heel is used to unload pulling force of the *quadriceps* muscle.

Degenerative Joint Disease

Degenerative joint disease also known as osteoarthritis is caused by breakdown and loss of the cartilage surface of the joints. It usually affects hands and large weight-bearing joints such as hips and knees.

When manifested on knees, orthopedic insoles with pronation can be applied [6]. The reason for pronation is the fact that medial joint space of the knee joint is narrower than lateral joint space and cartilage wears away more on the inner side (Fig. 8.36).

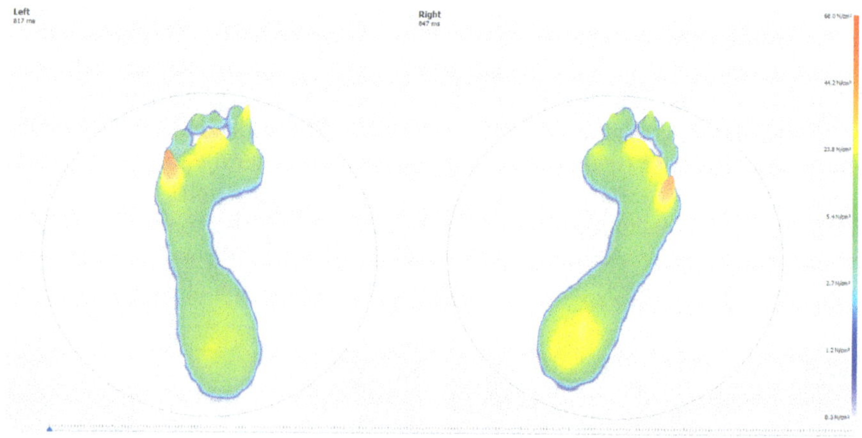

Fig. 8.36 Pedobarograms of patient with osteoarthritis of knees: varus position of feet-increased pressure of lateral part of both feet (from Janković et al. [30]. With permission)

Supporting foot arches and correction of irregular position of foot can contribute in reduction of symptoms. Leg length inequality should be considered, especially when sciatica occurs [6].

Groin Pain

Painful groin syndrome is generally considered the most frequent overuse syndrome in some athletic activities, e.g. soccer. The term groin pain, itself, clearly indicates the site and the principal symptom of the syndrome. The term syndrome is fully justified; indeed the symptoms are numerous and so are the causes of pain in the groin region. The modern medical literature abounds with terms defining pain in the groin region: necrotic osteitis pubis, anterior pelvic joint syndrome, traumatic pubic osteitis, gracilis muscle syndrome, pubic stress symphysitis, pubic symphysis osteoarthropathy, symphysitis, pubic chondritis, post-traumatic necrosis of the pubic bone, athletic pubalgia, sports hernia. It most often occurs during sports that require sudden changes of directions or intense twisted movements, such as soccer or ice hockey. Most frequently affected are oblique abdominal muscles and *adductor longus* and *gracilis* muscles at the attachment to the pubic bone (Figs. 8.37 and 8.38).

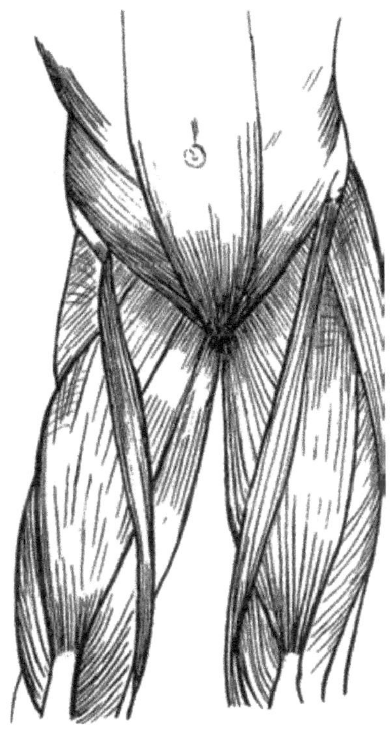

Fig. 8.37 The groin area is the crossroads of the trunk and lower extremity muscles. (from Janković et al. [30]. With permission)

Treatment of painful groin syndrome is a complex as the causes of its development. The best possible treatment is based on elimination of causes and prevention of development of the syndrome. The following are some common principles of non-operative treatment of painful groin syndrome regardless of its cause: (a) alleviate pain and control inflammation in the myotendinous apparatus, (b) hasten healing of the myotendinous apparatus; and (c) control further activities. The most important is to commence treatment as early as possible, i.e. when the first symptoms occur.

As inequality in leg length can be one of the causes of the groin syndrome (Fig. 8.39), orthopedic insole with leg length discrepancy correction can contribute to the treatment of this overuse syndrome [6].

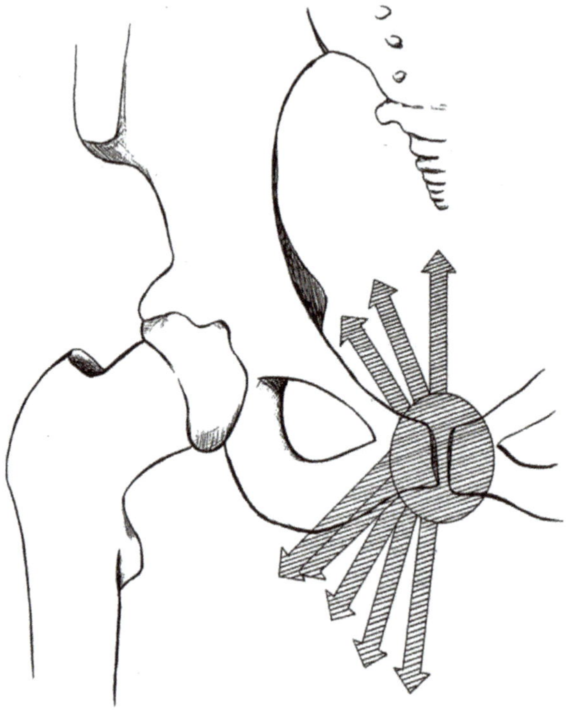

Fig. 8.38 Graphic representation of the imbalance of strength between the abdominal muscles and muscles of the lower extremity. (Figs. 8.37 and 8.38 from Janković et al. [30]. With permission)

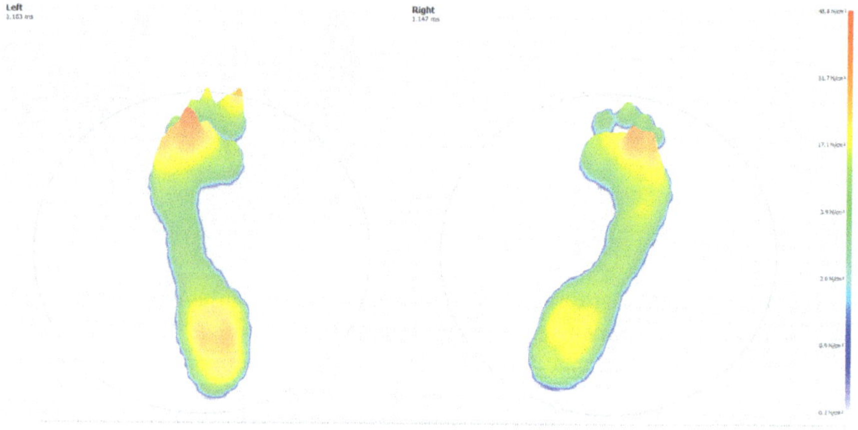

Fig. 8.39 Leg length inequality of soccer player with groin pain: less pressure beneath heel of the shorter right leg

Operative treatment of painful groin syndrome, to a certain extent, depends on the attitude in approaching this complex problem. Laparoscopic preperitoneal hernia repair should be considered as a treatment modality in athletes presenting with chronic groin pain and so-called sports hernia.

Diabetic Foot

Most foot problems that people with diabetes face arise from two serious complications of the disease: nerve damage and poor circulation. They contribute to formation of ulcerations. The area of high pressure of soles is predilective area for development of ulceration.

Pedobarographic analysis can detect dangerous areas (Fig. 8.40), and orthopedic insoles with special corrections can reduce plantar pressures and prevent ulcers. When ulceration already exists, a properly made orthopedic insole can help healing by redistribution of plantar pressure and elimination of mechanical cause of ulcer.

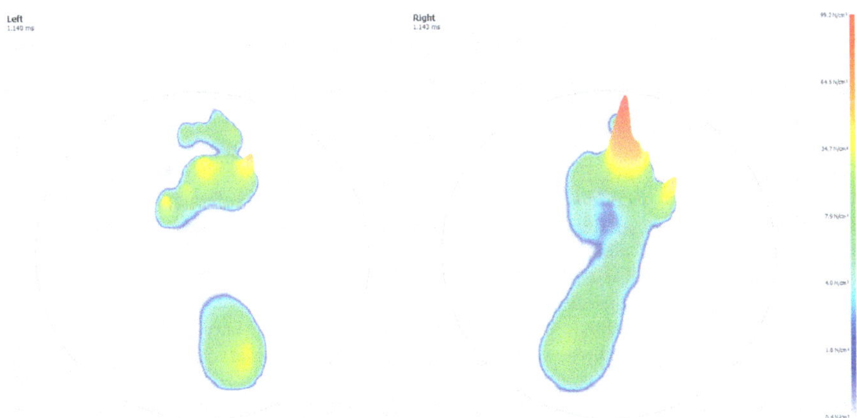

Fig. 8.40 Diabetic feet: pedobarograms of left foot after amputation of little toe and right foot after amputation of big toe and fourth toe with ulcer beneath third metatarsal head: increased pressure beneath head of the third metatarsal bone of the right foot

References

1. Pedobarography DT (1988). In: Helal B, Wilson D (eds) The foot. Churchill Livingstone, Edinburgh, London, Melbourne, New York, pp 108–130
2. Heikki U, Edgardo B (2004) Gait Analysis. In: Cuccurullo SJ (ed) Physical medicine and rehabilitation board review. Demod Medical Publishing, New York, pp 409–416
3. Burić M, Antičević D (2004) Nožni zglob i stopalo. In: Pećina M. i sur. Ortopedija. Zagreb: Naklada Ljevak
4. Keros P, Pećina M (2019) Funkcijska anatomija lokomotornoga sustava, 2nd edn. Naklada Ljevak, Zagreb
5. Ruszkowski I (1981) Normalan i poremećen hod čovjeka. Jumena, Zagreb
6. Pećina M, Bojanić I (2004) Overuse injuries of the musculoskeletal system, 2nd edn. CRC Press, Boca Raton-London-New York-Washington, D.C.
7. Kelly JL, Valier AR (2017) The use of orthotic insoles to prevent lower limb overuse injuries: a critically appraised topic. J Sport Rehabil 27:1–16
8. Bonanno DR, Murley GS, Munteanu SE, Landorf KB, Menz HB (2018) Effectiveness of foot orthoses for the prevention of lower limb overuse injuries in naval recruits: a randomized controlled trial. Brit J Sports Med 52(5):298–302
9. Medved V (2001) Measurement of kinetic variables. Medved V: Measurement of human locomotion. CRC Press, Boca Raton, Fl, pp 154–168

10. Pećina M, Obrovac K, Pećina HI, Jelić M, Obrovac-Vuković J (1998) Kompjuterska dijagnostika deformacija stopala i robotska kompjuterski vođena izrada ortopedskih uložaka. Hrvat Športskomed Vjesn 13:9–14
11. Pećina M, Obrovac K, Pećina HI, Jelić M (1998) Electronic measurment system for recording and evaluating dynamic plantar pressure distribution. In: Biomedical measurment and instrumentation: proceedings/8th International IMEKO Conference on measurment in clinical medicine & 12th international symposium on biomedical engineering, Dubrovnik, editor Ratko Magjarević-Zagreb: KoREMA, str. 8-56-8-59
12. Pećina M (1998) Robotska izrada ortopedskih uložaka na osnovu statičke i dinamičke kompjutorske analize hoda. Medix 4:64–67
13. Mirković M, Pećina M, Cicvara-Pećina T, Mađarević M, Klobučar H (2019) Primjena ortopedskih uložaka u prevenciji i liječenju sundroma prenaprezanja na stopalu i donjem ekstremitetu (poglavlje 35). In: Pećina M et al. Sportska medicina. Zagreb, Medicinska Naklada, pp 355–362
14. Lucijanić I, Bičanić G, Sonicki Z, Mirković M, Pećina M (2009) Treatment of hallux valgus with three-dimensional modification of mitchell's osteotomy. J Am Podiatr Med Assoc 99(2):162–172
15. Lucijanić I, Mirković M, Cicvara-Pećina T, Mirković M, Pećina M (2010) Kirurško liječenje hallux limitus/rigidus u mlade športašice. Hrvat Športskomed Vjesn 25:52–57
16. Hennessey WJ (2016) Urustal Heikki: Lower limb Orthosess. In: Braddom's physical medicine and rehabilitation, 5 edn, Elsevier, Philadelphia, pp 249–272
17. Smerdelj M, Mađarević M (2019) Sindromi prenaprezanja na potkoljenici i stopalu (poglavlje 34). In: Pećina M et al. Sportska medicina. Zagreb, Medicinska Naklada, pp 343–354
18. Nadari A, Degens H, Sakinepoor A (2018) Arch-support foot-orthoses normalize dynamic in-shoe foot pressure distribution in medial tibial stress syndrome. Eur J Sport Sci 7:1–11. https://doi.org/10.1080/17461391.2018.1503337
19. Borić I, Pećina M, Mirković M, Cicvara Pećina T, Plečko M, Matoković D, Čulić-Dumić I (2019) Hallux Sesamoiditis—radiological diagnostics and conservative management. RAD CASA, pp 48–49
20. Matoković D, Pećina M, Hašpl M (2019) Ortopedska propedeutika. Medicinska Naklada, Zagreb
21. Bojanić I, Pećina HI, Pećina M (2001) Prijelomi zamora. Arh Hig Rad Toksikol 52:471–483
22. Madjarevic M, Kolundzic R, Trkulja V, Mirkovic M, Pecina M (2009) Biomechanical analysis of functional adaptation of metatarsal bones in statically deformed feet. Int Orthop (SICOT) 33:157–163
23. Pećina M, Bojanić I, Smoljanović T, Ivković A, Mirković M, Jelić M (2011) Surgical treatment of diaphyseal stress fractures of the fifth metatarsal in competitive athletes. Long-term follow-up and computerized pedobarographic analysis. J Am Podiatr Med Assoc 101(6):517–522
24. Dubravčić-Šimunjak S, Pećina HI, Janković S (1999) Sindromi prenaprezanja sustava za kretanje. Hrvat Športskomed Vjesn 14:82–89
25. Milošević M, Jelić M, Vondra Sedlaček J, Pećina M (2002) Pedobarografija u nogometaša mlađe životne dobi. Hrvat Športskomed Vjesn 17:3–7
26. Mađarević M, Mirković M, Cicvara-Pećina T, Klobučar H, Mahečić K, Jelić M, Pećina M (2007) Ortopedski ulošci u prevenciji i liječenju sindroma prenaprezanja na stopalu i gležnju. Hrvat Škomed Vjportsesn 22:3–9
27. House C, Reece A, de Roiz SD (2013) Shock-absorbing insoles reduce the incidence of lower limb overuse injuries sustained during royal marine training. Mil Med 178(6):683–689. https://doi.org/10.7205/MILMED-D-12-00361
28. Seungbum K, Sangho Ch, Kyoung ML, Byung CC, Young-Jun K, Dong-Wan K, Moon SP (2018) Sex differences in pedobarographic findings and relationship between radiographic and pedobarographic measurements in young healthy adults. Clin Orthop Surg 10:216–224
29. Pećina M, Bojanić I, Ivković A, Brčić L, Smoljanović T, Seiwerth S (2010) Patellar tendinopathy: histopathological examination and follow-up of surgical treatment. Acta Chir Orthop Traumatol Cech 77(4):277–283
30. Janković S, Pećina M, Matoković D (2019) Sindrom bolne prepone (poglavlje 31). In: Pećina M et al. Sportska medicina. Zagreb: Medicinska Naklada, pp 311–328

Chapter 9
Kinesiological Electromyography

Mario Cifrek, Igor Gruić, and Vladimir Medved

Abstract Neuromuscular level of treatise is pursued, meaning that the motor control aspect comes into focus. In the chain of information transmission—neural impulses serving as information carriers—signal conduction takes place via neural pathways, followed by transmission of excitation to muscle fibers. Skeletal muscle function is presented as seen through its bioelectrical manifestation; electromyographic (or myoelectric) signals (EMG), the signal being modeled in a form of an interference pattern. The method of electromyography is succinctly explained next including technical aspects of signal detection, amplification and registering. In kinesiology, primarily surface electromyography (sEMG) is used. A classical repertoire of signal processing methods in time and in frequency domain is presented next, with interpretations commonly used in biomechanics and kinesiology. It is known as kinesiological electromyography, with principal applications in the evaluation of movement skill and of local muscle fatigue.

9.1 Introduction

Since the beginnings of modern consolidation of the field of biomechanics electromyography was recognized as an important factor [1]. It is precisely surface EMG (sEMG) being a variable to monitor functioning of the neuro-muscular system in in vivo conditions, so it may be considered a sort of non-invasive window into its action, which, combined with other available measurement quantities, provides a unique, second to none, information. There is a high degree of sensitivity and specificity offered by this signal in studying, specifically, movement skill acquisition [2]. The present chapter is therefore focused to, first, shortly explain how this signal is

M. Cifrek (✉)
Faculty of Electrical Engineering and Computing, University of Zagreb,
Unska 3, 10000 Zagreb, Croatia
e-mail: mario.cifrek@fer.hr

I. Gruić · V. Medved
Faculty of Kinesiology, University of Zagreb, Horvaćanski zavoj 15, 10000 Zagreb, Croatia

V. Medved (ed.), *Measurement and Analysis of Human Locomotion*,
Series in Biomedical Engineering, https://doi.org/10.1007/978-3-030-79685-3_9

being generated in the organism and then, how it is being detected, processed and interpreted to serve in human movement analysis.

9.2 Origins, Formation, and Properties of Myoelectric Signals

This subchapter provides a condensed explanation of the subject matter. It refers primarily to [3], and also to [4] and [5].

As adequately depicted in the scheme by Winters (Chap. 2, Fig. 2.18) musculoskeletal system of human body is controlled by the nervous system, and distinct directions of information transmission may be identified in this kind of a plant. Focusing to the transmission of information in the efferent direction, the following processes are identified: conduction of information via the neural pathways, transmission of information from the neural to the muscular system, and realization of muscular contraction.

Nervous system provides the basis for conduction of information by means of neural action potentials. A neural action potential, being a change of equilibrium potential at the cell membrane spreading along an axon, manifests an impulse waveform (spike) of 0.2–1 ms duration and with an amplitude of about 100 mV (Fig. 9.1).

As carrier signals of control (or sensory) information in the nervous system, neural action potentials are equal in shape, while the information to be transmitted is contained in time intervals between the successive action potentials, i.e. using engineering terminology, a pulse-frequency modulation of sorts occurs [3]. Propagation speed of neural action potentials falls within a 0.5–120 m/s interval [9]. The term

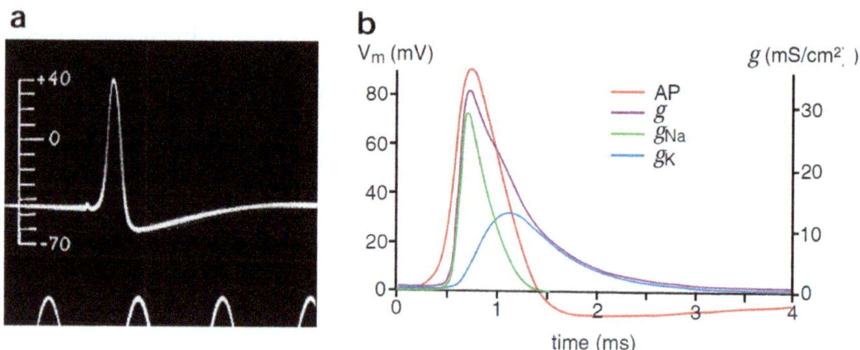

Fig. 9.1 The action potential: from experiment to theory. **a** The first intracellular recording of an action potential, from squid axon. Time calibration, 2 ms. Modified from Hodgkin and Huxley [6]. **b** Separation of ionic conductances underlying the action potential (AP) in the H–H model. Modified from Hodgkin and Huxley [7]. (from Häusser [8]. With permission)

pulse-frequency modulation, borrowed from the engineering field of communications, seems appropriate in characterizing this biological mechanism, however, in a simplified and schematized manner.

Neural axons possess cable-like properties. It is interesting to note that the Englishman Arthur A. Baines, who was the first to postulate the analogy between the spreading of the impulse through neural pathways and an electric cable (1918), is considered by Basmajian and De Luca the first biomedical engineer [10]. He modeled the nervous system on the theory of lines and circuits, which in a way preceded the idea of systems theory and cybernetics, as pursued by Norbert Wiener, Lotfi A. Zadeh, and others. In general, axon manifests its cable properties through two physical modes of information transmission: passive and active conduction. The active conduction happens by virtue of myelinated fibers (axon's membrane (myelin) acting as electrical insulator), enabling saltatory conduction (in "jumps").

With a nerve impulse arriving at a motor plate (neuromuscular plate, motor point) a process of transmission of excitation to a muscle fiber is initiated. The motor plate is a specially formed synapse, a site of innervation of a muscle fiber by a motoneuron. An incoming nerve impulse causes an electric change in the neural terminal leading to the release of a transmitter, acetylcholine, which binds itself to muscle cell receptors. Depolarization follows in both directions along a muscle fiber. Consequently, an electromagnetic field is generated in its environment. Because the final global effect is our concern, the underlying electrochemical mechanism shall not be discussed in further detail here however.

Muscle action potentials are the "companions" of chemical processes by means of which mechanical and thermal energy is released. According to Katz [11]: "In muscle, the action potential, traveling at a speed of a few meters per second, serves to produce sufficiently quick "mobilization" of the contractile apparatus in the interior of the cell." Muscle action potentials propagate at speeds ranging from 2 to 6 m/s, while impulse width ranges from 1 to 5 ms.

While itself a unity within a neuromusculoskeletal system, a unity within one muscle is the motor unit. It is a structure consisting of one efferent motoneuron (alpha) and a group of muscle fibers innervated by this neuron. The number of innervated fibers may vary from a few, as in some muscles active in "fine" movements, to several thousands, as for example in the lower extremity muscles. Muscle fibers of the same motor unit are anatomically grouped into subgroups that are spatially dispersed throughout the muscle. Therefore, motor neuron action potentials reach them with different latencies, contributing to the spatial gradation of muscular contraction.

The electrical processes of propagation of the action potential in a neuron and then in a muscle fiber lead to the realization of a muscle twitch with delay of a few milliseconds, called a latency time. A twitch may be defined as a mechanical response of a muscle as a whole to one stimulus, manifested as a time change of isometric force. The twitch phenomenon lasts an order of magnitude longer than the electric action potential. The total bioelectrical signal of a muscle is a result of the spatiotemporal summation of activity of a large number of motor units, producing what is called an interference pattern (to be discussed later). Assuming a systems analysis standpoint, the electrophysiological process of stimulation and contraction of a skeletal muscle [11] can be considered a transformation of information contained in a train of electrical impulses, i.e., pulse-coded information traveling along the neural fiber, into a mechanical force. Aspects of temporal summation of muscle twitches leading eventually—when stimulation frequency reaches approximately 50/s—to the state of tetanus will not be further discussed here. For a more detailed description of muscle (bio)mechanical functioning, interested reader is directed to a short summary [3], while a profound explanation may be found in [12].

Apart from the mechanical effect, as mentioned, an electrical change occurs in the muscle's state through the propagation of electrical impulses in the motor unit. The parallelism in these two processes, where an electrical precedes a mechanical, provides a basis for the kinesiological importance of myoelectric signals which are to be found at the muscle surface and are therefore accessible for detection and measurement. Technical issues of electrodes and other stages in a measurement chain, including signal processing and signal interpretation aspects, will be dealt upon in Sects. 9.3.

Basmajian and De Luca presented a mathematical model of the myoelectric signal [10]. Originating from De Luca's previous study and research on the physiology of muscular activity as seen and interpreted through myoelectric indices [13, 14], a mathematical/statistical approach is pursued, where generation and formation of myoelectrical signals by summing the activity of multiple motor units is comprehensively modeled as a so-called "interference pattern". The complete chain from the motoneuron to the final amplified myoelectric signal at the output of the measurement device is encompassed. The goal of this—as they named it, a "structured"—approach was to interpret information content of the EMG signal. Beginning with the elementary physiological phenomena, over the nerve and muscle action potentials, all higher anatomical levels of integration were taken into consideration which finally results in a total EMG signal. Tissue filtering effects, as well as the influence of measurement instrumentation, were taken into account. (The model of myoelectric signal shortly presented here assumes a standard bipolar detection technique, followed by conventional signal amplification and filtering. As already mentioned, measurement aspects are a subject of the next Sect. 9.3).

A situation in which there are more muscle fibers active is illustrated in Fig. 9.2, where the emergence of the motor unit action potential is presented schematically (MUAP goes for motor unit action potential).

A contracting motor unit changes its length, so a waveform of a MUAP changes accordingly. This anatomically complex situation is well described in Loeb and Gans [15]. It can be considered, in fact, a sort of mapping of a spatial (3D) process into a 1D signal. Individual potentials recorded may be monophasic, biphasic, etc., their total sum representing MUAP, denoted by h(t). Potentials may attain a triphasic form at best, while the appearance of more than four phases is indicative for pathology and a prolonged duration indicates the organism's aging.

A train of motor unit action potentials represents a MUAPT (motor unit action potential train). Basmajian and De Luca develop the approach further, so as to finally encompass a muscle as a whole, with large number of active motor units. They

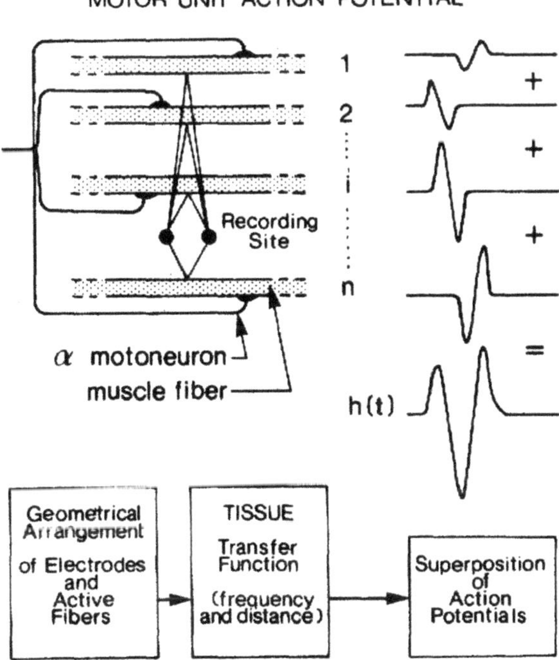

Fig. 9.2 Schematic representation of the physiological model of motor unit action potential generation (from Basmajian and De Luca [10]. With permission)

develop the model of EMG signal by synthesizing and linearly summing up the MUAPTs:

$$m(t, F) = \sum_{i=1}^{p} u_i(t, F) \tag{9.1}$$

while a complementary spectral representation is:

$$S_m(\omega) = \sum_{i=1}^{p} S_{ui}(\omega) + \sum_{\substack{i,j \\ i \neq i}}^{q} S_{uiuj}(\omega) \tag{9.2}$$

Schematic representation of the generation and formation of a global myoelectric signal is given in Fig. 9.3.

The Fig. 9.4 shows, further, electromyographic signals in several grades of isometric contraction of m. *biceps brachii*. It is nicely shown how a muscle force is developed, starting from an initial activity of a single motor unit, to two, three and more units, into a full interference pattern.

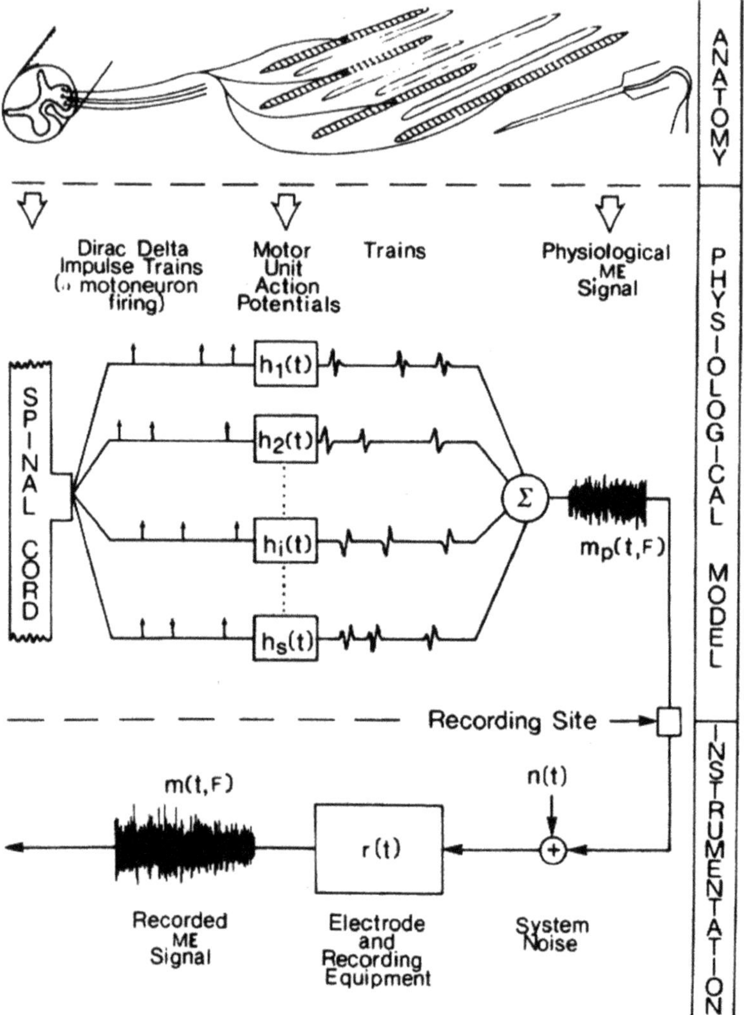

Fig. 9.3 Schematic representation of the generation and formation of a global myoelectric signal (from Basmajian and De Luca [10]. With permission)

Finally, Fig. 9.5. shows a raw surface EMG recording for three successive contractions at the example of m. *vastus medialis* [17].

A characteristic quasi-stochastic appearance of the signal may be observed. Raw sEMG signal can range between ±5 mV (maximum usually achieved in athletes) and typically the frequency contents ranges between 6 and 500 Hz, showing most power between ~20 and 250 Hz.

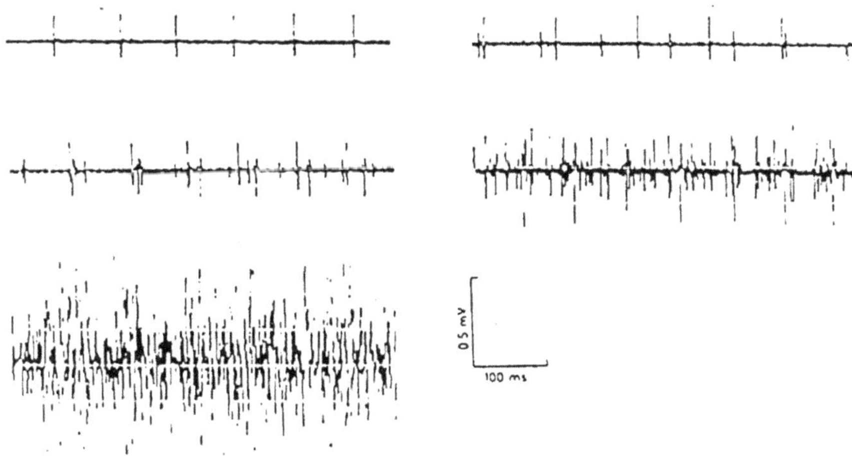

Fig. 9.4 Bioelectric muscle potentials—EMG signals—in several grades of isometric contraction of m. *biceps brachii* (from Basmajian et al. [16]. All resonable efforts were made to ask permission, any entitled party can contact us. In addition, Nancy Basmajian, daughter of J.V. Basmajian, had no objections.)

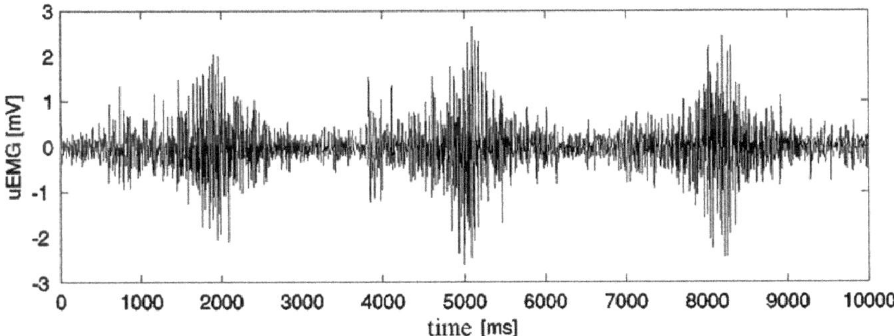

Fig. 9.5 Raw surface EMG recording for three successive contractions of m. *vastus medialis* while exercising lower leg extension (from Cifrek [17]. With permission)

For a more profound elaboration of the genesis, formation and properties of myoelectric signals interested reader is referred in the first place to [10], being a premier reference for the subject, and also to [3–5, 18, 19].

9.3 Surface Electromyography: Measurement Technique, Signal Processing and Interpretation

9.3.1 *Measurement Technique*

Electrical signals originating from skeletal muscle are rather complex because they depend on the anatomical and physiological properties of the muscle, the organization of the peripheral nervous system that innervates the muscle, the measurement conditions and characteristics of the measuring instrumentation. They can be measured in two significantly different ways: using subcutaneous electrodes or, alternatively, using surface electrodes.

Subcutaneous electrodes enable detecting the activity of individual motor units. (In fact, the waveform, amplitude, and duration of the measured myoelectric signal detected in this manner are considered as diagnostically significant parameters). Surface electrodes, on the other hand, are used to record a global myoelectric signal generated by the superposition of a series of MUAPs of a large number of motor units in the area covered by the configuration of the electrodes used (Figs. 9.2 and 9.3). Further treatise will proceed to describing surface detection, and occasionally also subcutaneous techniques will be commented upon.

Physiological myoelectric signal undergoes several modifications in its path from muscle to the registration site, which alters its properties (Fig. 9.6):

- Tissue has the property of a spatial low-pass filter whose upper cut-off frequency depends on the distance between electrodes and the activemotor unit,
- Skin–electrode interface has the property of a high-pass filter and polarization potential is present,

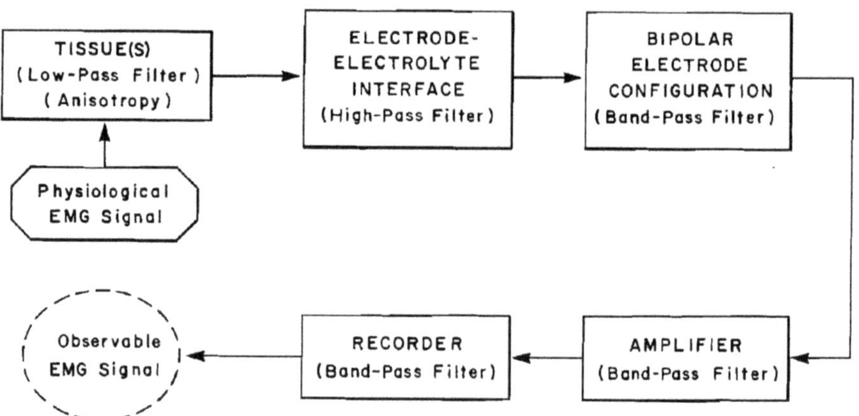

Fig. 9.6 Principal aspects of surface EMG signal acquisition (from Basmajian and De Luca [10]. With permission)

- Frequency properties of the bipolar configuration of surface electrodes depend on the distance between the electrodes,
- EMG amplifier with its bandwidth also affects the frequency composition of the measured signal.

9.3.1.1 Influence of Tissue as a Volume conductor—Tissue Filtering Effect

On its path from the source (muscle fiber) to the electrodes on the skin surface, muscle action potentials propagate through tissues (muscles, subcutaneous fat, skin) that alter their properties.

Figure 9.7 shows electromyogram recorded simultaneously by subcutaneous (FINE WIRE) and surface (SURFACE) electrodes. There is clearly a decrease in the amplitude (attenuation) as well as attenuation of the high-frequency components of the signal recorded by surface electrodes relative to the signal recorded by subcutaneous electrodes.

This is due to the low permeability of tissue as a volume conductor. In the graphical representation of the filtering property of tissue in Fig. 9.8, we can observe the indicated effects of the tissue on muscle fiber action potential. The first effect is the reduction of signal amplitude with the distance from the membrane of the observed muscle fiber. This means that muscle fibers located close to the electrode will make a greater contribution to the overall amplitude of the measured signal than fibers located in the deeper parts of the muscle. When recording with surface electrodes, a thicker layer of subcutaneous adipose tissue will further reduce the amplitude of the measured signal. Furthermore, muscle tissue acts as a low-pass filter whose upper cut-off frequency also depends on the distance of the muscle fiber from the electrodes (Fig. 9.8).

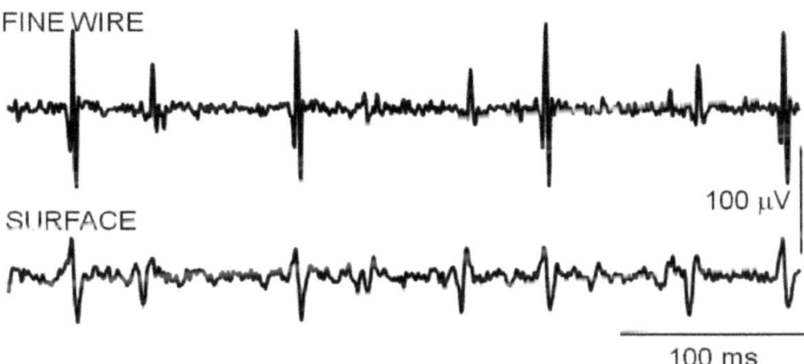

Fig. 9.7 Electromyogram recorded by subcutaneous (FINE WIRE) and surface (SURFACE) electrodes during muscle contraction of m. *brachioradialis* in a healthy subject. Surface electrode diameter 2.5 mm. A low-pass filter effect of subcutaneous tissue can be observed (from Merlo and Campanini [20]. With permission)

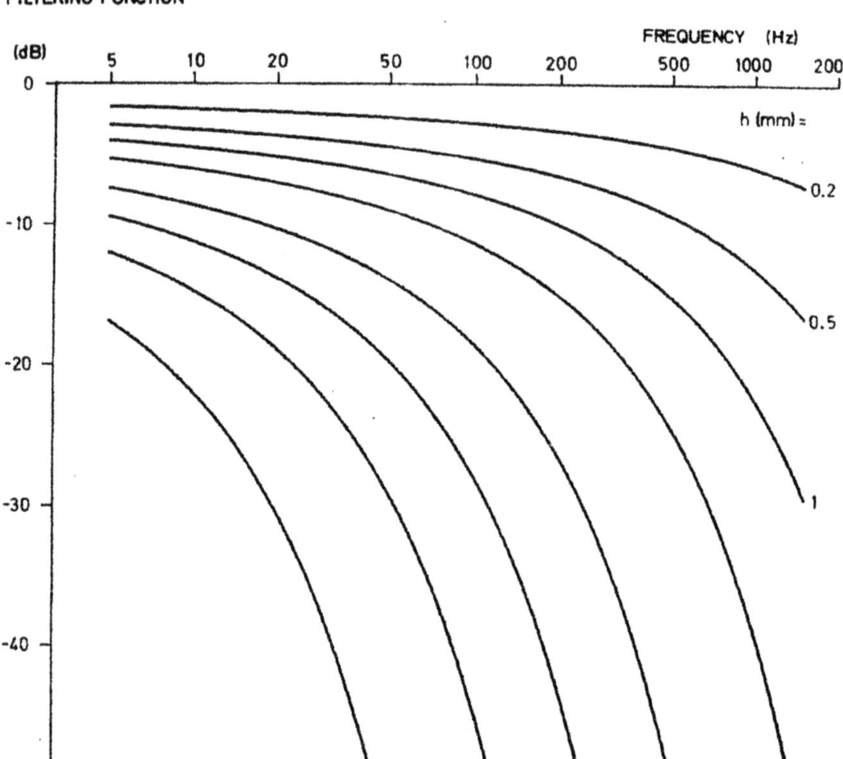

Fig. 9.8 Effect of tissue filtering properties on the amplitude and frequency content of the action potential of a single 50 μm muscle fiber, with an assumed of 4 m/s. The distance of the active muscle fiber from the electrodes is indicated by *h* (from Lindström [21]. With permission of the author)

9.3.1.2 Electrodes

Myoelectric signals, as mentioned, can be detected in two significantly different ways: by surface electrodes (Fig. 9.9) or by subcutaneous electrodes.

Two types of surface electrodes are used: passive and active. Passive electrodes are usually made as single-use self-adhesive electrodes and have a conductive gel that is used to reduce the resistance of the surface layer of the skin (stratum corneum). Electrodes with decentralized cable connection (Fig. 9.10) are used if increased pressure on electrodes is expected (for example if the subject is sitting on electrodes). Solid gel electrodes (Fig. 9.11) are used recently.

With the surface electrodes, it is possible to monitor only the activity of the muscles just under the skin, with the present problem of cross-talk (to be discussed

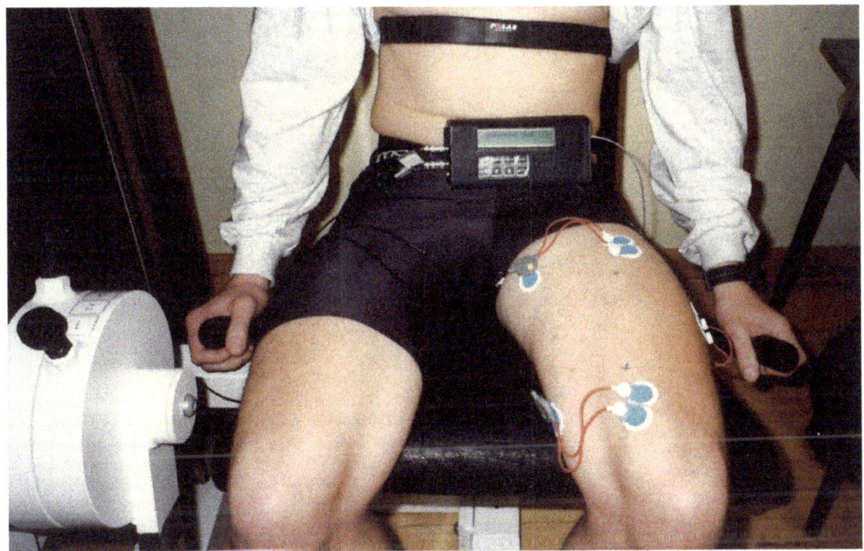

Fig. 9.9 For recording the EMG in biomechanical experiments, the non-invasive EMG electrodes applied to the skin of the subject are the best choice. Experimental setup for upper leg muscle fatigue investigation during the leg extension exercise (from Cifrek [17]. With permisison)

Fig. 9.10 Self-adhesive surface electrodes Ambu® BlueSensor P (from Ambu® Cardiology catalog 2015 [22]. With permission)

Fig. 9.11 Ambu® WhiteSensor 0215 M solid gel electrode (from Ambu® Cardiology catalog 2015 [22]. With permission)

later). Deeper muscle activity can be monitored by subcutaneous electrodes, with much smaller detection volume and less cross-talk. Signal recorded under the skin has a wider frequency spectrum due to the smaller influence of tissue filtering effect. Either needle electrodes or fine wire electrodes are used (Fig. 9.12).

Thin wire electrodes are, due to their dimensions, less invasive than EMG needle electrodes and are sometimes used in kinesiological electromyography because they do not interfere with movement. In addition to invasiveness and discomfort during placement, the major drawback of subcutaneous electrodes is the problem of repeatability of placing the electrodes in the same place for repeated measurements. But despite all the drawbacks, some muscles can only be monitored using subcutaneous electrodes.

The active surface electrodes (Fig. 9.13) have a built-in preamplifier with a very high input impedance allowing measurement on dry skin without the use of conductive gel. They are less sensitive to motion artifacts and have a better signal-to-noise ratio than passive ones.

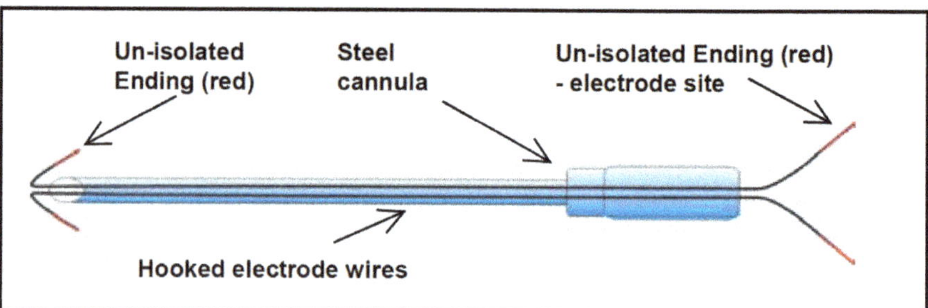

Fig. 9.12 Schematic illustration of the fine wire electrodes: two fine wires with un-isolated endings are located with a steel canula (from Konrad [23]. With permission)

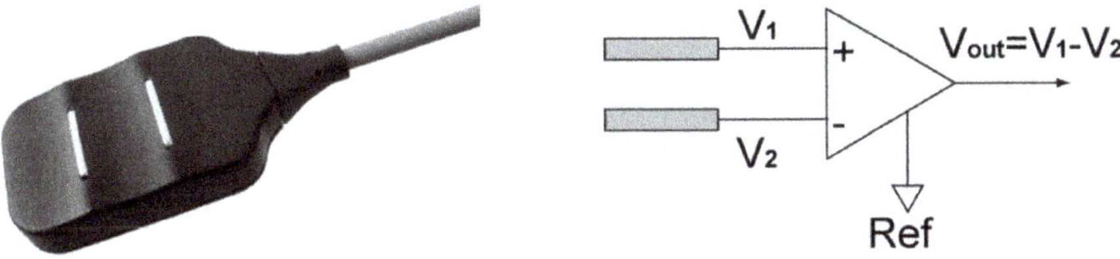

Fig. 9.13 Example of an active surface electrode (from Bagnoli™ EMG System User'sss Guide [24]. With permission)

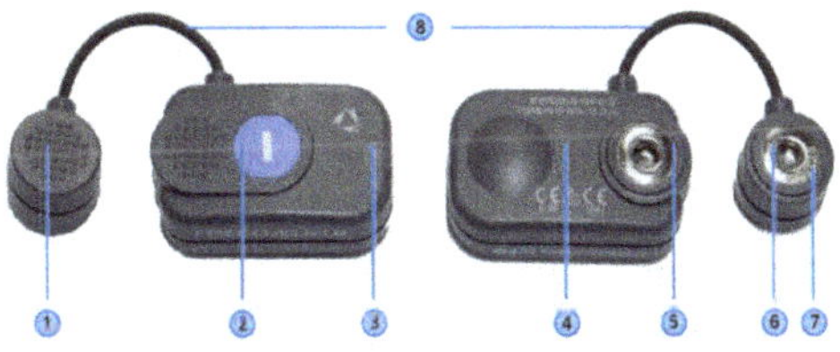

Fig. 9.14 Example of an active wireless EMG sensor (from BTS FREEEMG 1000 User Manual [25]. With permission)

Technological development and miniaturization enable the creation of wireless EMG sensors which are especially suitable in sports applications where connecting cables are often a problem. As an example, Fig. 9.14 shows the wireless EMG sensors of Biotechnology Systems (BTS) SpA.

9.3.1.3 Skin–Electrode Interface

There is a relatively high resistance between the surface electrode and the conductive interior of the body to which the stratum corneum, the superficial layer of the skin, contributes most. Since this layer of skin also has a certain dielectric constant, there is an electrical capacity between the electrode and the inside of the body.

To reduce resistance, the skin has to be cleaned before the electrodes are placed. In most cases, simple alcohol cleaning may be sufficient. In the case of stronger movement artifacts, abrasive pastes can be used to mechanically remove the upper corneal layer of the skin, thereby significantly reducing resistance. A conductive gel can be rubbed into the skin after cleaning. Commercially available disposable self-adhesive electrodes already have a conductive gel.

At the junction of the metal electrode and the electrolyte, an electric bilayer is created that separates the positive and negative charges, and they create a potential difference (half-cell voltage) that depends on the metal from which the electrode is made. An additional problem is the instability of this voltage. In the case of the bipolar electrode configuration, although the two electrodes are identical, there is a potential difference of about 10 mV between the electrodes, which is both time- and temperature-varying, and especially changes especially due to electrode displacement, which is pronounced when measuring dynamic contractions, especially in fast movements. To reduce artifacts due to fast movements, the lower limit frequency of the measurement amplifier is set to 10 Hz and, in the case of very strong movement artifacts, to 20 Hz. Silver/silver chloride (Ag/AgCl) is predominantly used as the electrode material, primarily because of the stability of the polarization potential [3].

At the point where an electrical bilayer is formed between the metal electrode and the electrolyte, a very large specific capacitance (about 30 $\mu F/cm^2$) occurs because of the very thin dielectric layer. In addition to this capacity, there is also the resistance of the bilayer. The schematics of the electrode–skin interface is shown in Fig. 9.15 [26].

Hary et al. [27] conducted impedance measurements of several surface and subcutaneous EMG electrodes in the frequency range of 50 Hz to 2 kHz. The measurement results are shown in Fig. 9.16. A higher impedance of the surface electrodes is observed as well as a decrease in the impedance by increasing the electrode surface.

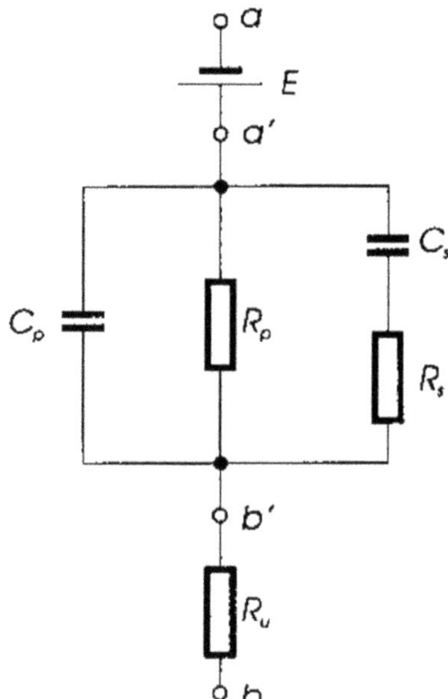

Fig. 9.15 Schematics of the skin–electrode interface. R_p corneal skin resistance, C_p skin capacity, C_s bilayer capacity, R_s bilayer resistance, E half-cell voltage, R_u inner resistance od the body (from Šantić [26]. With permission)

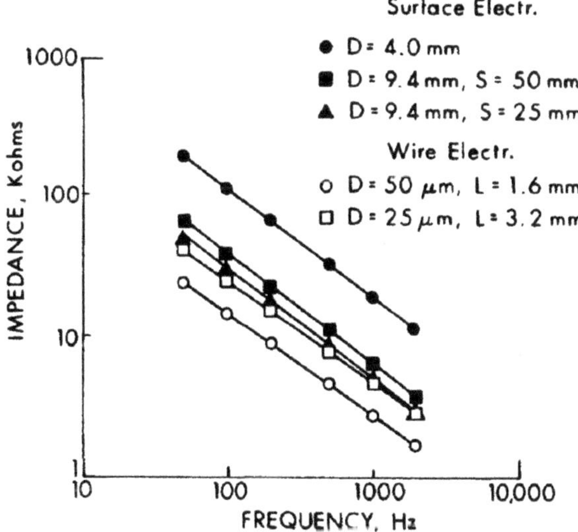

Fig. 9.16 The impedance of surface and subcutaneous EMG electrodes as a function of frequency, electrode diameter (D), electrode spacing (S), length of active electrode part (L) (from Hary et al. [27]. With permission)

9.3.1.4 Configuration of Bipolar Electrodes and Differential Amplifier

When measuring myoelectric signals using electrodes connected to a differential amplifier, the situation is shown in Fig. 9.17. Assuming that the bipolar electrodes are arranged parallel to the muscle fibers, and if the action potential wavelength is equal to the distance between the electrodes, the amplifier output voltage is zero (Signal 1 in Fig. 9.17). Signal 2, whose wavelength is twice the distance between the electrodes, will be amplified in full.

Lindström [21] derived the expression for the frequency response of a bipolar electrodes configuration whose shape is shown graphically in Fig. 9.18. He showed that the frequencies f_G at which the attenuation is greatest can be calculated according to expression (9.3)

$$f_G = \frac{nv}{d}, n = 1, 2, 3, \ldots \tag{9.3}$$

In the expression (9.3), v denotes the muscle fiber, and d is the distance between the electrodes.

It is apparent from the expression (9.3) that the band-pass of the described electrode and amplifier configuration is widened by reducing the distance between the electrodes. But it should also be noted that by reducing the distance between the electrodes, the amplitude of the recorded signal is reduced. Both effects can be seen in Fig. 9.19, showing the influence of three different electrode spacings on the power spectrum of myoelectric signals measured on m. *biceps brachii* [28].

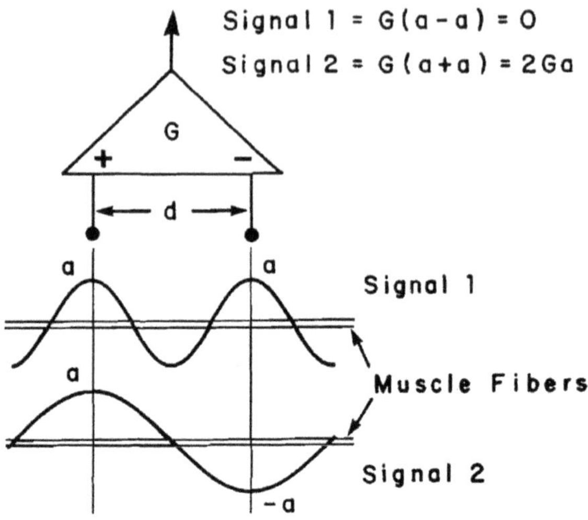

Fig. 9.17 Influence of bipolar configuration of surface electrodes and differential amplifier on the measurement of myoelectric signals of different wavelengths. The signals are presented in spatial coordinates (from Basmajian and De Luca [10]. With permission)

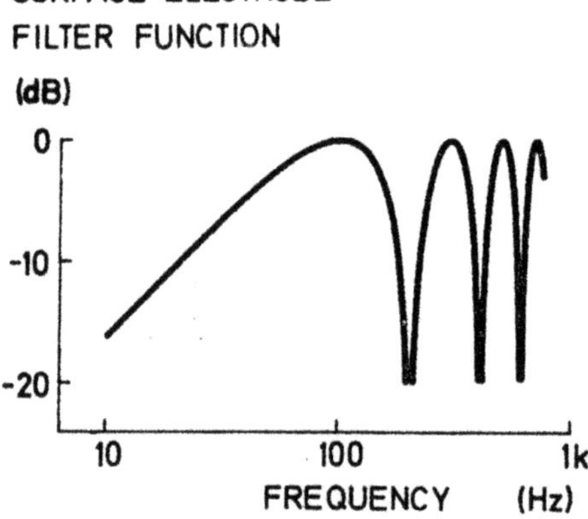

Fig. 9.18 The frequency response of bipolar configuration of surface electrodes and a differential amplifier. The assumed of a muscle fiber is $v = 4$ m/s and the distance between the electrodes is $d = 20$ mm (from Lindström [21]. With permission of the author)

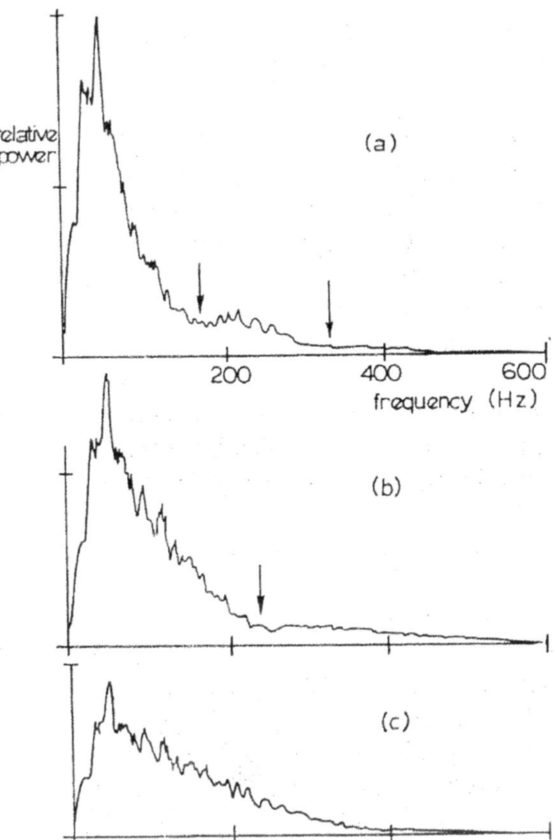

Fig. 9.19 Power spectrum of a myoelectric signal measured on m. *biceps brachii* of one subject, for three different electrode spacings: **a** 30 mm, **b** 20 mm and **c** 10 mm. The arrows indicate the frequencies at which the zeros should be in the spectrum due to the filtering property of the bipolar electrodes. All three graphs are drawn to the same horizontal and vertical scales (from Lynn et al. [28]. With permission)

9.3.1.5 Position of the Electrodes Relative to the Innervation Zone

In order to obtain myoelectric signals that reflect the true state of the contracting muscle, great care should be taken in the placement of surface electrodes. It is recommended that the electrodes be placed between the motor point (innervation zone) and the tendon so that the longitudinal axis (the axis passing through the middle of the electrodes) is parallel to the muscle fibers of the observed muscle. The issue of electrode positioning is commented in rather detail in [3].

The effect of placing the electrodes in the area of the muscle innervation zone is shown schematically in Fig. 9.20. In the case where the electrodes are placed in the area of the innervation zone, the displacement of the electrodes (which is unavoidable in dynamic contractions), e.g. from position A to position B, results in a significant change in shape of the measured MUAP, while in the latter case this influence is much less pronounced. It is clear that changes in the measured signal due to changes in the relative position of the electrodes and muscle do not reflect the physiological activity of the muscle. The influence of electrode positioning at different locations

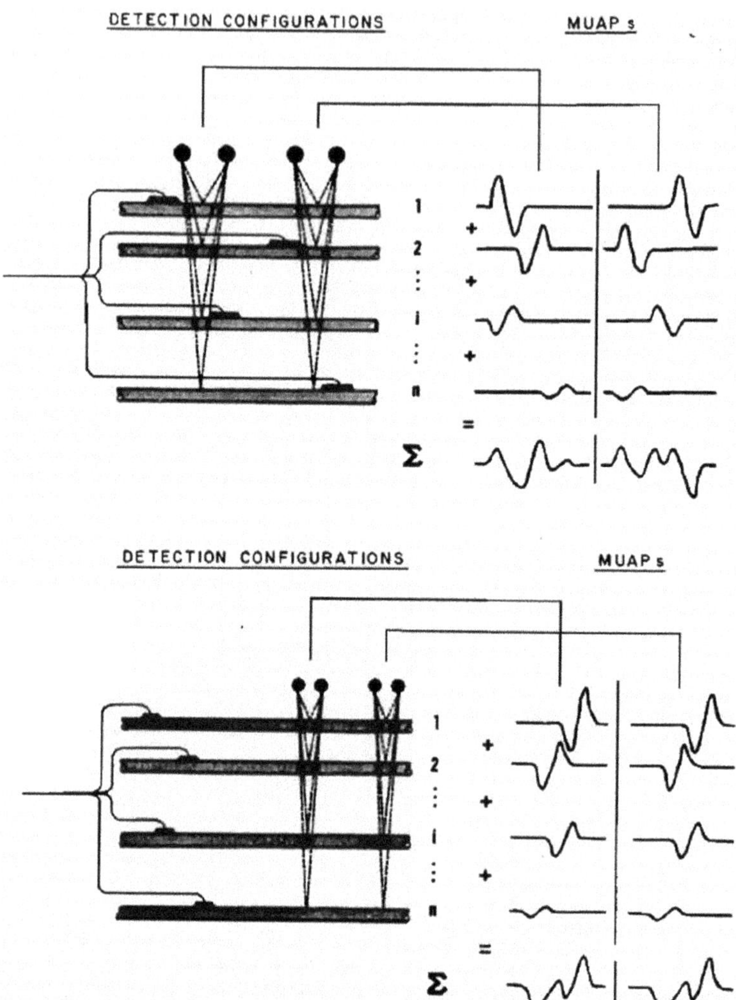

Fig. 9.20 Schematics showing the influence of the position of the electrodes on the waveform of the action potential. The electrodes may be located in the area of the innervation zone (upper scheme), and far enough from the innervation zone (lower scheme) (from Basmajian and De Luca [10]. With permission)

on the muscle can also be seen from the signal spectrum (Fig. 9.21). Positions of the motor points on each individual muscle can be determined using an electrostimulator prior to placement of the electrodes. The position of motor points is determined by the location on the skin above the muscle that requires the smallest amplitude of the electrical stimulus to induce contraction.

Technological development has made it possible to produce fields of miniature electrodes enabling measurement of myoelectric activity at high spatial resolution

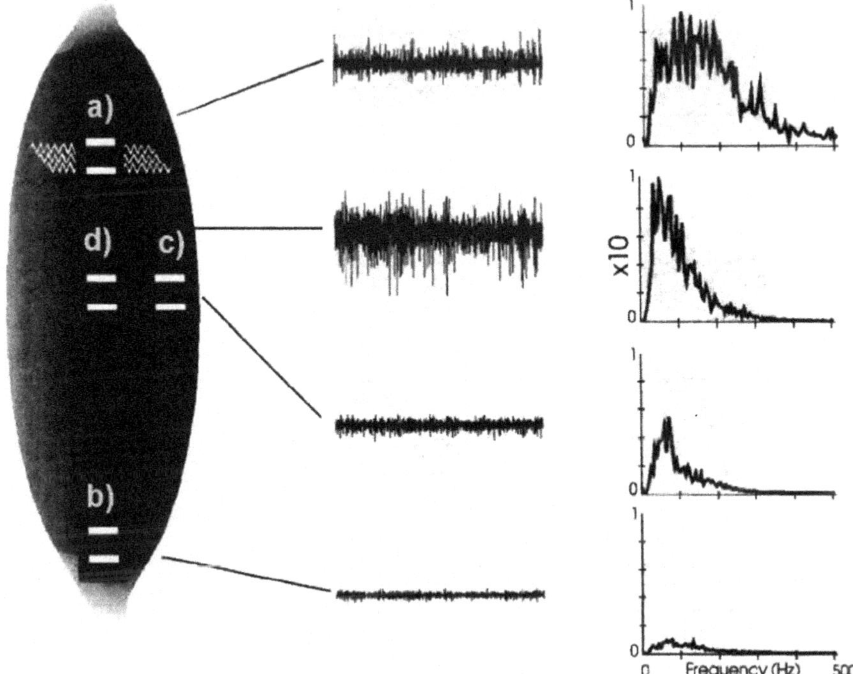

Fig. 9.21 Influence of electrode position on the appearance of the signal spectrum of measured myoelectric signals with respect to the innervation zone (**a**), tendon (**b**), and muscle edge (**c**). The recommended electrode placement is in the middle of the muscle, between the innervation zone and the tendon (**d**). At this point, the myoelectric signals of highest amplitude will be measured (from De Luca [29]. With permission)

across the entire muscle surface. Masuda et al. [30] were the first to suggest measuring multiple-point myoelectric signals along the muscle. Subsequently, a number of research groups were involved in the development of EMG mapping [31].

Figure 9.22 shows the result of the measurement using a linear electrode array. The position of the motor point as the location from which the myoelectric signals propagate in two opposite directions to the tendons is obvious.

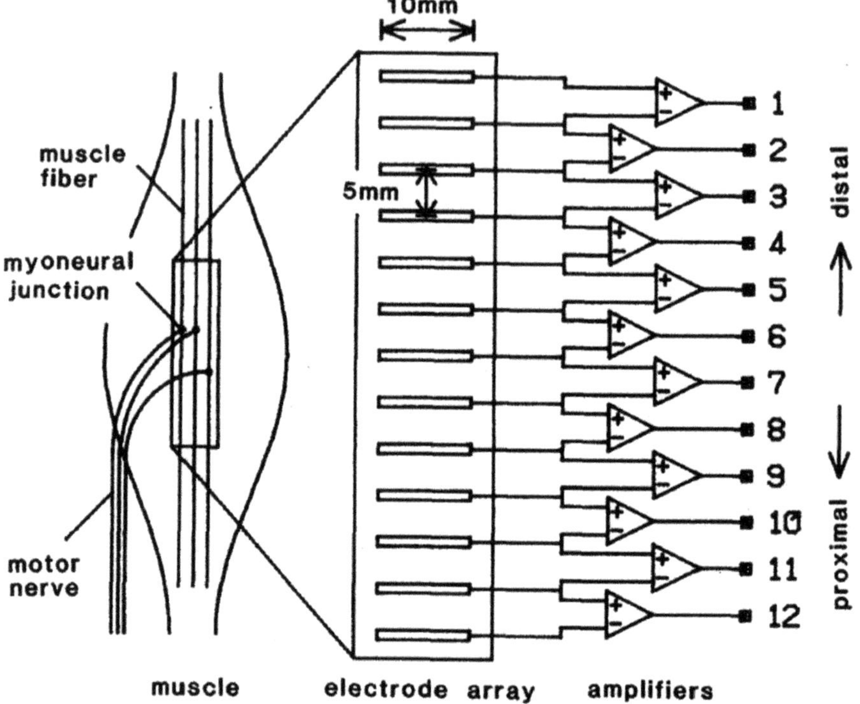

Fig. 9.22 Example of measuring a surface myoelectric signal using a linear electrode array. The myoelectric signals were received bipolarly from adjacent stainless-steel wire contacts, which were place parallel to each other with interelectrode spacing of 5.0 mm nad perpendicular to the underlying muscle fibers (from Masuda et al. [30]. With permission)

9.3.1.6 Position of the Electrodes Relative to the Direction of Muscle Fibers

Surface electrodes should be positioned so that the longitudinal axis (the axis passing through the centers of both electrodes) is parallel to the muscle fibers of the observed muscle. When this condition is not satisfied, the number of muscle fibers covered by both electrodes decreases. Figure 9.23 shows that when the longitudinal axis of the electrode closes with the muscle fiber angle α, the number of muscle fibers covered by both electrodes decreases by a factor $\cos\alpha$.

Figure 9.24 shows the position of the surface electrodes set up to measure myoelectric signals from three muscles from the quadriceps femoris group.

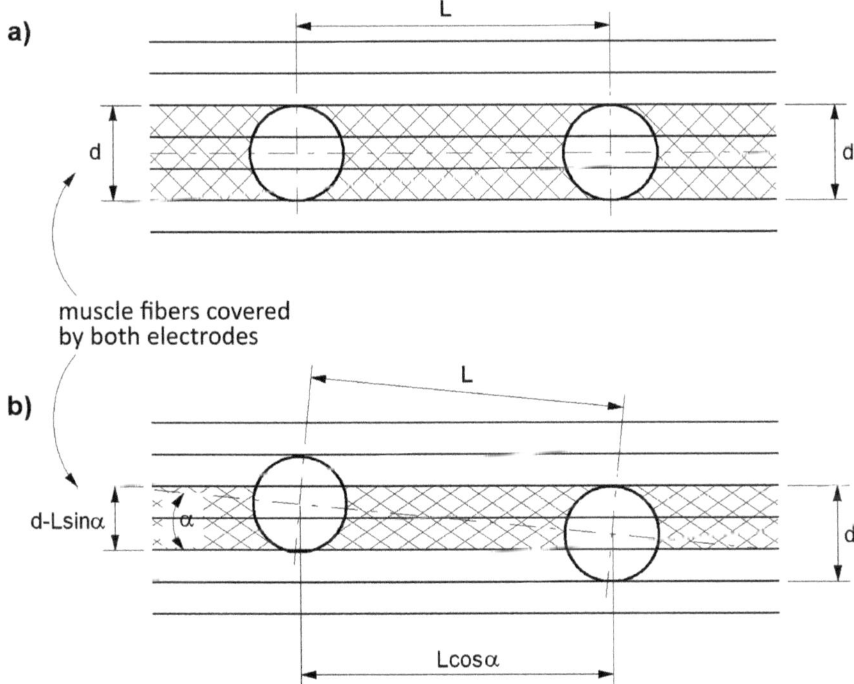

Fig. 9.23 Effect of electrode position relative to the direction of the muscle fibers: **a** the longitudinal axis of the electrodes is parallel to the muscle fibers, **b** the longitudinal axis of the electrodes closes with the muscle fibers angle α. The diameter of the electrodes is denoted by d and the distance between the electrodes by L (adapted from Cifrek [17]. With permission)

9.3.1.7 Electromyographic Amplifier and Analog-to-Digital Conversion

Amplitudes of myoelectric signals that can be measured by surface electrodes range from a few μV (when measuring resting tone) to a few mV (in athletes, with very strong contractions), so they need to be amplified before further processing. Also, the level of the input signal places a requirement on the input noise of the amplifier as well, and in modern EMG amplifiers, it ranges from 1 to 5 μV RMS.

EMG amplifiers must also meet certain requirements regarding the rejection of artifacts that may occur during measurement. The source of the artifacts during the recording of bioelectric signals can generally be divided into external and internal. The most prominent external artifact is caused by capacitive coupling between a power line (50 Hz or 60 Hz) and the input of the amplifier. Internal artifacts are biological signals that we do not want to measure (e.g. other muscle signals, ECG) and movement artifacts present during dynamic contractions. Rejection of interference signal, assuming that this signal appears at the input of the amplifier as a common signal, is achieved by using a differential amplifier with a sufficiently high CMRR (Common Mode Rejection Ratio). Today's EMG amplifiers have a CMRR >100 dB.

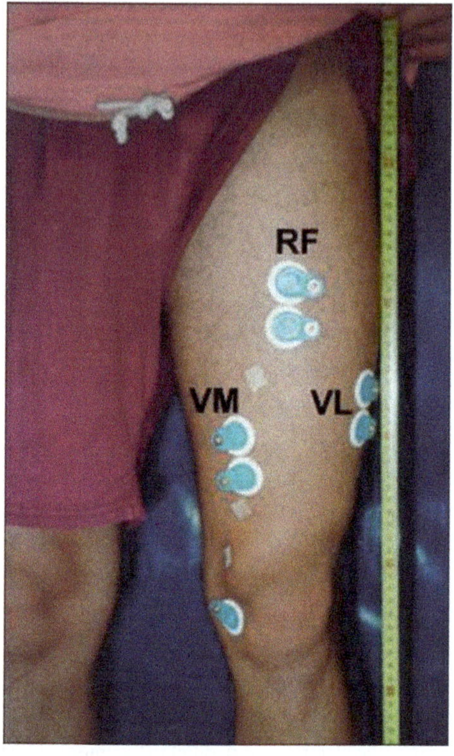

Fig. 9.24 Example of placement of surface electrodes above the muscles *rectus femoris* (RF), *vastus lateralis* (VL) and *vastus medialis* (VM). Motor points were determined using electrostimulator, and electrodes were positioned on a half way between a motor point and a distal tendon (from Cifrek [17]. With permisison)

Avoiding cross-talk from other muscles is possible by applying a double differential configuration (see the section on cross-talk), while displacement artifacts can be reduced by setting the lower cutoff frequency to 10 Hz or, in the case of stronger movement artifacts, to 20 Hz [3].

When using surface electrodes, an upper cutoff frequency of 500 Hz is sufficient, while for subcutaneous electrode measurements the upper cutoff frequency should be 1000 Hz. These values also determine the sampling frequency for analog-to-digital conversion. According to the Nyquist theorem, the sampling frequency should be at least twice the highest frequency component in the signal, which is 1000 Hz for surface electromyography and 2000 Hz for measurements using subcutaneous electrodes. Of course, if a detailed reconstruction of the myoelectric signal in the time domain is required, the sampling frequencies should be higher than the specified minimum.

9.3.1.8 Repeatability of Electromyographic Measurements

Repeated measurements on the same subject raise the problem of reliability and comparability of results. Any possible significant differences between the two experiments (not physiological differences) can be classified into three categories:

1. Bad electrodes or incorrectly placed electrodes,
2. Different location of electrode placement,
3. Different distance between electrodes.

The problems in the first category are solved by careful selection and fixing of the electrodes. A major problem is the location of the electrodes because it is very difficult to place the electrodes anatomically identically in two or more subjects and to determine whether a registered electromyogram represents the effect of the same muscle groups. Another problem is the repetition of experiments with the same subject when, in order to compare the results, we have to ensure the repeatability of the electrode placement. Specifically, changing the electrode distance has an effect on the properties of the registered myoelectric signal, since it affects both the amplitude and the frequency properties of the signal. The location of the electrodes relative to the muscle is also important, as biological tissue acts as a low-pass filter as a volume conductor between the muscle as a source of action potential and the electrode.

Due to all of the above, it is necessary to standardize the measurements, to document them in detail and to take into account all the above mentioned influencing factors, which could reduce the possibility of comparing the signals. Sometimes the photograph is taken as a document of actual measurement (as in Fig. 9.24).

9.3.1.9 Amplitude Normalization of Myoelectric Signals

It has already been stated earlier that the properties of the measured myoelectric signal are influenced by the tissues through which the signal passes on the path from the muscle fiber to the electrodes. In addition, the thickness of the subcutaneous adipose tissue has a great influence on the amplitude of the recorded surface myoelectric signal (Fig. 9.25).

Therefore, the signal amplitude expressed in mV or μV cannot be used to compare the contraction intensity of the same muscle in two different subjects. As a solution, the normalization of the amplitude to a reference value is used and the amplitude is shown in percentage of the maximum value. The reference value is often the amplitude of the observed muscle signal measured under conditions of maximum voluntary isometric contraction (MVIC) and with defined biomechanical conditions. Other ways of normalization are possible, such as, for instance, the maximum signal amplitude recorded during the measurement [18].

Strictly speaking, the subject of amplitude normalization of myoelectric signal belongs into Sects. 9.3.2: „myoelectric signal processing“. We have included it here, however, to accompany Fig. 9.25, illustrating relevant measurement aspects.

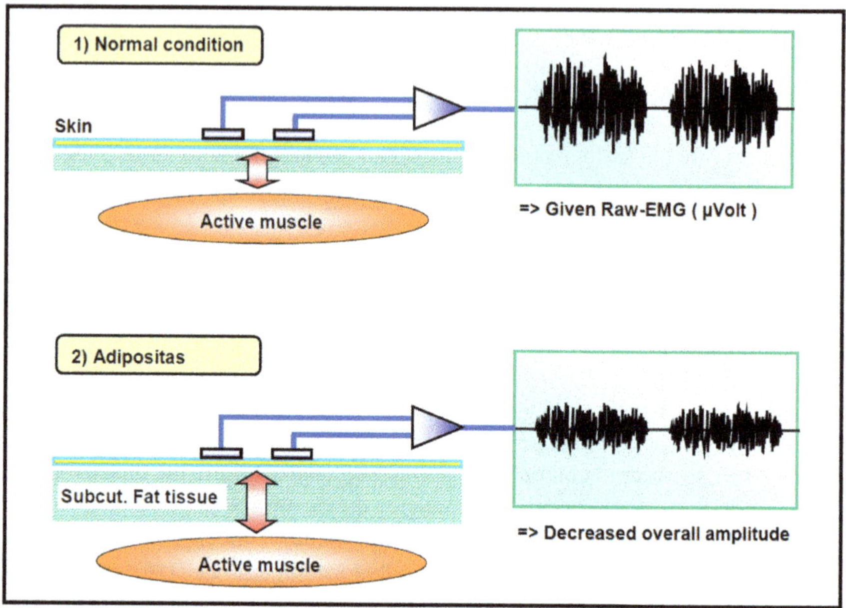

Fig. 9.25 Influence of subcutaneous adipose tissue on the amplitude of the measured myoelectric signal (from Konrad [23]. With permission)

9.3.1.10 The Cross-Talk Problem

Cross-talk in surface electromyography is the contribution of signals from adjacent muscles or some other electrically active organs to the myoelectric signal of the muscle above which the surface electrodes are positioned. It is a consequence of the conductivity of the interior of the body, through which the signals propagates. Due to the properties of the tissue as a volume conductor, it is obvious that the contribution of signals from adjacent muscles or other organs will decrease with distance, and that this influence will be more significant at lower frequencies (the tissue has the properties of a low-pass filter).

Based on all of the above, it is obvious that the cross-talk problem will be more pronounced when measuring signals from adjacent small size muscles that contribute synergistically to the same function, such for example as in the forearm muscles.

A certain amount of cross-talk is always present, but it can be reduced by properly positioning the electrodes on individual muscles (in the middle of the muscle, between the motor point and the tendon) while keeping the maximum distance between the pairs of electrodes on the individual muscles.

If the contribution to cross-talk is substantial but cannot be reduced by changing the position of the electrodes, it can be reduced by using a multi-electrode recording. Double differential configuration and muscle cartography will be shown as an example.

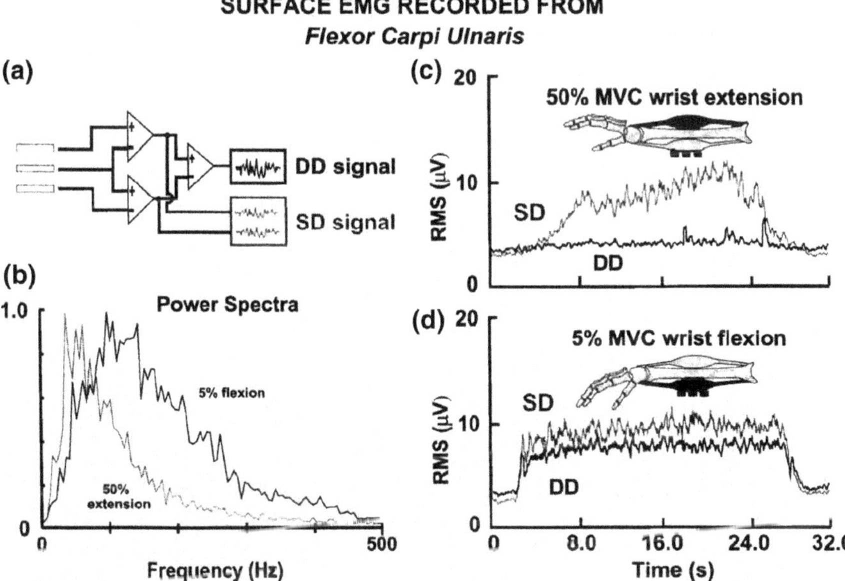

Fig. 9.26 Using a double differential method to reduce crosstalk **a** Scheme **b** spectra of SD signals recorded on flexor muscle (electrodes are on it) under 5% MVC (maximal voluntary contraction) conditions and on arm extensor muscle under 50% MVC conditions. The effect of the filter property of the tissue on the signal from the arm extensor muscle away from the electrodes is visible—the frequency spectrum is shifted to lower frequencies **c** SD and DD signals recorded during extension—it can be seen that the SD signal would give the wrong information about the flexor muscle activity due to the very strong signal coming from the extensor. The DD signal provides information that better delineates the condition of the observed muscle **d** SD and DD signals recorded during flexion. Since only the observed muscle signal is present on the electrodes, the signals are practically indistinguishable (from De Luca [29]. With permission)

9.3.1.11 Double Differential Method

Unlike the usual bipolar configuration of electrodes connected to a single differential amplifier (SD), the double differential method (DD) requires three electrodes per measuring point (Fig. 9.26). The method is based on the fact that the contribution of the myoelectric signal of the distant muscle is the same on both differential (bipolar) electrodes, so at the second differential stage, it is canceled, leaving only the contribution of the signal from the measured muscle.

9.3.1.12 Muscle Cartography

One of the significant advances of the last decades is the technology of two-dimensional EMG (2D-EMG or high-density EMG, HDEMG) or EMG imaging, based on electrode grids [32].

Already mentioned were contributions to measuring multiple-point myoelectric signals along the muscle [30], as well as development of EMG mapping [31]. Figure 9.27 shows examples of mono- and bi-dimensional electrode arrays [33]. Muscle cartography allows the application of spatial filtering [18] other than double differential, and in addition to determining the position of motor points, it can also serve to separate myoelectric signals of the measured muscle from those of adjacent muscles.

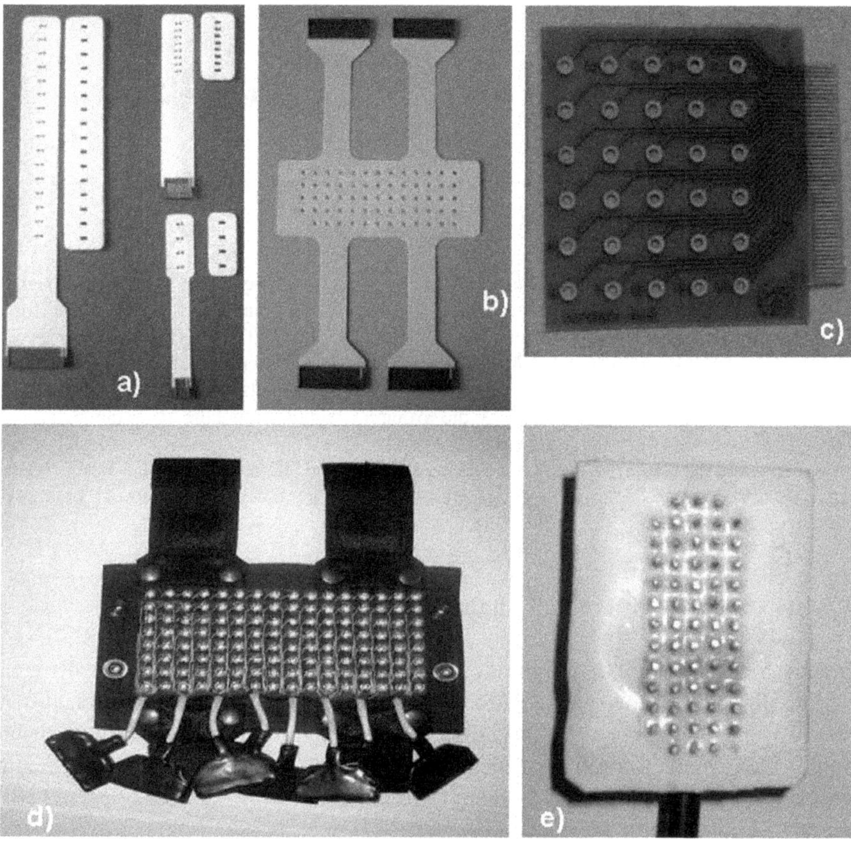

Fig. 9.27 **a** Examples of linear electrode arrays with different electrode numbers and interelectrode distances. The cavities in the double adhesive foam are filled with gel, through the holes in the arrays, after positioning. **b** Example of a two-dimensional flexible electrode grid (LISiN-Spes Medica, Italy) with 64 electrodes. **c** Printed circuit bidimensional electrode array of 6 × 5 electrodes on Mylar. **d** 128 silver coated eyelets on a textile support, with eight columns and 16 rows of electrodes. **e** Bidimensional pin electrode array of 61 silver electrodes, with five columns and 13 rows of electrodes (1-mm diameter, 3-mm interelectrode distance in both directions) without the four corner electrodes (from Merletti et al. [33]. With permission)

As an example, effect of muscle shift under a pair of electrodes during m. *biceps brachii* elbow flexion is shown in Fig. 9.28. The effect is already mentioned in the Sect. 9.3.1.5.

Fig. 9.28 Example of the effect of muscle shift under a pair of electrodes. *Biceps brachii* sEMG RMS map (grid of 64 electrodes, IED = 10 mm) during elbow flexion. The images are interpolated. As the muscle shortens the innervation zone shifts upward and the white electrode pair reads a progressively greater signal whereas the black electrode pair reads a progressively smaller one. These opposite changes of amplitude are not necessarily reflecting changes of muscle activity. They mostly reflect a variation of the muscle-electrode geometry (muscle shortening under the electrodes) and may be misleading and misinterpreted. The use of a sEMG map reduces the possibility of misinterpretation. [34] Reproduced from www.robertomerletti.it (from Merletti and Muceli. Open Access, Creative Commons CC-BY license)

9.3.2 *Myoelectric Signal Processing*

9.3.2.1 Time Domain Signal Processing

Raw surface myoelectric signal is of irregular shape but exhibits characteristic properties, the most obvious of which is the increase in amplitude with increasing contraction intensity (Fig. 9.5). Historically, the first method of monitoring and analyzing muscle activity was precisely the visual control of a raw myoelectric signal. Furthermore, on some electromyographs, the myoelectric signal could be connected to the loudspeaker.

As in kinesiology, a primary concern is the study of human movement, sEMG during movement is often called dynamic EMG. Development of this measurement technique has contributed to the establishment of the field sometimes called electrophysiological kinesiology.

Averaging (Smoothing) of the Full-wave Rectified Signals

Some of the averaging methods used are shown in Fig. 9.29.

The mean value of a fully rectified myoelectric signal (Average Rectified Value, ARV) within the time interval (t_2–t_1) is calculated according to expression (9.4).

$$\bar{x} = \frac{1}{t_2 - t_1} \int_{t_1}^{t_2} |x(t)| dt \tag{9.4}$$

In the above expression, the myoelectric signal is denoted by $x(t)$. Considering the mean value of the signal, the duration of the time interval (t_2–t_1) within which the mean value is calculated should be specified. This is usually the duration of a contraction or of a movement.

The root mean square (RMS) value of the myoelectric signal within the time interval (t_2–t_1) is calculated according to expression (9.5).

$$x_{RMS} = \sqrt{\frac{1}{t_2 - t_1} \int_{t_1}^{t_2} x^2(t) dt} \tag{9.5}$$

Moving average (MA) allows monitoring the change in the mean amplitude of a myoelectric signal. The moving averaging procedure consists of two steps: (i) full-wave rectification, (ii) low-pass filtering (expression (9.6)). The combination of a full-wave rectifier and a low-pass filter is also called a linear envelope detector.

$$x_{MA}(t) = \frac{1}{T_{MA}} \int_{\tau=0}^{T_{MA}} |x(t - \tau)| d\tau \tag{9.6}$$

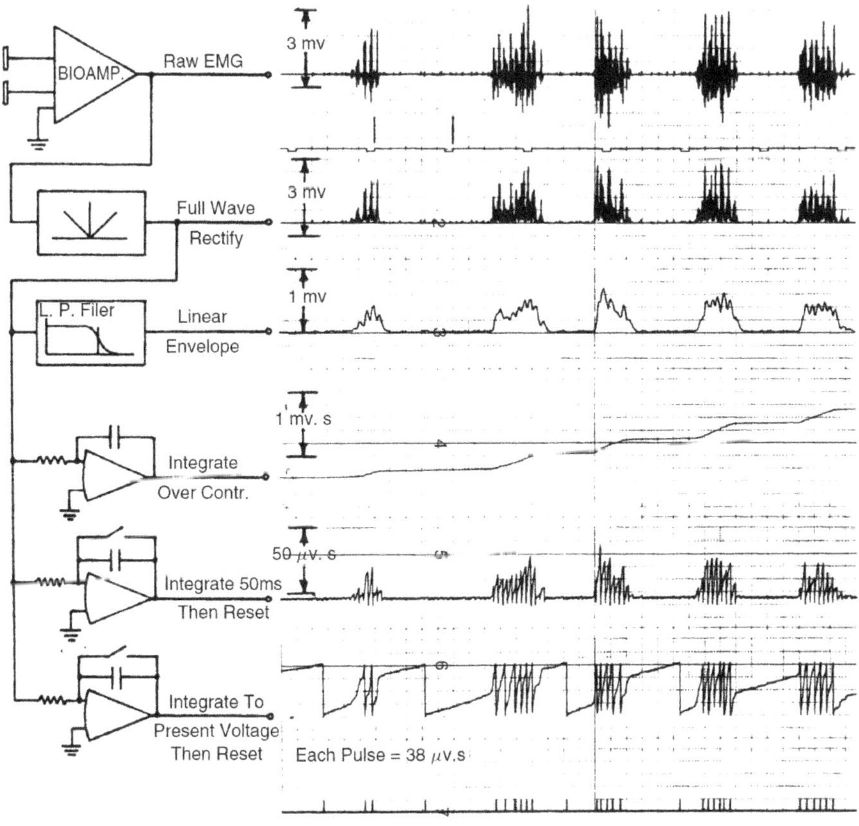

Fig. 9.29 Common surface EMG signal processing methods in the time domain. Signal processing operators in an analog (physical) form are symbolically depicted, however, in modern systems digital signal processing is performed (A/D conversion of measured and amplified signals prior to processing is assumed) (from Winter [35]. With permission)

Moving average operation can also be performed by computing the mean squared value of the signal, as shown by expression (9.7).

$$x_{MA\_RMS}(t) = \sqrt{\frac{1}{T_{MA}} \int_{\tau=0}^{T_{MA}} x^2(t-\tau)d\tau} \qquad (9.7)$$

Figure 9.30 shows how the averaging interval (T_{MA}) in the moving averaging process affects the appearance of the averaged signal. The effect of low-pass filtering on the amplitude of the averaged signal and on the estimation of contraction duration is evident.

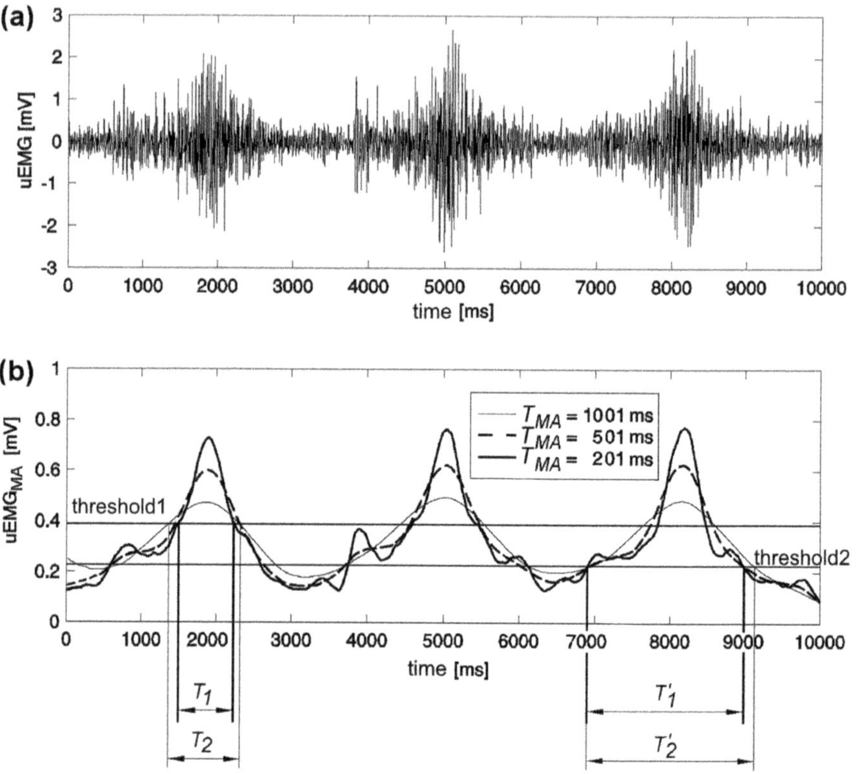

Fig. 9.30 Effect of averaging interval (TMA) on the amplitude of the averaged signal and on the estimation of contraction duration (from Cifrek [17]. With permission)

By definition, the integral of myoelectric signals represents the area under the curve of a full-wave rectified myoelectric signal and is expressed in unit Vs. The integration of a continuous myoelectric signal is described by expression (9.8).

$$x_I(t) = \int_0^T |x(t)|dt \tag{9.8}$$

When reporting the results obtained by integrating a myoelectric signal, the time period (T) within which the integration was performed should be specified. In practice, two approaches are usually used for quantification of the integrated myoelectric signal. The first method integrates a fully rectified myoelectric signal over a period of time (e.g. 50 ms) and after which the integrator output is reset. As a result, a series of pulses of equal duration but different amplitudes are obtained, depending on the intensity of the muscular contraction. In the second method, the fully rectified myoelectric signal is integrated to a certain voltage level, followed by an integrator reset. As a result, a series of pulses is obtained whose frequency is proportional to the intensity of the muscular contraction (Fig. 9.29).

Calculation of Coefficient of Correlation Between Averaged Signals

Multichannel EMG may serve in studies of muscular coordination. Cross-correlation between averaged (smoothed) signals may provide a method for objectively comparing the timing and shape of EMG signals.

Cross-correlation coefficient R between pair of averaged EMG signals x_i and y_i is calculated according to Eq. 9.9:

$$R = \frac{\sum x_i y_i}{\sqrt{\sum x_i^2 \sum y_i^2}} \tag{9.9}$$

The cross-correlation measures the similarity in shape between two curves as a scalar between 0 and 1 (two curves with exactly the same shape will have a cross-correlation of 1). Changing the amplitude of the curve without changing its shape does not affect the cross-correlation results.

9.3.2.2 Signal Processing It the Frequency Domain

Power Spectrum of the Myoelectric Signal

The power spectrum and the discrete parameters derived from it are used as quantitative indicators of the characteristics of myoelectric signals.

The power spectrum, or more precisely, the power spectrum density $P(\omega)$ of the myoelectric signal $x(t)$ is defined as the square of the modulus of its Fourier transform $X(\omega)$ according to expression (9.10).

$$P(\omega) = X(\omega) \cdot X^*(\omega) = |X(\omega)|^2 \tag{9.10}$$

From the obtained power spectrum, the parameters that serve to characterize and monitor changes are calculated, most commonly due to muscle fatigue, which will be demonstrated later at the example. These are usually the median frequency (f_m) and the mean frequency of the power spectrum.

The median frequency is the frequency that divides the area under the power spectrum density curve into two equal parts, calculated according to (9.11).

$$\int_0^{f_m} P(f)df = \int_{f_m}^{\infty} P(f)df \tag{9.11}$$

The mean frequency of the power spectrum is calculated as the ratio of first- and zero-order spectral moments to expression (9.12).

$$\bar{f} = \frac{m_1}{m_0} = \frac{\int_0^f fP(f)df}{\int_0^f P(f)df} \tag{9.12}$$

In addition to the above parameters, mode (f_{mod}) is used, the frequency corresponding to the maximum of the power spectrum and the bandwidth of the power spectrum defined as the frequency range in which the power spectrum density curve has a value greater than 50% of the maximum amount.

The mentioned characteristic frequencies are indicated on the idealized representation of the power spectrum density function in Fig. 9.31.

In addition to the above parameters derived from the calculated power spectrum, a single parameter of the myoelectric signal, calculated in the time domain, is used, which can be related to the frequency content of the signal. This is the so-called zero-crossing rate. This parameter is interesting because of its simple method of determination, but it should be used with caution because of the high dependence of the results on the signal-to-noise ratio of the measured signal.

In addition to the Fourier transform, newer methods for estimating the power spectral density of myoelectric signals based on wavelet analysis and Hilbert-Huang transformation are used [36].

When calculating the power spectrum by frequency analysis methods, care should be taken of the fact that the parameters of the analysis affect the shape of the power spectrum and therefore the calculated characteristic frequency. As an example, Fig. 9.32 shows the influence of the windowing function on the appearance of the power spectrum calculated by the Fourier transform.

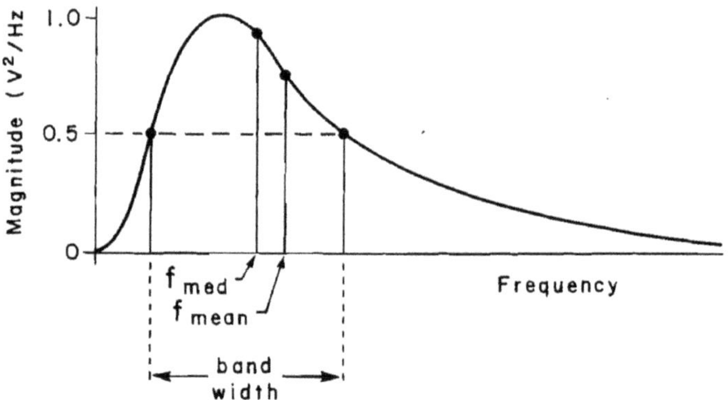

Fig. 9.31 Idealized representation of the normalized power density function of the surface power of a surface myoelectric signal with characteristic frequencies (from Basmajian and De Luca [10]. With permission)

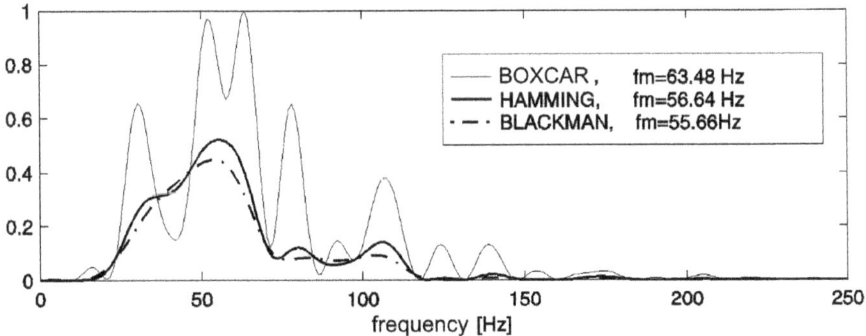

Fig. 9.32 Influence of the windowing functions on the power spectrum of a myoelectric signal

9.3.2.3 Recommendations and Standards for Surface Electromyography

At the very beginning of the application of surface electromyography, research was conducted in separate laboratories, using different equipment (most often polygraphs), with different measurement protocols, which resulted in different terminology, methods of measurement and the way of presenting the results. In order to ensure reproducibility of results and comparability between different authors and laboratories, recommendations and standards have been introduced. Back in 1980, The International Society of Electrophysiology and Kinesiology (ISEK) published recommendations aimed at harmonizing terminology, measurement procedures, measurement instrumentation parameters, processing and analysis methods, and methods of reporting surface electromyography results [37]. The development of the area has also resulted in the revision and amendment of the above recommendations, and the current version of the document "ISEK Standards of Reporting EMG Data" can be found at the following link:

https://isek.org/wp-content/uploads/2015/05/Standards-for-Reporting-EMG-Data.pdf.

As part of the Biomedical Health and Research Program (BIOMED II), the European Commission funded the Surface Electromyography for Non-Invasive Assessment of Muscles project. The project began in 1996 and has resulted in a series of recommendations for conducting electromyographic measurements (electrodes, electrode placement, analysis and signal modeling. The final results of the project have been published in [38].

Journal of Electromyography and Kinesiology introduced Standards for Reporting EMG Data in April 2014 [39].

In [40] fundamental concepts in EMG signal acquisition are presented.

9.3.3 Interpretation of Surface EMG Signals

The interpretation of surface myoelectric signals is shown through five examples. Firstly, it is reflected shortly on the relationship between the amplitude of the sEMG signal and the developed muscle force, a longyear subject of research by professionals practicing the technique of electromyography. Remaining examples report concisely on research conducted by the authors of the chapter. The sport of alpine skiing is addressed first. A comparison of the performance of wedge turn on standard and carving skis, by measuring the activity of the four upper leg muscles and ground reaction force, and analysing these data later, is shown. Then, a sports biomechanics method for evaluating locomotor skill using myoelectric and kinetic indicators is reported, at the example of backward somersault performance in artistic gimnastics. Application of multichannel sEMG in tennis follows then, whereby a structured approach to the game of tennis is developed aimed for the teaching school of the sport of tennis. Finally, the use of electromyographic indicators of muscle fatigue in the diagnostics of back pain is shown as an example of the interpretation of sEMG in the frequency domain.

9.3.3.1 Relationship Between the Amplitude of the Myoelectric Signals and Developed Force

A large number of researchers have investigated the quantitative relationship between the amplitude of the myoelectric signal and the force developed [3, 29, 41]. Research results have shown that this dependence is generally nonlinear (Fig. 9.33), except

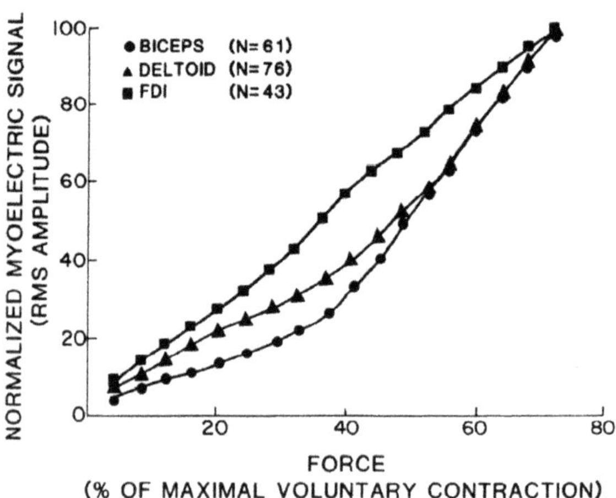

Fig. 9.33 Normalized representation of the correlation between the force and amplitude of a myoelectric signal for three different muscles. There is a noticeable difference in linearity (from Lawrence and De Luca [42]. With permission)

under conditions of a very controlled contraction in certain muscles [42]. Specifically, in order to accurately determine this relationship, it is necessary to estimate the contribution of synergistic, agonistic and antagonistic muscles to the total force developed. Also, the question remains whether changes in the distribution of force between the individual muscles involved in a contraction occur during the measurement. Thanks to this positive correlation of an EMG signal and muscle force, and in the special case of isometric muscle contraction in particular, a good measure of proportion has been established between the averaged or smothed (full wave rectified and low pass filtered) EMG signal (Sect. 9.3.2.1) and muscle force, a feature taking advantage of in the following two examples.

9.3.3.2 Biomechanical Measurement and Testing Method for Turns in Alpine Skiing

Pursuing an outdoor type kinesiological research, Rađenović et al. [43] attempted to experimentally investigate into how sports equipment may influence relevant sports technique and, consequently, induce new traumatologic factors at a large scale. In alpine skiing, a novel type of skies was introduced having a so-called carving cut (as opposed to standard, classical cut) and, being shorter, enabled faster acquiring of a proper skiing technique by beginners and also significant changes in technique in skilled skiers and masters. This was particularly interesting for numerous ski schools. In cooperation with colleagues from Ljubljana, we have measured surface EMG signals of four left leg muscles: m *rectus femoris*, m. *vastus lateralis*, m. *vastus medialis*, m. *biceps femoris*, and ground reaction force below a ski boot, during a performance of wedge turns, standard elements of alpine skiing technique. Standard (classical) and carving skis were used and two types of wedge turns were performed, known as the Croatian and the Austrian variant. Video recordings were used to later monitor kinematics for control and classification (Croatian variant—Austrian variant of the turn) purposes. EMG signals and ground reaction force signals were compared for both types of skis and both types of turns. To enable comparisons, measured sEMG signals were smoothed (full wave rectified and low pass filtered), followed by parameter extraction from smoothed signals. Parameters were also extracted from the force signals recorded. Then, sEMG and force signal parameters were statistically compared.

Video records are illustrative of differences between Croatian and Austrian variants of the wedge turn (subtle differences, recognizable to alpine ski teachers); see Figs. 9.34 and 9.35.

When performing the Croatian variant, in the second phase of the turn, the outside (dominant) leg is mostly loaded, in contrast to the Austrian turn, where during the same phase both legs are loaded almost equally. The results of the quantitative comparison and analysis showed that the least strain develops while performing the Austrian variant of the wedge turn using carving skis [43]. These results may

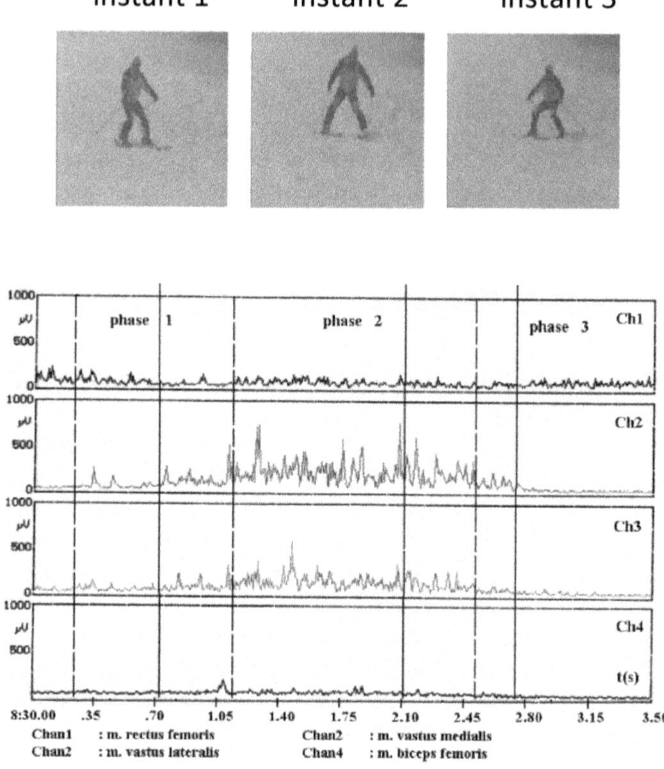

Fig. 9.34 Croatian variant of the turn with EMG signals for each phase (from Rađenović et al. [43]. With permission). phase 1. the entrance, phase 2. the steering and phase 3. the exit; Ch1: m. *rectus femoris*, Ch2: m. *vastus medialis*, Ch3: m. *vastus lateralis*, Ch4: m. *biceps femoris*. SEMG signals are in a full-wave rectified and low-pass filtered (smoothed) form

have implications for the use of this type of skis in teaching basic skiing techniques. They are therefore relevant for ski schools. Besides, they contribute some evidence to the field of sports traumatology.

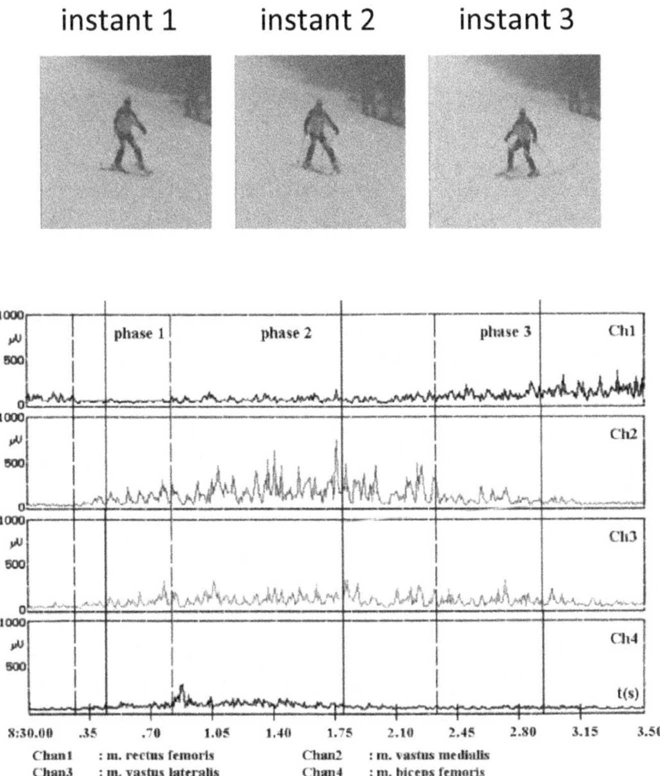

Fig. 9.35 Austrian variant of the turn with EMG signals for each phase (from Rađenović et al. [43]. With permission). phase 1. the entrance, phase 2. the steering and phase 3. the exit; Ch1: m. *rectus femoris*, Ch2: m. *vastus medialis*, Ch3: m. *vastus lateralis*, Ch4: m. *biceps femoris*. SEMG signals are in a full-wave rectified and low-pass filtered (smoothed) form

9.3.3.3 Skill of Performance Evaluation—An Example of Backward Somersault

Medved et al. [44] have studied skilled human locomotions in artistic gymnastics by measurement and analysis of ground reaction force and myoelectric signals. The research was a part of monitoring progress in young gymasts in a Zagreb club, over a period of two years. Specifically, elements of take-off in several artistic gymnastics movement tasks were studied. Take-off is a very important element of many a movement pattern in artistic gymnastics as it reflects the degree of explosive power manifested by lower extremity extensor musculature, a prerequisite of successful airborn performances. In particular, the movement pattern of backward somersault from a standing position was studied (Fig. 9.36). Besides kinetic parameters (ground reaction force) also myoelectric signals were recorded from major

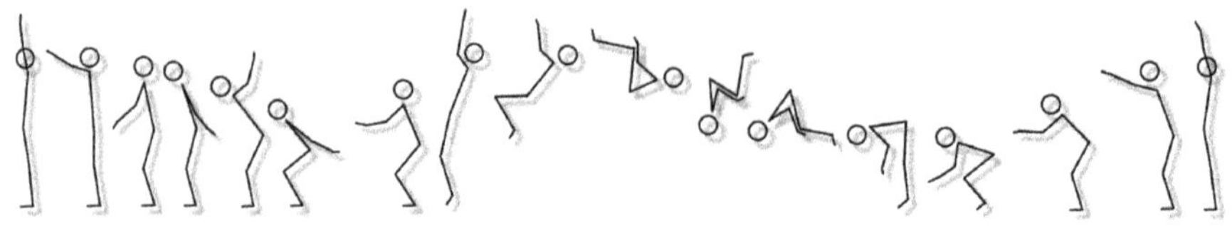

Fig. 9.36 Stick figure representation of the backward somersault kinematics

lower extremity muscles (m. *gastrocnemius*, m. *tibialis anterior*, m. *rectus femoris*, m. *biceps femoris*).

Quality of performance of each particular somersault was estimated by experts, certified gymnastics judges, through their visual judgment in real time. Assuming the quality of performance to be a skill level criterion acquired signals were mathematically analyzed via basic signal processing and cross-correlation methods. The study of backward somersault yielded quantitative criteria of performance, proposing an original measure—an inter-muscular cross-correlation function (Eq. 9.13).

$$H(j) = \frac{\sum_{i=1}^{N}\left[A(i+j)-\bar{A}\right]\cdot\left[B(i+j)-\bar{B}\right]}{\sqrt{\sum_{i=1}^{N}\left[A(i+j)-\bar{A}\right]^2\cdot\sum_{i=1}^{N}\left[B(i+j)-\bar{B}\right]^2}} \quad (9.13)$$

The function $H(j)$, is a collection of scaled correlation coefficients, calculated one by one for each j shows the correlation between two selected averaged (smoothed) EMG signals, A and B, with particular focus to bilaterally measured muscles. It is calculated by moving an N point window.

Figure 9.37 shows calculated moving correlation functions between left (LGa) and right (RGa) *gastrocnemius* muscle for the "top-level" and "poor" performer, that is, for performances by a top-level performer "at his best" and "deliberately poor", respectively.

A good discriminability feature is observed in the procedure for the evaluation of skill level realized in this way; EMG signals have thus shown to be rather sensitive measures of neuromuscular performance.

Besides motor learning, the quantification of movement skill also relates to possible applications in biocybernetics, robotics, and rehabilitation medicine.

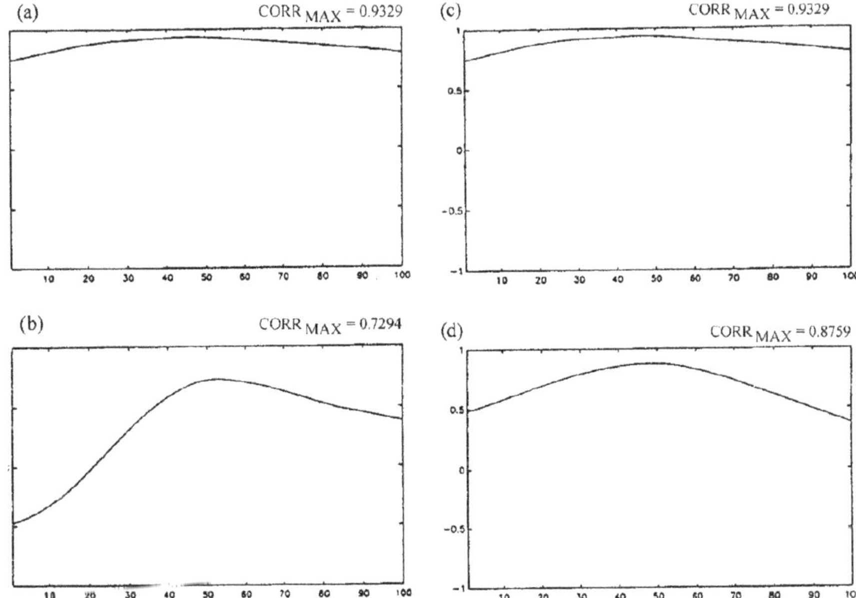

Fig. 9.37 Left column: correlation (LGa vs. RGa) functions $H(j)$ of “top-level” performer (**a**) and “poor” performer (**b**). Right column: correlation (LGa vs. RGa) functions $H(j)$ of a top-level gymnast performing backward somersault “at his best” (**c**) and “deliberately poorly” (**d**) (from, Medved and Cifrek [19]. Open access)

9.3.3.4 sEMG Patterns in Tennis

A laboratory research was undertaken with the goal to study several standard tennis strikes. One subject participated, a tennis expert and certified trainer, age 58. He was instrumented with electrodes on eight muscles (Figs. 9.38 and 9.39: see legend) and has performed multiple times several strikes: backhand, forehand, forehand spin, serve. Strikes were performed without hitting a ball (simulated strikes). Typical multi-channel sEMG recordings of tennis serve and of forehand top spin are shown as an illustration (Figs. 9.38 and 9.39, respectively). This kind of presentation of signals can be visually evaluated, compared, and analysed, and can give an indication of both timing and amplitudes of activation of the set of muscles measured. Findings of this sort were incorporated into the book devoted to teaching the sport of tennis [45]. It represents an original structured approach to this popular sports game, systematically describing and defining tennis strikes according to their constituent parts and elements. Besides covering the elements of technique, the book also explains playing strategy and tactics, as well as necessary physical conditioning elements.

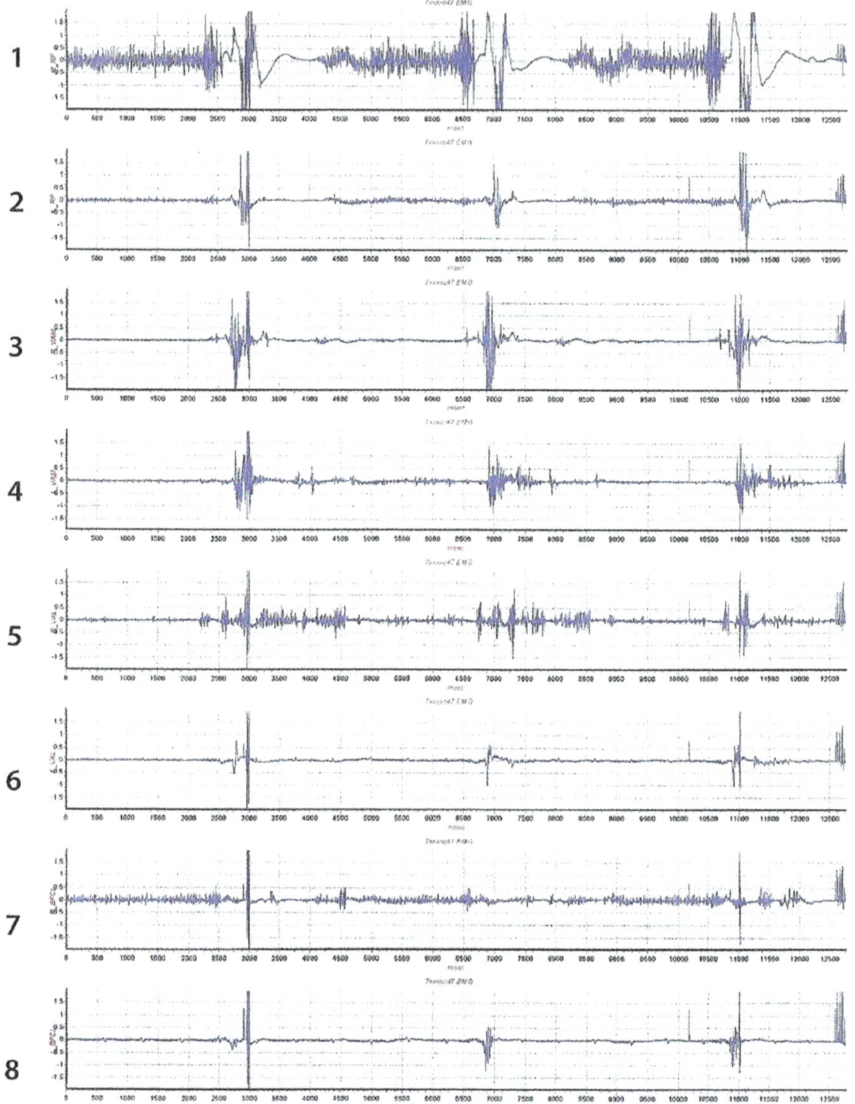

Fig. 9.38 sEMG record of tennis serve. Channels denote: (1) *m. deltoideus*, (2) *m. biceps brachii*, (3) *m. triceps brachii*, (4) *m. brachio-radialis*, (5) *m. pronator teres*, (6) *m. pectoralis*, (7) *m. rectus abdominis and* (8) *m. obliqus sinister* (from Rupić [45]. With permission)

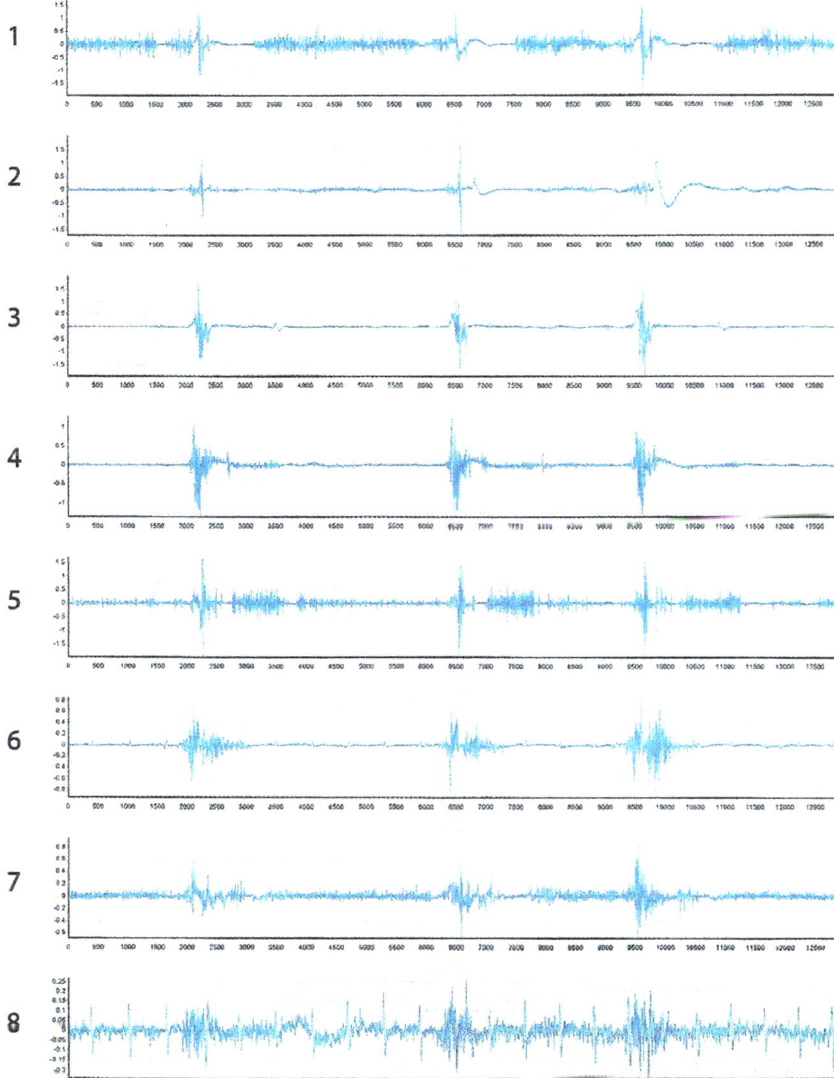

Fig. 9.39 sEMG record of forehand top spin. Channels denote: (1) *m. deltoideus*, (2) *m. biceps brachii*, (3) *m. triceps brachii*, (4) *m. brachio-radialis*, (5) *m. pronator teres*, (6) *m. pectoralis*, (7) *m. rectus abdominis and* (8) *m. obliqus sinister* (from, Rupić [45]. With permission)

9.3.3.5 Frequency Domain Processing Example—Muscle Fatigue

Generally About EMG Indicators of Muscle Fatigue

Muscle fatigue is a very complex phenomenon that encompasses various causes, mechanisms, and forms of manifestation. It occurs as a result of a series of metabolic, structural and energy changes in the muscles, as well as a lack of supply of oxygen and nutrients through the bloodstream, and due to changes in the efficiency of the nervous system.

The method of surface electromyography enables continuous monitoring of the local fatigue of individual muscles during the performance of certain work due to the fact that biochemical and physiological changes in the muscles during fatigue also reflect the properties of myoelectric signals. One of the major influencing factors that alter the properties of the surface myoelectric signal during fatigue is the decrease in muscle fiber . This phenomenon causes the power spectrum of the surface electromyogram to shift to lower frequencies and, as a second-order effect, an increase in amplitude, as shown in Fig. 9.40. In addition, there are minor changes in the shape of the power spectrum [46].

Changes in EMG signal parameters showing fatigue development during isometric muscle contraction of m. *tibialis anterior* are shown in Fig. 9.41. Although the subject was able to maintain the given contraction force for the first 60 s, the development of

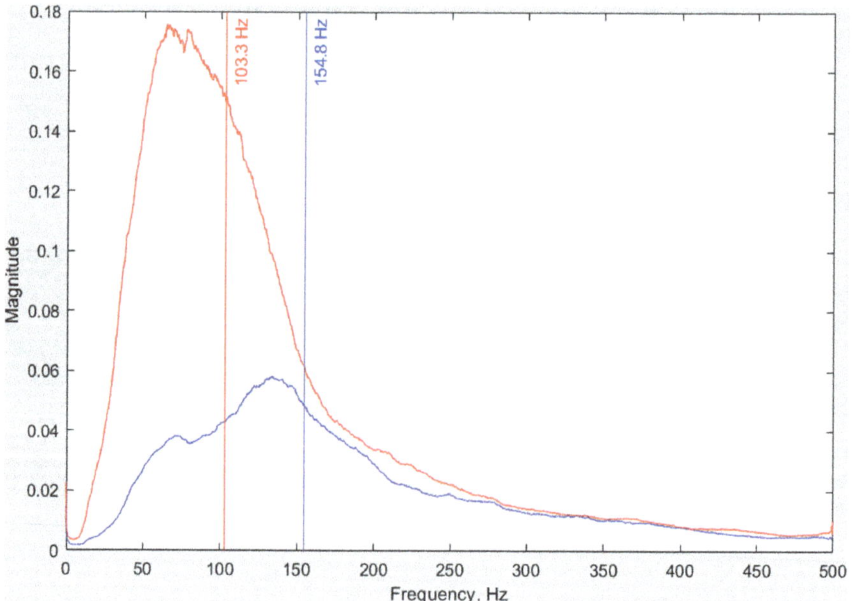

Fig. 9.40 Power spectrum of the myoelectric signal at the beginning (blue) and end (red) of the 60 s, 80% MVC static contraction of the muscle *triceps brachii*. Healthy female subject, 23 years. The corresponding median frequencies are also shown

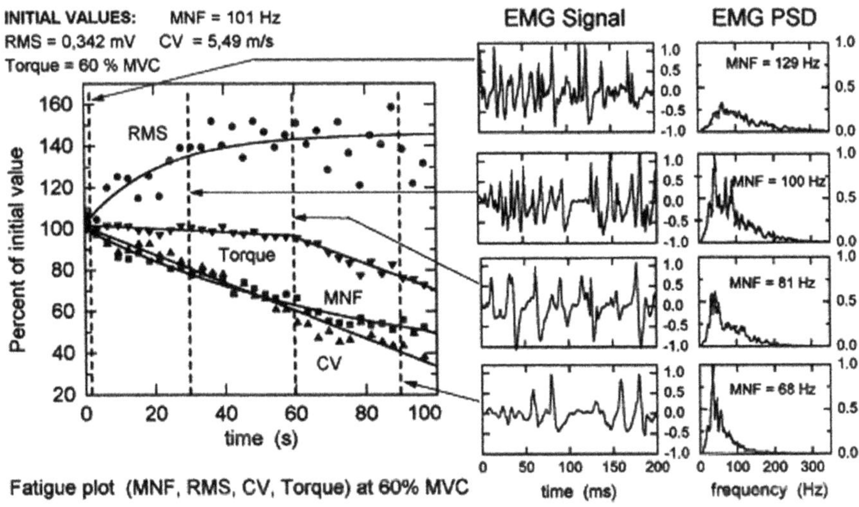

Fig. 9.41 Changing the characteristic parameters of the surface myoelectric signal of the m. *tibialis anterior* muscle during a sustained isometric contraction for 100 s. The subject was able to maintain a given force level for 60 s, after which this force decreased. In addition, it is evident that the indicators of fatigue (median frequency, mean frequency) change from the beginning of the contraction. On the right is EMG signal and its power spectral density (PSD) during specific time windows (from Merletti and Lo Conte [47]. With permission)

muscle fatigue from the electromyographic indicators (in both time and frequency domains) is visible from the very beginning of contraction.

Merletti and Parker [18] have edited a book providing broad coverage of modern modeling and signal processing issues in the area of sEMG, among other also fatigue influences and means of quantification of this phenomenon. Cifrek et al. [48], however, have presented a state of the art summary on the issue of sEMG based muscle fatigue evaluation. An overview was given of classical and modern signal processing methods and techniques from the standpoint of applicability to sEMG signals in fatigue-inducing situations relevant to the broad field of biomechanics. Time domain, frequency domain, time–frequency, and time-scale representations, and other methods such as fractal analysis and recurrence quantification analysis were described succinctly and were illustrated with their biomechanical applications, research or clinical alike.

sEMG recordings during dynamic contractions are particularly characterized by non-stationary (and non-linear) features. Standard signal processing methods using Fourier and wavelet-based procedures demonstrate well-known restrictions on time–frequency resolution and the ability to process non-stationary and/or non-linear time series, thus aggravating the estimation of the spectral parameters. The Hilbert–Huang transform (HHT), comprising of the empirical mode decomposition (EMD) and Hilbert spectral analysis (HSA), provided a possible new approach to overcome these issues (Srhoj-Egekher et al. [36]). The time-dependent median frequency estimate has been used as a muscle fatigue indicator, and linear regression parameters were derived as fatigue quantifiers.

Moreover, emerging methods based on nonlinear signal analysis are being applied. These techniques, known as recurrence quantification analysis (RQA), are based on detecting deterministic structures in the signals that repeat throughout a contraction [49].

Example: Differentiating Patients with Radiculopathy from Chronic Power spectrum Patients by Single Surface EMG Parameter

In [50] a classification model based on single surface EMG parameter—regression line slope of power spectrum median frequency—is introduced to differentiate low back pain patients with radiculopathy from chronic low back pain patients and control subjects. A variant of the Roman chair (Fig. 9.42) was used to perform static contractions, where the subject's own upper body weight was used to induce muscle fatigue in low back muscles. Surface EMG signals were recorded over the paraspinal muscles at L1–L2 and L4–L5 interspace level (Fig. 9.43).

As a descriptor of spectral changes, the median frequency of the power spectrum was estimated by use of Hilbert–Huang transform [36]. Student's t–test detected that the regression line slope of the median frequency is significantly different ($p < 0.05$) only between low back pain patients with radiculopathy and the other two groups. There was no significant difference between chronic low back pain patients and control subjects. The achieved overall accuracy of the implemented decision tree classification model was at best 86.8%. The results suggest the possibility of differentiating low back pain patients to subgroups depending on clinical symptoms.

sEMG is used extensively in researching (loco)motor patterns, both healthy and pathological, and makes an indispensible tool of biomechanical experimental accessory at disposal to modern student of human movement. In these endavors it is allways important to adequately document both the measurement procedure applied as well as signal processing method used. (It is a general practice to acquire raw

Fig. 9.42 The tilting device used for testing with subject placed in starting position (from Ostojić et al. [50]. Open access)

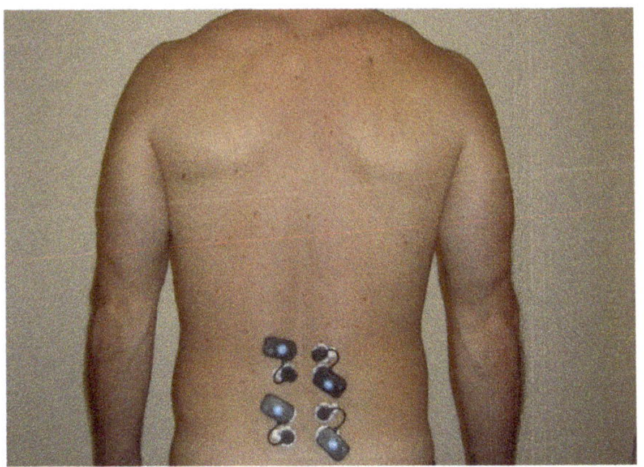

Fig. 9.43 Placement of the electrodes over the paraspinal muscles. Two electrode pairs are placed over m. *erector spinae* at L1-L2 interspace level and two electrode pairs are placed over m. *erector spinae* at L4-L5 interspace level (from Ostojić et al. [50]. Open access)

EMG signals, subject them to A/D conversion and store into computer memory, after which they can be mathematically process by the method chosen, either in time or in frequency domain.) This approach has been used extensively in biomechanical research. Numerous applications are on a brink of clinical applications. Besides monitoring and diagnostics-type applications of sEMG, shown here, this physiological variable plays important role also in control and in biofeedback applications, such as are for instance control of myoelectric prostheses of extremities and biofeedback application in the field of rehabilitation. A critical appraisal of sEMG clinical value in this context, in fields of medicine and of kinesiology, has been attempted recently [51].

Research on EMG features characterizing various movement patterns with regard to skill of performance and/or muscle fatigue, such as reported in references [36, 43–45, 48], as well as in [50], typically is presented in journals and at conferences in realms of the field of biomedical engineering, where it occupies a significant portion of biosignal analysis and processing tracks. As mentioned in Chap. 1, the Croatian Biomedical Engineering and Medical Physics Society (CroBEMPS) has been active for decades and has been organizing numerous scientific symposia and workshops. Figure 9.44 shows several members of the Society together with an international guest at the international scientific conference (see figure caption for details).

9.3.4 Conclusion

Surface electromyography as a simple and non-invasive method of monitoring the electro-physiological activity of muscle has long been applied in biomechanics and kinesiology. However, while simple to apply, there are a number of factors that can influence the properties of the recorded signal and, therefore, the diagnostic conclusions drawn from the results of the analysis of these signals. The standardization that has evolved over time in this field, as well as technological developments,

Fig. 9.44 Members of the Croatian Medical and Biological Engineering Society (CroMBES), renamed in 2017 to Croatian Biomedical Engineering and Medical Physics Society (CroBEMPS), from left to right: Velimir Išgum, Ante Šantić, Stanko Tonković, Stanislav Peharec and Vladimir Medved, and Håkan Lanshammar of Uppsala University, Sweden, keynote speaker at the MEDICON 2001: IX Mediterranean Conference on Medical and Biological Engineering and Computing, June 12–15, 2001, Pula, Croatia

have resulted in specialized systems for surface electromyography that significantly facilitate the application of the method in research and clinical practice.

References

1. Waterland JC (1968) Integration of movement. In: Wartenweiler J, Jokl E, Hebbelnick M (eds) Biomechanics I: 1st international seminar, Zürich, 1967.S. Karger, Basel, pp 178–187
2. Medved V (2009) Surface EMG applications in clinical biomechanics. Editorial. Clin Biom 24(2):121. https://doi.org/10.1016/j.clinbiomech.2008.12.011
3. Medved V (2001) Measurement of myoelectric variables. Measurement of human locomotion. CRC Press, Boca Raton, Fl, pp 169–213
4. Enoka RM, Duchateau J (2016) Physiology of muscle activation and force generation. In: Roberto M, Dario F (eds) Surface electromyography: physiology, engineering, and applications, 1st edn. 2016 by the Institute of Electrical and Electronics Engineers, Inc. Published 2016 by Wiley Inc., pp 1–29
5. Farina D, Stegman DF, Merletti R (2016) Biophysics of the generation of EMG signals. In: Roberto M, Dario F (eds) Surface electromyography: physiology, engineering, and applications,

1st edn. 2016 by the Institute of Electrical and Electronics Engineers, Inc. Published 2016 by Wiley, Inc., pp 30–53
6. Hodgkin AL, Huxley AF (1939) Action potentials recorded from inside a nerve fibre. Nature 144:710–711
7. Hodgkin AL, Huxley AF (1952) A quantitative description of membrane current and its application to conduction and excitation in nerve. J Physiol 117:500–544
8. Häusser H (2000) The Hodgkin-Huxley theory of the action potential. Nat Neurosci Suppl 3:1165
9. Guyton AC (1987) Basic neuroscience—anatomy and physiology. W.B.Saunders, Philadelphia
10. Basmajian JV, De Luca CJ (1985) Muscles alive: their functions revealed by electromyography, 5th edn. Williams & Wilkins, Baltimore, Md
11. Katz B (1979) Nerv, Muskel und Synapse. Georg Thieme Verlag, Stuttgart
12. McMahon TA (1984) Muscles, reflexes, and locomotion. Princeton University Press, New Jersey
13. De Luca CJ (1979) Physiology and mathematics of myoelectric signals. IEEE Trans BME 26(6):313–326
14. De Luca CJ (1984) Myoelectrical manifestations of localized muscular fatigue in humans. CRC Crit Rev BME 11:251–279
15. Loeb GE, Gans C (1986) Electromyography for experimentalists, The University of Chicago Press, Chicago & London
16. Basmajian JV, Clifford HC, McLeod WD, Nunnally HN (1975) Computers in electromyography, Butthcrworths
17. Cifrek M (1997) Analiza mioelektričkih signala tijekom dinamičkog umaranja (Myoelectric signal analysis during dynamic fatigue). Dissertation, University of Zagreb
18. Merletti R, Parker PA (eds) (2004) Electromyography—physiology, engineering, and noninvasive applications. IEEE Press, Wiley, Hoboken, New Jersey
19. Medved V, Cifrek M (2011) Kinesiological electromyography. In: Vaclav K (ed) Biomechanics in applications. InTech, pp 349–366 http://www.intechopen.com/articles/show/title/kinesiological-electromyography (as reprint with added abstract) Medved V, Cifrek M (2015) Kinesiological electromyography. In: Andročec V (ed) Annual 2015 of the Croatian Academy of Engineering. pp 279–299
20. Merlo A, Campanini I (2010) Technical aspects of surface electromyography for clinicians. Open Rehab J 3:98–109
21. Lindström L (1973) On the frequency spectrum of EMG signals, 2nd edn. Research Laboratory of Medical Electronics, Chalmers University of Technology, Göteborg
22. Ambu® Cardiology catalog 2015—Ambu® Cardiology Sensors, Ambu A/S, Ballerup, Denmark
23. Konrad P (2006) The ABC of EMG. A practical introduction to kinesiological electromyography, Version 1.4. Noraxon INC USA
24. Bagnoli™ EMG system user's guide. Delsys Incorporated, Natick, MA
25. BTS FREEEMG 1000 User Manual (2018) BTS SpA: Milano
26. Šantić A (1995) Biomedicinska elektronika (Biomedical electronics). Školska knjiga, Zagreb
27. Hary D, Bekey GA, Antonelli DJ (1987) Circuit models and simulation analysis of electromyographic signal sources-I: the impedance of EMG electrodes. IEEE Trans Biomed Eng 34:91–97
28. Lynn PA, Bettles ND, Hughes AD, Johnson SW (1978) Influences of electrode geometry on bipolar recordings of the surface electromyogram. Med Biol Eng Comput 16:651–660
29. Masuda T, Miyano H, Sadoyama T (1985) The position of innervation zones in the biceps brachii investigated by surface electromyography. IEEE Trans Biomed Eng 32:36–42
30. Merletti R, Aventaggiato M, Botter A, Holobar A, Marateb H, Vieira TM (2010) Advances in surface EMG: recent progress in detection and processing techniques. Crit Rev Biomed Eng 38:305–345
31. De Luca CJ (1997) The use of surface electromyography in biomechanics. J Appl Biomech 13:135–163

32. Merletti R, Vieira TM, Farina D (2016) Techniques for information extraction from the surface EMG signal: highdensity surface EMG. In: Merletti R, Farina D (eds) (2016) Surface electromyography: physiology, engineering, and applications. First edition. IEEE, Inc. John Wiley & Sons, Inc., pp 126–157
33. Merletti R, Botter A, Troiano A, Merlo E, Minetto MA (2009) Technology and instrumentation for detection and conditioning of the surface electromyographic signal: state of the art. Clin Biom 24:122–134
34. Merletti R, Muceli S (2019) Tutorial. Surface EMG detection in space and time: Best practices. JEMG Kinesiol 49, 102363
35. Winter DA (2009) Biomechanics and motor control of human movement, 4th edn. Wiley, Inc., Hoboken, New Jersey
36. Srhoj-Egekher V, Cifrek M, Medved V (2011) The application of Hilbert-Huang transform in the analysis of muscle fatigue during cyclic dynamic contractions. Med Biol Eng Comput 49:659–669
37. Winter DA, Rau G, Kadefors R, Broman H, De Luca CJ (1980) Units, terms and standards in the reporting of EMG research; Report by the Ad hoc committee of the International Society of Electrophysiological Kinesiology
38. Hermens HJ, Freriks B, Merletti R, Stegeman DF, Blok J, Rau G, Disselhorst-Klug C, Hägg G (1999) SENIAM 8: European recommendations for surface electromyography. Roessingh Research and Development
39. Standards for Reporting EMG Data (2014) JEMG Kinesiol, 24(2): I-II
40. De Luca G (2003) Fundamental concepts in EMG signal acquisition. Copyright DelSys Inc
41. Disselhorst-Klug C, Schmitz-Rode T, Rau G (2009) Surface electromyography and muscle force: limits in sEMG–force relationship and new approaches for applications. Clin Biom 24:225–235
42. Lawrence JH, De Luca CJ (1983) Myoelectric signal versus force relationship in different human muscles. J Appl Physiol 54(6):1653–1659
43. Rađenović O, Nemec B, Medved V (2001) A new biomechanical measurement and testing method for turns in alpine skiing. IFMBE Proc MEDICON 2001:624–627
44. Medved V, Tonković S, Cifrek M (1995) Simple neuro-mechanical measure of the locomotor skill: an example of backward somersault. Med Prog Technol 21(2):77–84
45. Rupić S (2008) Energizirana teniska loptica (The energized tennis ball). Print centar Krapina
46. Merletti R, Lo Conte LR (1995) Advances in processing of surface myoelectric signals: Part 1. Med Biol Eng Compu 33(3):362–372
47. Merletti R, Lo Conte LR (1997) Surface EMG signal processing during isometric contractions. JEMG Kinesiol 7(4):241–250
48. Cifrek M, Medved V, Tonković S, Ostojić S (2009) Surface EMG based muscle fatigue evaluation in biomechanics. Clin Biom 24:327–340
49. Farina D, Fattorini L, Felici F, Filligoi G (2002) Nonlinear surface EMG analysis to detect changes of motor unit conduction velocity and synchronization. J Appl Physiol 93(5):1753–1763
50. Ostojić S, Peharec S, Srhoj-Egekher V, Cifrek M (2018) Differentiating patients with radiculopathy from chronic low back pain patients by single surface EMG parameter. Automatika, 59(3–4)
51. Medved V, Medved S, Kovač I (2020) Critical appraisal of surface electromyography (sEMG) as a taught subject and clinical tool in medicine and kinesiology. Front Neurol 11:560363. https://doi.org/10.3389/fneur.2020.560363

Chapter 10
Gait Analysis

Vladimir Medved, Rodolfo Vastola, Daniele Albano, and Marko Pećina

Abstract Gait analysis can be considered a backbone of modern clinical locomotion biomechanics. Basic methodology, integrating kinematic, kinetic and myoelectric measurements in course of a subject's walking and subsequent interpretation of results is summarized and illustrated mainly with findings from our Zagreb and Salerno situated laboratories. In addition, pedobarography and portable oxygen consumption measurement system make desirable components of an equipment inventory for gait analysis. Physical examination and observational analysis precede measurement of several gait trials in a laboratory. Interpretation of measurement findings is the next step. Modern tendencies, pursued worldwide, incorporate individualized (subject-specific) neuro-musculo-skeletal modeling into a clinical procedure. Non-linear analysis of gait data emerges as yet another methodology of gait analysis. Alternative modern tendencies are finally pointed to whereby classical measurement instrumentation setup confined to laboratory environment is substituted by portable, so-called pervasive multisensor measurement solutions offering freedom of movement and aiming to replace classical approach, albeit with no success yet.

10.1 Establishment of Gait Analysis as a Clinical Vehicle

Benno Nigg has named the nineteenth century the gait century [1], however, a real standardization of gait analysis in a clinical context in modern means has begun to be crystallized in the twentieth century. Robert Plato Schwartz's work was of importance. He devised measurement methods, along the lines of Marey's and

V. Medved (✉)
Faculty of Kinesiology, University of Zagreb, Horvaćanski zavoj 15, 10000 Zagreb, Croatia
e-mail: vladimir.medved@kif.hr

R. Vastola · D. Albano
Department of Human, Philosophical and Educational Sciences, University of Salerno, Via Giovanni Paolo II, 132 - 84084, Fisciano (SA), Italy

M. Pećina
School of Medicine, University of Zagreb, Šalata 2, 10000 Zagreb, Croatia

V. Medved (ed.), *Measurement and Analysis of Human Locomotion*,
Series in Biomedical Engineering, https://doi.org/10.1007/978-3-030-79685-3_10

Carlet's systems, and applied them clinically. He first postulated requirements for gait measurement method with a clinical purpose. The achievements of The Berkeley Group, focused to measurements of both healthy and of pathological gait including gait of amputees equipped with prostheses, have provided further impetus and have set the stage for gait analysis in modern means. Perry and Sutherland, successors to The Berkeley Group, developed standardized observational analysis protocol for human gait, complementing it with their measurements (Chap. 2).

In this Chapter, under the term *gait analysis* we factually consider *instrumented gait analysis* (or *biomechanical gait analysis*). At the beginning of 1980es in Newington, Connecticut, USA, first clinically oriented instrumented gait laboratory was installed by the United Technologies Corporation, and besides, laboratories were also initiated in Boston, Glasgow and Dundee [2], as well as in realms of NIH (the National Institutes of Health), Department of Rehabilitation Medicine [3]. Since then, advances in measurement technology and computers and accumulation of clinical knowledge have contributed to the standardization of the field, so that at the dorn of the twenty-first-century one witnessed the presence of a clinical method of gait analysis [2, 4, 5]. Brand has modified Plato Schwartz's arguments for performing clinical gait analysis (Brand 1987, Brand and Crowninshield 1981, citations after 2), while Baker defined them into: 1. Diagnosis between disease entities, 2. Assessment of the severity, extent or nature of a disease or injury, 3. Monitoring progress in the presence or absence of intervention and 4. Prediction of the outcome of intervention (or the absence of intervention) [2, 5].

An example of how gait analysis can provide differential diagnostics, through kinematic variables, comes from Gage et al. [6] (Fig. 10.1). Identifying subtypes of hemiplegia may help in the development of specific treatment protocols.

A concise description of a typical approach in measuring gait follows, illustrated with examples from our Zagreb and Salerno situated laboratories. As in previous chapters the components of measurement instrumentation (parts of equipment in a laboratory), and relevant signal and data processing methods have been elaborated in some detail (kinematics, kinetics, sEMG—Chaps. 5–9), here we build upon this information, referring to it where appropriate.

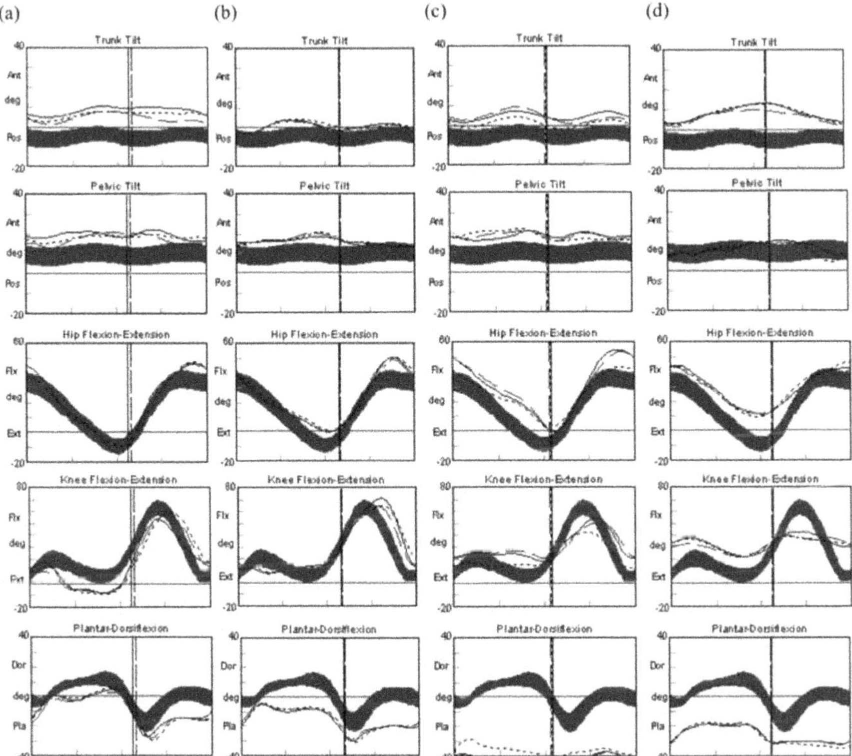

Fig. 10.1 Sagittal plane patterns for hemiplegia type 1 (a), 2 (b), 3 (c) and 4 (d) for the trunk, pelvis, hip, knee and ankle. Multiple gait cycles from an individual child for each type are plotted in relation to typical motion (grey band) (from Gage et al. [6]. With permission)

10.2 Gait Analysis—Components

10.2.1 Gait Cycle

From a kinematics standpoint human gait is viewed as a quasi-periodic phenomenon comprising, in a stationary regimen, a series of approximately repetitious periods (cycles). To describe what happens during walking it is useful to divide one cycle into phases. The simplest division consists in dividing the cycle of each leg in the stance phase, when the foot is in contact with the ground, and the swing phase, when the foot is off the ground. Usually, the stance phase lasts about 60% of the cycle. The scheme to subdivide gait cycle has been proposed by Perry [7], and has been adopted universally in gait community [8] (Fig. 10.2). It also takes into account the contralateral limb and therefore distinguishes two phases of double support and a single support phase, the latter divided into initial, intermediate and terminal phase; the same subdivision is used for the swing phase.

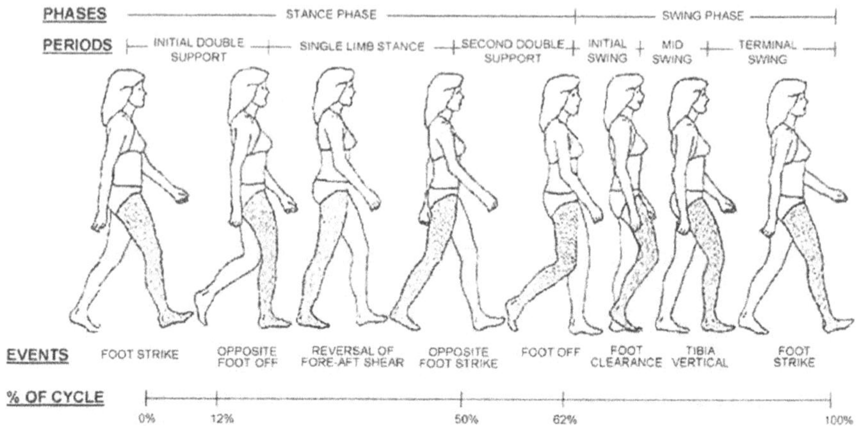

Fig. 10.2 Normal walk cycle illustrating the events of gait. The time axis is normalized to % of the gait cycle (from Kaufman and Sutherland [8]. With permission)

While this subdivision is easily proposed in healthy subjects (normals), it is often not feasible in subjects with particular pathologies, for example, the terminal support phase begins when the heel rises from the ground, but in the case of patients who never rest their heels on the ground, the identification of this phase is impossible, the same thing happens for the intermediate swing phase, identified with the vertical tibia, a situation that may not occur in subjects that walk with a crouch gait pattern. This is the reason why it is important to normalize data about gait cycle. Each particular gait pattern has, in principle, a different timing, so they are not directly comparable. Normalization of gait cycle consists in defining the initial contact (0% of the cycle) and the subsequent contact of the same foot with the ground (100% of the cycle). This sort of time normalization is shown on the horizontal axis. Baker points to limitations to the conventional subdivision of the gait cycle, giving several critical comments/suggestions [4, pp. 216, 217].

10.2.2 Standard Technical Requirements for Gait Analysis

10.2.2.1 Instrumentation

Among the tools used for gait analysis and evaluation, one can distinguish in the first place tools that provide data regarding the kinematic aspects of the movement, and those concerning the dynamics. Among the systems for kinematic measurement and analysis, the gold standard is represented by optoelectronic systems that allow the acquisition of the spatial coordinates of particular markers positioned on anatomical body landmarks. Being very accurate, these systems are quite expensive, they require expert staff for their use and are related to the laboratory environment to enable close-range photogrammetry. Furthermore, the fact is that these systems require quite time consuming marker positioning procedures, and the same markers can be invasive or annoying for the execution of the gesture or movement (Chaps. 5 and 6).

Among the most used tools in the quantitative assessment of gait, there are also the electrogoniometers for measurement of joint angles, however, these systems are rather cumbersome and can alter the normal course of movement (speaking strictly, whereas modern lightweight technical solutions strive to minimize encumberance), moreover, they provide limited information. A digression to one of early experiences with using electrogoniometers of that time to measure gait is warranted at this point, however. The CARS-UBC Electrogoniometer system has been used: a joint angle measuring device providing a spatial measurement of relative hip, knee and ankle angles [9–11]. Movements of each of these joints are followed by three mutually orthogonal torsional potentiometers, mounted so that the angle arms of potentiometers track flexion–extension, abduction–adduction and rotation angles, respectively, thus providing a 3D analog measure of joint angles' changes in time. In the CARS-UBC System, this task has been accomplished by incorporating a so-called parallelogram chain mechanism. It allowed translational movements up to 2.5 cm in radius in the sagittal plane and 5 cm in radius in the frontal and horizontal planes, without causing misalignment between potentiometer arms and corresponding axes defining angles being measured. This feature also made proper positioning of the system on the subject's body less critical. Completed with the switch-type foot contact detection device and the Kistler piezoelectric platform, the equipment was intended to be used for recording walking in experiments testing the ballistic walking model. One limitation of the instrument was its inappropriateness for measuring fast movements (run, etc.) due to its mass. It was concluded that with some slight mechanical modifications the instrument has proven to be useful and reliable in providing a means of monitoring and recording human joint angles during walking (only 2D, sagittal, information was used in this research). Its main advantages were relative simplicity, satisfactory accuracy, ease of use and acceptable price [11]. Surely, today there are even better, and less invasive, goniometers available for research application and/or clinical use.

On the other side, in recent years, thanks to the ever-increasing development of wearable technologies, it has been a growing use of inertial sensors in motion analysis (Sect. 5.2.3). These devices are in general less expensive compared to traditional systems (opto-electronic), portable, versatile, and can supply a lot of kinematic information in real time for gait analysis both indoors and outdoors, thus suggesting that in the future they could displace the systems currently used. However, as well as goniometers, they don't offer inverse dynamics calculation possibility, although there have been literature reports claiming inverse dynamics realization with only flexion/extension kinematics in hip, knee and ankle joints at disposal combined with some optimization techniques [12]. This issue will not be commented further here, but classical complete 3D kinematics measurement setting will be assumed.

Considering the dynamic aspect of the movement, the measurement possibility comes in a form of force platforms for measurement of forces exchanged with the surrounding environment; these forces are referred to as "ground reaction forces" (GRF). For the analysis of dynamics of the body as a whole, inverse dynamics approach is to be provided for which, given the kinematics of a body—the human body is represented as a multi-body system of rigid segments linked by joints—and the forces exchanged with the environment, forces and moments (couples) that act on a system are estimated (Sect. 2.2; Chap. 6). (Again, some authors claim approximate realization possibility of inverse dynamics when in-shoe pressure measurement insoles are at disposal instead of force platform(s) [13]).

In general, a gait analysis laboratory is equipped with an integrated system composed mainly of an optoelectronic system for kinematic measurement and evaluation, a force platform(s), and a multichannel telemetric surface electromyograph (sEMG) for dynamic muscle action monitoring and measurement. Further, pedobarography and a portable oxygen consumption measurement system are also desirable parts of equipment to embrace more completely the process of gait quantification and diagnostics. A commercial video camera may also be used for documentation and later qualitative visual analysis of movements recorded.

Motion Analysis Laboratory in Salerno possesses an integrated system composed by six infrared cameras (BTS DX600), 6 force platforms (BTS P6000), two cameras (BTS Vixta) for qualitative evaluation of movement, an 8 channel waterproof electromyograph (BTS FreeEMG H2O) and an IMU (BTS Gsensor) for basic Gait and Jump analysis out of the laboratory (Fig. 10.3: Some elements of instrumentation).

Fig. 10.3 Elements of instrumentation in the Salerno Motion Analysis Laboratory (Handicap Laboratory), Department of Human, Philosophical and Educational Sciences, University of Salerno (With permission of BTS Bioengineering)

BTS DX600

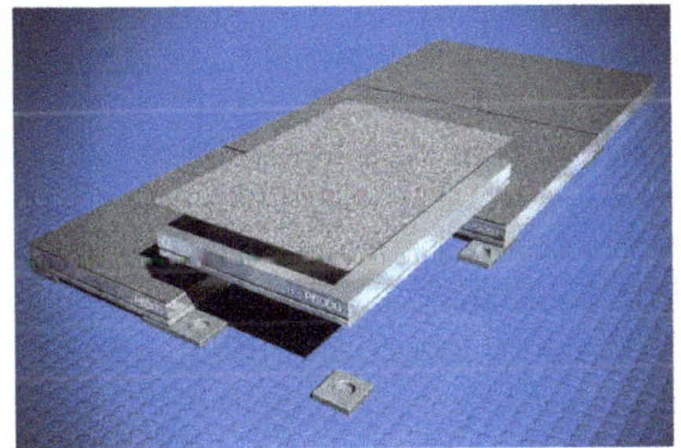

BTS P6000

BTS FreeEMG

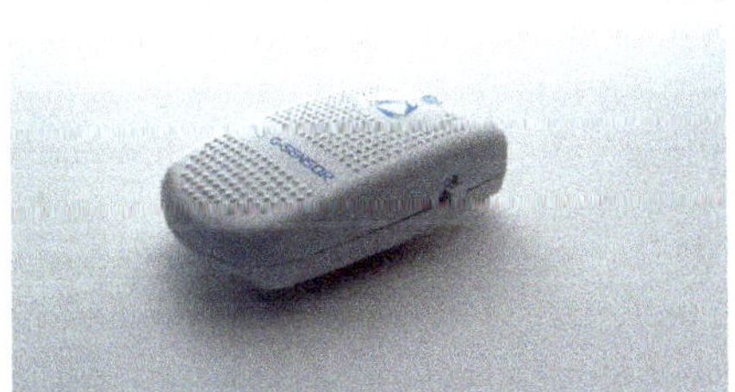

BTS Gsensor

10.2.2.2 Measurement Procedure

Upon arrival to the laboratory, subjects undergo a standard physical examination to document the status of the locomotor system. Chambers and Sutherland gave a concise „recipe“ of how to approach this task, and also the task of preliminary observational gait inspection, and we follow their description closely [14]. Physical examination including measuring the range of motion of at least the hip, knee, and ankle joints should be performed on all patients having gait problems. Noted are: the presence of muscle or joint contractures, spasticity, possible extrapyramidal motions, muscle weakness, reported pain, etc. Abnormal neurologic signs also should be documented as they may contribute to gait abnormalities. Possible radiographically documented abnormalities should be noted of the lumbar spine, pelvis, or lower extremities. Then, a patient's gait is to be evaluated first by visual observation. The patient should be asked to walk both toward and away from the observer. In this manner, coronal plane abnormalities such as trunk sway, pelvic obliquity, hip adduction/abduction, and possibly rotation can be assessed. Each segment's (trunk, thigh, leg, and foot) movement should be observed while the patient walks each way, and any possible abnormalities have to be noted. Walking back and forth in front of the observer allows evaluation of sagittal plane abnormalities reflected in pelvic tilt and flexion and extension of the hip, knee, and ankle. Axial or rotational abnormalities are difficult to detect by visual observation. It is useful to videotape the patient while walking from the front and the side, and later study the record.

One component of a physical examination may be the Thomas Test (TT), a standard test in physiotherapy aimed at examining hip joint function. Compared to TT which is performed in full supine on the testing table, a variant called Modified Thomas Test (MTT) is performed in supine on the table, with pelvis and hips near the edge, to allow movement of the examined upper leg below the horizontal line [15]. Consequently, MTT provides an opportunity to evaluate not only one-joint, but also two-joint hip flexors. Today, in research and clinical practice, MTT is applied more often than TT. MTT is a frequently used clinical test for posture, postural adaptation, and lower extremity kinematics. Indirectly, it supplies information on the flexibility of hip and knee musculature, primarily *m. iliopsoas*, *m. rectus femoris*, *m. tensor fasciae latae* and *m. sartorius*. It is often used for clinical evaluation of the passive range of motion (ROM) while extending the hip and flexing the knee. Most often the evaluation is rather subjective, based on observation: it is evaluated as positive or negative, for commonly accepted norms. Assessment of length of lower extremity muscles is provided indirectly, by measuring hip and knee ROM and results are compared bilaterally, and with normative values. (Normative values for the length of lower extremity muscles, i.e. hip and knee ROM, may be important for injury prevention through detection of impaired flexibility).

We have attempted to realize this test using an optoelectronic system, and also manually [15]. Fifty-four male subjects participated, healthy and manifesting high level of physical activity as assessed by IPAQ (International Physical Activity Questionnaire—Croatian variant), age spanning from 19 to 27, mean age 21.9. They were

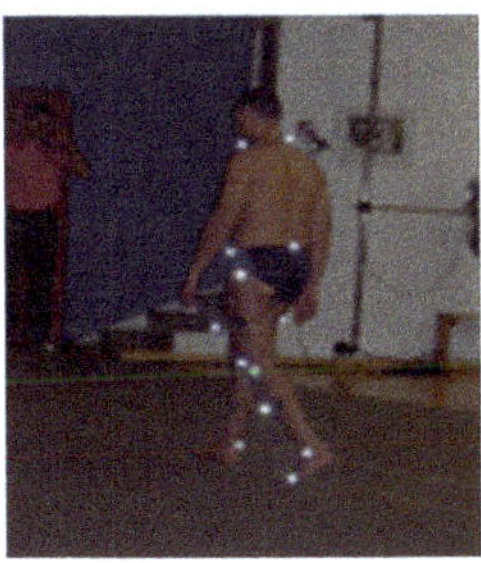
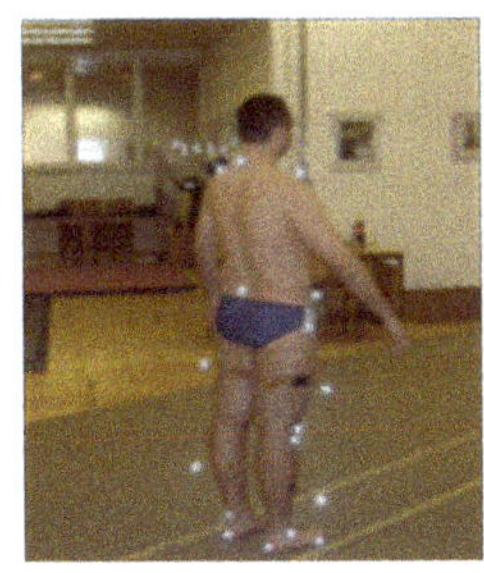
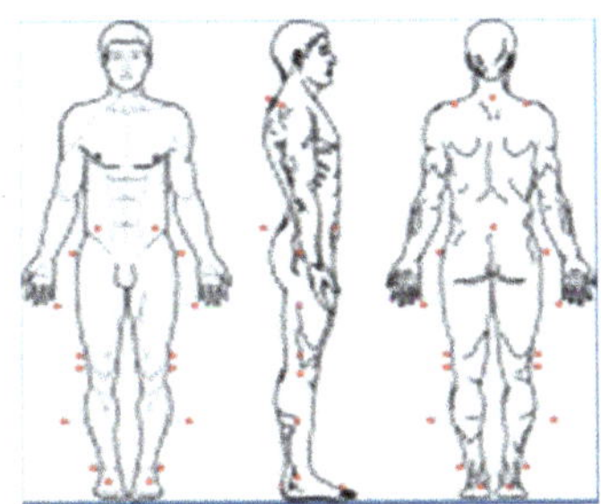

Fig. 10.4 Kinematics and ground reaction force (GRF) measurement of normal gait. Davis protocol for marker positioning. Biomechanics Laboratory, Faculty of Kinesiology, Zagreb (from Heimer [18]. With permission)

measured bilaterally. The optoelectronic kinematic 3D evaluation proved to be superior to 2D goniometry due to the possibility of simultaneous measurement of all angles and automatic data processing with fast feedback, which makes it a relevant modality for accurate and comprehensive measurement of ROM. The results have demonstrated the validity of using optoelectronic kinematic evaluation of MTT, not only for research purposes, but also for use in clinical practice. Normative values for four angles of ROM have been established for MTT. This research showed a high correlation between the two methods of measurements, suggesting that both, 3D and 2D method, can be used for evaluating passive ROM in MTT.

Prior to the measurement of gait in a laboratory, a subject has to be prepared in accordance with the analysis protocol to be used. Focusing to kinematic and kinetic measurement first, the protocol includes standardized procedures to be implemented to carry out the analysis, and among the most widespread protocols for gait analysis are the Davis [16] and Helen Hayes [17] protocol. Markers are to be positioned at defined anatomical landmark points (Fig. 10.4). In addition to these well-known protocols, there are several other that attempt to reduce the number of markers needed. Reducing the number of markers is necessary to make subject preparation operations more feasible, especially in clinical applications, in which subjects may suffer for a prolonged stay in the lab. A protocol developed to reduce the number of markers on the subject is the calibrated anatomical systems technique (CAST) [19].

The Davis protocol foresees acquisition of anthropometric parameters of the subject such as: body mass, body height, lower limb length, pelvis width, pelvis height, knee diameter, and ankle diameter. Markers are positioned respectively on: C7, right and left acromion, right and left ASIS, sacrum, right and left great trochanter, right and left femoral condyle, right and left fibula head, right and left lateral malleolus, right and left 5th metatarsal head, right and left heel, also bars are used on the thigh and leg. The lateral position of the markers allows for optimal visibility during walking. This protocol is applied in both our Zagreb and Salerno laboratories (Fig. 10.4).

The CAST consists in recording the position of an anatomical landmark relative to three technical markers that belong to the same anatomical segment (assumed as rigid) in order to capture full 3D segment's kinematics [19]. This type of protocol makes possible the creation of a local reference system of the segment from which one can reconstruct the position of the other points belonging to the same segment. This type of protocol foresees an initial acquisition phase in which—through a particular stick whose dimensions are known—it is possible to mark the various landmarks and then reconstruct their positions later on the coordinates of the clusters. This type of procedure thus offers the advantage of decreasing the number of markers present on the subject's body.

Based on this technique several protocols have been developed.

The LAMB (Laboratory of Movement Analysis in the Child) protocol was created at the University of Milan (Italy). It includes a marker set for the lower limbs and pelvis and includes 15 anatomical and 4 technical markers. The protocol uses other 8 anatomical markers positioned on the subject only during the static calibration, then they are removed and reconstructed thanks to the anatomical reference systems built thanks to the combination of anatomical and technical markers [20].

The Total 3D Gait developed by Aurion (Aurion Srl, Milan, Italy) consists in the application of 20 markers and uses as a basis the CAST technique to reconstruct some marker such as the medial condyle of the femur, the medial malleolus and the head of the second metatarsal, using a pointer mounting two markers in known positions with respect to the tip. The average total time for a single subject preparation and data collection was approximately 30 min [21].

The SAFLo is a protocol for clinical applications of gait analysis developed at the Bioengineering Centre in Milan (Italy). This protocol has been used in a clinical setting at both the Bioengineering Centre in Milan (laboratory S.A.F.Lo.-Servizio di Analisi della Funzionalita Locomotoria) and the Spaulding Rehabilitation Hospital in Boston. This protocol uses only 9 anatomical markers plus three additional markers on each lower limb attached to the ends of the rods rigidly fixed on the lateral femoral condyles, the anterior tibial shaft and the forefoot. These "extended markers" are used to create reference systems for the lower limb segments. From the combination of the position of the markers, the reference systems and the anthropometric measurements it is possible to estimate the position of the joint centers of the hip, knee and ankle [22].

If electromyography is also used, it is necessary to position the electrodes for acquiring the electromyographic signals. Usually, surface electromyography (sEMG) is used. For the positioning of the electrodes the guidelines of the SENIAM project related to the signal acquisition procedures are usually followed (www.seniam.org). In this case, the subject's skin at the electrode application points must be free of hair and oil, therefore it must be cleaned with alcohol and treated with a particular conductive solution (see Sect. 9.3 for details on preparing and providing a measurement of dynamic sEMG).

Once the preparation procedure is completed, measurements of gait may begin.

Usually, two types of acquisitions are performed for the measurement, a static one and a dynamic series. Dynamic walking trials are usually performed in a straight line, at a speed chosen by the subject and considered comfortable by him, and the data acquisition begins after the subject has carried out a couple of trials to become familiar with the laboratory environment. If data are also acquired utilizing force platform(s), the subject must hit the surface of the platform with his whole foot during walking. To realize this, it is usually necessary to make adjustments to the starting point; it is useful in this sense to place marks on the floor. In any case, particular attention must be paid so that the subject is not influenced by the position of the platform(s), as attempting to hit it(them) with his foot alters his normal way of walking. His/her walk has to be spontaneous, a requirement not easily attained by children or the elderly. Following remark is waranted here: special marking on the floor along an intended walking path which is supposed be followed is to be avoided, as this kind of „guided walking attempts“ might alter subtle properities of walking pattern and introduce deviations from the spontaneous walking performance. Usually, three successfully recorded trials by each hitting foot are stored. Baker and collaborators recommend at least ten walking trials during the appointment [5]. Video camera recording is also desirable.

Once measurements are finished, the data is stored in computer memory to be processed and analyzed.

10.2.2.3 Data Processing, Presentation, and Analysis

Regarding kinematics, it is possible, by applying the laws of rigid body mechanics, to provide information on linear and angular displacement, velocity and acceleration, useful for describing global and relative movements. As human anatomy is too complex to provide a detailed description of how the body moves, the use of biomechanical models of the human body is customary. The most used models consider the body as a set of rigid segments linked together by constraints represented by the joints, for each segment, a reference system is established that allows describing the orientation in space or in relation to another segment, so it is possible to calculate joint angles (Sect. 2.2; Chaps. 5 and 6).

Regarding dynamics, forces and moments (couples) exchanged with the ground, measured with force platforms, can be used on their own or to estimate internal forces. The dynamics problem can be faced in two ways: one called the direct dynamics problem for which, given the forces acting on a body, its movement is predicted, and another called the inverse dynamics problem for which, given the kinematics of a body and the forces exchanged with the environment, the forces and moments acting on it are predicted. The second approach, called inverse dynamics approach (mentioned in 10.2.2.1), is used more often in motion analysis. The measured mechanical quantities are used as an input to a musculoskeletal model. The outputs of the model are the joint moments (Sect. 2.2; Chaps. 5 and 6, [23]).

A further contribution to the description and estimation of forces acting is provided by measuring muscle activity during the execution of movement, thanks to multichannel sEMG technique. Using electromyographic data it is possible to determine when a muscle is active and the activation sequence of a muscle group. To better interpret and use the electromyographic data, it is necessary to carry out a series of operations on the acquired signal. First of all, it is necessary to filter the signal to eliminate unwanted elements, noise, generally due to the movement of the cables, electric current (power line), etc. Subsequently, to perform further mathematical operations on the signal (calculation of average and peak values) and make it more readable, smoothing techniques are applied which help to better describe the signal and to identify, for example, the activity peaks. Another procedure that may be applied to the electromyographic data is the calculation of the activation threshold, which allows determining the onset/offset of the muscle under examination (details of sEMG signal processing in Sect. 9.3.2).

All the measurement data obtained then require normalization procedures, since each individual has different anatomical characteristics, such as mass and height, therefore any measure to be compared between different subjects needs to be related to these characteristics. For example, force measurements are related to the subject's mass, so they will not be expressed in N, but in N/kg, lengths are generally related to the subject's height, so measures such as the length of the step will be expressed as a percentage of the body height. The electromyographic data is generally normalized as a ratio with respect to a reference value obtained through the execution of a maximum voluntary isometric contraction (MVIC), or with respect to the maximum value recorded during the task, thus a value expressed as a percentage is obtained.

All measurement data (kinematics, kinetics, EMG) conveniently are presented/plotted as a function of time to enable visual insight and analysis.

From the statistical point of view, data may be analyzed to quantify the intra- and inter-subject variability. The indices used include the coefficient of variation (CV), which is given by the standard deviation of a measurement divided by the average value and expressed as a percentage, the multiple coefficient of variation (CMC), given by the square root of the coefficient of multiple determination (CMD) which equals 1 minus the ratio between variance between cycles and total variance, and finally the interclass correlation coefficient (ICC), given by the ratio between variance between individuals and the total variance (including measurement variability). A rather popular view is that these coefficients overestimate the variability and in some cases lead to misinterpreting the data, and the standard deviation, on the contrary, underestimates the variability, as by definition over a third of the measurements deviate by more than one standard deviation from the average value.

The above mentioned variability measures assume normal distribution of variables, so that, strictly speaking, normalcy test(s) should be implemented to decide between parametric or non-parametric statistics.

10.3 Examples

Typical examples of measuring gait in our laboratories are presented as an illustration. Mostly, these were for purpose of research, but sometimes clinical contribution has been attampted as well.

Example 1: Measurement of Normal Gait and Introduction of Colored Visual Tool of Signal Characterization (Zagreb) In the master thesis by Heimer [18] the aim has been to standardize kinematic and kinetic gait measurement protocol. 14 healthy males participated with following characteristics: Age: 29–44 (37.50 ± 5.25); Mass: 74–103 kg (86.4 ± 10.00) and Height: 170–188 cm (177.9 ± 5.1). Kinematic measurements were realized using the ELITE BTS System. GRF has been acquired by means of the Kistler platform. Standard measurement procedure was followed comprising three successful trials by each contacting foot (photographies of measurement make part of Fig. 10.4).

Figure 10.5 shows kinematic curves for all 14 participants superimposed in

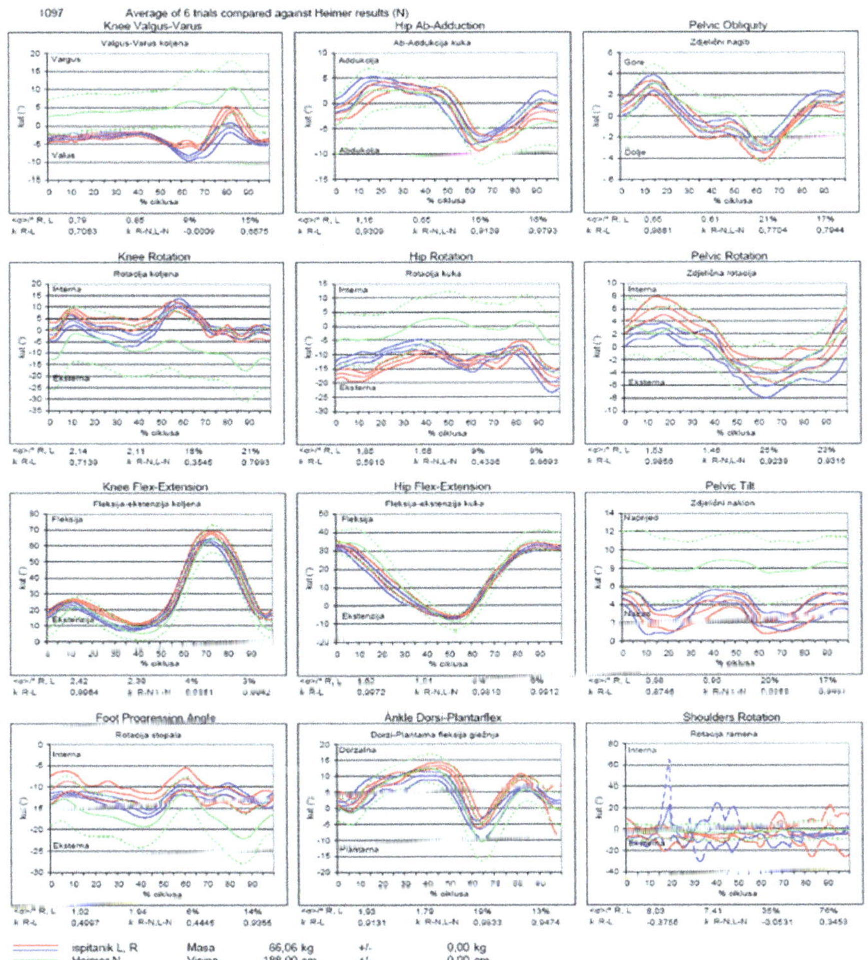

Fig. 10.5 Kinematic curves of normal gait of 14 participants (from Heimer [18]. With permission)

the same diagrams. A high degree of similarity between curves measured can be observed, characterizing normal, healthy gait.

Besides a convenient way to present measured signal curves, an ability of their visual analyses and comparisons, as well as of their statistical analyses and comparisons, a novel feature has been introduced; the Manal-Stanhope method has been implemented [18, 24, 25].

Foremost due to the complexity of the interpretation of the gait analysis finding and the absence of a standard approach in the presentation of these results at that time, biomechanical gait evaluation has not been regarded a routine clinical procedure yet, as elaborated in [26]. Gait analysis report was lengthy, its data were not well understood, it included a clinical interpretation, unlike other clinical tests [26]. Kinematic and kinetic motion curves represented rotation of certain human body segments in three planes, and the components of resultant (net) forces and moments in certain joints. Defining of the angles was not unambiguous; until the mid-1990's various authors might have used their own variants, contributing to the disarray in the area. This was the reason why the International Society of Biomechanics issued recommendations [27, 28] as the basis to be used by all the manufacturers of the biomechanical measurement systems, as well as in biomechanics laboratories all over the world. Further, the kinematic motion curves comprise typically of twelve function graphs, usually each containing separate curves for left and right leg measurement records, combined with the curves representing the standards (norms), while the kinetic curves comprise additional twelve graphs of forces, moments and powers. It is difficult in this assemblage of curves to identify clinically significant aberrations, i.e. deviations from the standards that may indicate a clinical disorder.

In an effort to devise a sort of display of gait measurement record that would aid in navigating the data and direct the clinician towards those findings where he should point his attention the most during the interpretation of the results, we have applied a specific solution. Method of this kind was proposed by Manal and Stanhope [24] and we have applied it to our data record. The principle proposed by Manal and Stanhope [24] comprises in the display of output data of the gait analysis system where the difference of a value of a measured curve in a certain point of time during the gait cycle y_i and a value of the standardized curve yi is displayed in relation towards the amount of the standard deviation of the standardized curve in that point (y_i):

$$d_i = \frac{y_i - \bar{y}_i}{\sigma(y_i)} \tag{10.1}$$

Let us name the value d_i the normalized deviation, since it actually signifies the deviation in the observed point measured by the standard deviation. The idea is to mark („code") the amount of this deviation with a continuous scale of different colors that would allow insight into the nature of the deviation with a quick glance.

Manal and Stanhope proposed the „color coding" i.e. the colored display such that the amounts of di falling within the zones less than one standard deviation would be coded with the shades of green, while the deviations towards negative would be coded with ever more red shades toward the full red part of the spectrum (through the shades of yellow) and the deviations toward positive would be coded with ever more blue shades toward the full blue part of the spectrum (through the shades of cyan), so that the extremes would be reached at the values of three standard deviations. The usual way of color coding in computerized systems is used according to the amount of red, green and blue component (i.e. the additive color blending) where the amount of each component is graded in the range between 0 (complete absence of the component) and 255 (maximal presence of the component). It is shown by Manal and Stanhope that the entire spectrum of colors may be used to code the range between −3 and +3 multiples of the standard deviation.

These authors proposed a mathematical function that may be used for the purpose of color coding composed of two equations for the two characteristic regions of the normalized deviation—linear one for the amount within the zone of one standard deviation, and biquadratic one for the region outside the zone:

$$\mathrm{C} = 85 \cdot \left(3d^2\right) \quad \text{for} - 1 \leq \mathrm{d} \leq 1 \text{ and} \tag{10.2}$$

$$C = 305.7 - 15.597 \cdot \left(3d^2\right) + 0.2324 \cdot \left(3d^2\right)^2 \quad \text{for} - 1 \leq d \leq 1. \tag{10.3}$$

This is shown in Fig. 10.6.

The value C denotes the amount of that component that is variable in the region, while the other two components take the constant values characteristic for the zone. For example, for d = -0.5 the value C is calculated according to the expression (9.2): C = 63.75; in practice, this is rounded to an integer. In this zone the red component is variable (cf. Figure 10.6: for -0.5 blue and green are constant), so R = C = 64, while for green and blue the values for constants are G = 255; B = 0.

Using this kind of color coding for each integer percentage of the gait cycle, a multicolored stripe is obtained corresponding to the set of curves consisting, as a rule, of at least four curves: the curve of the standardized values, two curves of

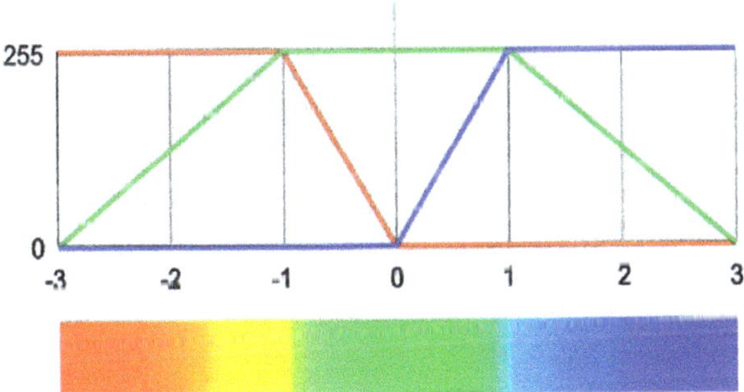

Fig. 10.6 Zones of change of the components of colors for the region of normalized deviation between −3 and +3 of the standard deviation (from Heimer [18]. With permission)

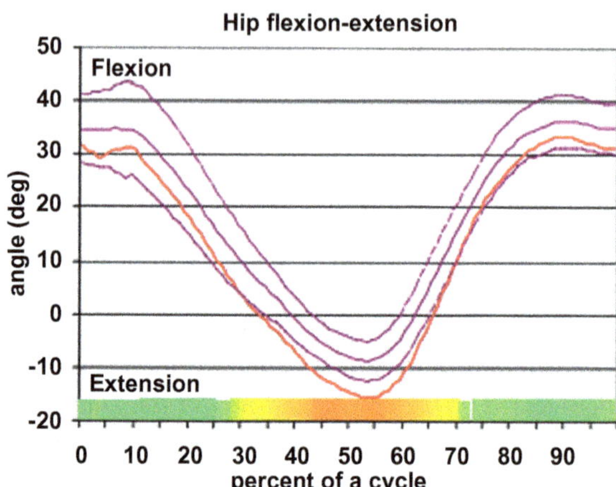

Fig. 10.7 The hip joint flexion–extension curve with superimposed Manal-Stanhope color representation. Normal curve obtained by measurement of sample in lilac. Curve for one particular subject in red (from Heimer [18]. With permission)

its standard deviations and the measured curve. The color coding stripes could be included together with the matching curves, which is shown at an example of the hip joint flexion–extension curve shown in Fig. 10.7 [18].

The curves of the full report itself, that follow, can then be analyzed only for the possible observed significant deviations (Fig. 10.8).

Naturally, it is assumed that also other clinical diagnostic tools are used and findings are taken into account if appropriate. Further, a certain curve may display a „normal" pattern as a result of compensation, therefore it is necessary to differentiate between primary and secondary adaptations.

The colored signal coding incorporated into the clinical gait analysis report is aimed at facilitating the readout of the output report, however the applicative value of this method should still be evaluated in clinical practice. Namely, it is the matter for a clinician, and expert in orthopedics and other disciplines that are using these reports to determine whether the color coding is performed in a way that is significant for their use. It may be assumed that the deviation larger that one standard deviation that is used as significant for this method of color-coding may in some cases (among the 45 curves in a standard report of the system) be of less significance, while in some other cases the deviation considerably less of one standard deviation from the healthy curve shape (i.e. from the standard) may have clinical implications. On the other hand, for some variables in the range between -3 and $+3$ of standard deviation may not be enough and the larger deviations that aren't yet considered pathological may require a larger range of coding [25].

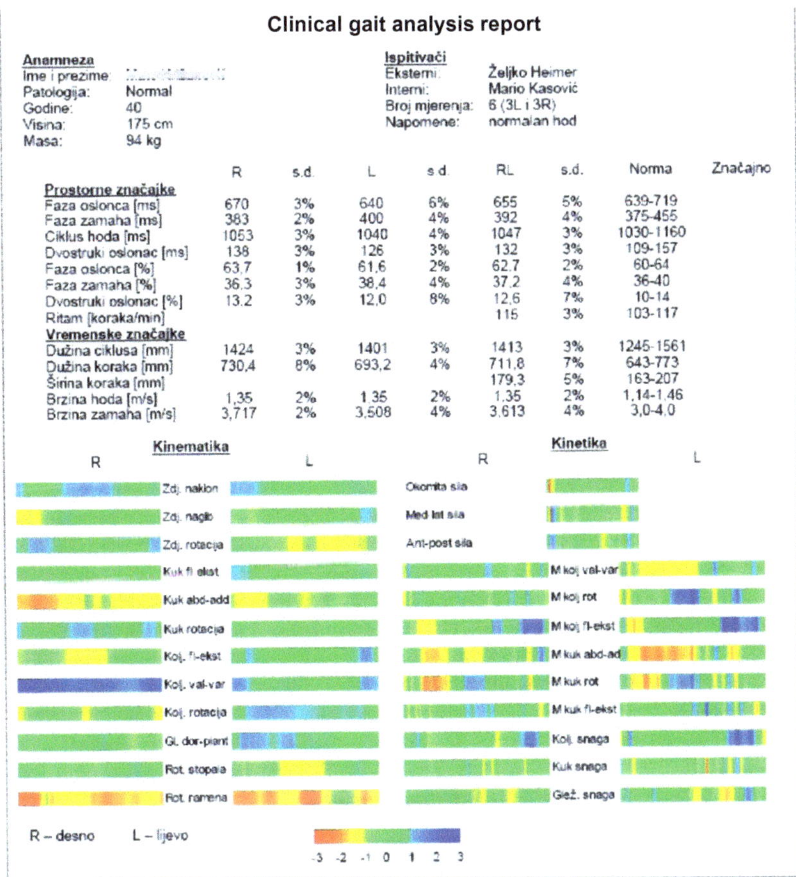

Clinical gait analysis report

Anamneza
Ime i prezime:
Patologija: Normal
Godine: 40
Visina: 175 cm
Masa: 94 kg

Ispitivači
Eksterni: Željko Heimer
Interni: Mario Kasović
Broj mjerenja: 6 (3L i 3R)
Napomene: normalan hod

	R	s.d.	L	s.d.	RL	s.d.	Norma	Značajno
Prostorne značajke								
Faza oslonca [ms]	670	3%	640	6%	655	5%	639-719	
Faza zamaha [ms]	383	2%	400	4%	392	4%	375-455	
Ciklus hoda [ms]	1053	3%	1040	4%	1047	3%	1030-1160	
Dvostruki oslonac [ms]	138	3%	126	3%	132	3%	109-157	
Faza oslonca [%]	63,7	1%	61,6	2%	62.7	2%	60-64	
Faza zamaha [%]	36,3	3%	38,4	4%	37,2	4%	36-40	
Dvostruki oslonac [%]	13.2	3%	12,0	8%	12,6	7%	10-14	
Ritam [koraka/min]					115	3%	103-117	
Vremenske značajke								
Dužina ciklusa [mm]	1424	3%	1401	3%	1413	3%	1245-1561	
Dužina koraka [mm]	730,4	8%	693,2	4%	711,8	7%	643-773	
Širina koraka [mm]					179,3	5%	163-207	
Brzina hoda [m/s]	1,35	2%	1,35	2%	1,35	2%	1,14-1,46	
Brzina zamaha [m/s]	3,717	2%	3,508	4%	3,613	4%	3,0-4,0	

Fig. 10.8 Cover page of report for one subject of the clinical system for gait analysis using Manal-Stanhope representation, serving as a compendium of all kinematic and kinetic signals (from Heimer [18]. With permission)

Example 2: A Glance on the Measurement of Gait in Dancers (Salerno) Gait patterns in dancers were the focus of attention in the Salerno laboratory. The subjects, while not being professional dancers, were dance practitioners for many years. Therefore, the purpose was to explore if there were differences in gait patterns between non-professional dancers and normals using an optoelectronic system and force platforms. The sample was composed of six female dancers: average age 22.83 years, average height 163.83 cm and average mass 56 kg. Markers were positioned following the Davis protocol [29]. The dancers performed 10 walking trials at their prefered speed. The kinematic and dynamic data have been recorded. SMART software has been used to process the tests and to create the final report [30]. Figure 10.9 gives an example of kinematic gait data.

The overall characteristics of data is that they differ from normality with a percentage less than 5%, but during double support phases (which is from $\approx$0% to $\approx$10% and $\approx$50% to $\approx$60%) they differ more (right cycle -9.85%, left cycle

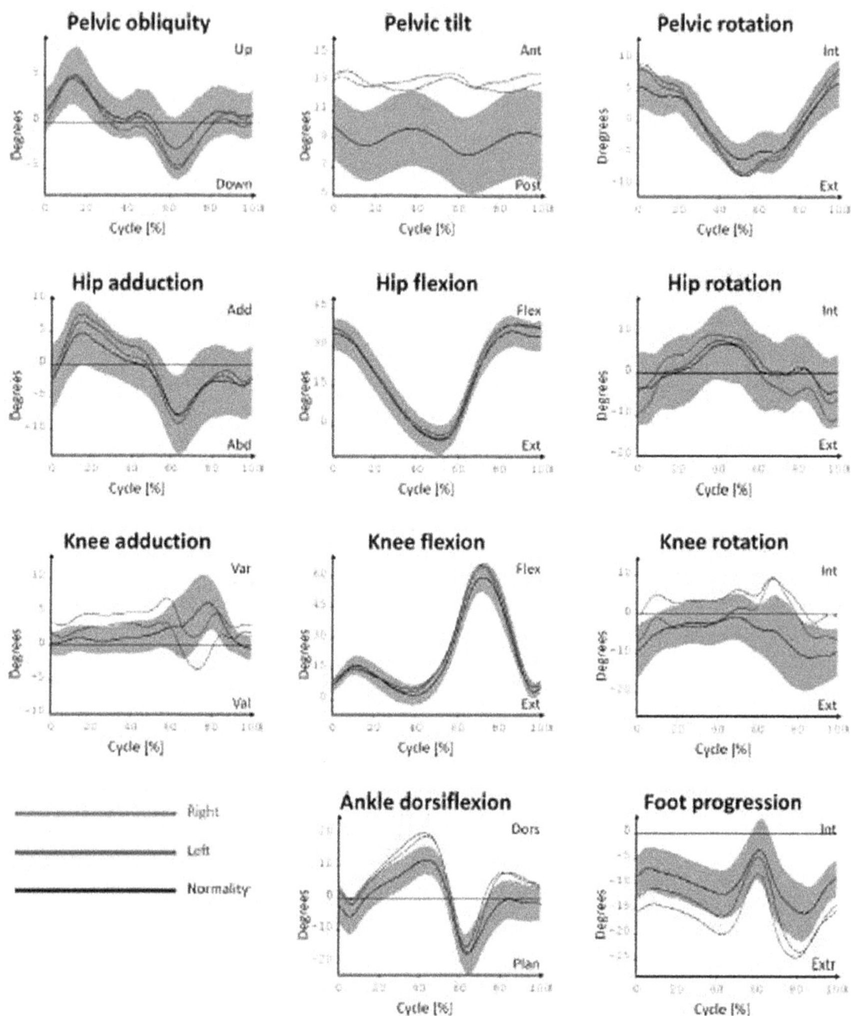

Fig. 10.9 Kinematic record of qualified dancer's gait (from Astone et al. [30]. With permission)

−7.22%), such as step width (−60.51%). (Step width is lateral distance from heel center of one footprint to the line of progression formed by two consecutive footprints of the opposite foot.) Regarding kinematics in the sagittal plane, we see differences in the pelvic tilt which moves forward for all the cycle, and in the ankle dorsiflexion, because it is more accentuated than normal at ≈40% and ≈80% of the cycle. We can see a little increase in internal rotation of the knee and an extra-rotation in foot progression.

A general kinematic report is presented in Fig. 10.10.

BTS MOTION ANALYSIS LAB
Integrated solutions for multifactorial clinical movement evaluation

Gait Profile Score	ARTO DX	ARTO SX	NORMALITA'
Gait Profile Score (grad):	4.2 ± .1	3.9 ± .2	< 7

Gait Variable Scores	ARTO DX	ARTO SX
Obliquità pelvica (grad):	2.6 ± .5	1.3 ± .2
Tilt pelvico (grad):	1.7 ± .3	1.6 ± .1
Rotazione pelvica (grad):	2.1 ± .6	2.2 ± .4
Abd-Adduzione anca (grad):	2.1 ± .4	2.9 ± .9
Flesso-Estens anca (grad):	5.1 ± .7	4.4 ± 1.1
Rotazione anca (grad):	2.8 ± .2	5.3 ± .2
Flesso-Estens ginocchio (grad):	7 ± .5	5.3 ± 1.5
Dorsi-Plantifless caviglia (grad):	6 ± .5	3.7 ± .1
Progressione del piede (grad):	4 ± .9	5.3 ± 1.1

Gait Deviation Index	ARTO DX	ARTO SX	NORMALITA'
Gait Deviation Index:	112.15 ± 2.7	111.9 ± 4.08	> 100

Fig. 10.10 General kinematic report of qualified dancer's gait. Gait variable scores refer to nine parameters considered in the calculation of the Gait Profile Score (GPS—see text). The report generated by the smart analyzer software doesn't provide respective normality values

The Gait Profile Score (GPS) is a clinical index originally published in [31]. This is a single index outcome measure that summarises the overall quality of a patient's gait kinematics. The GPS can be decomposed to provide the Gait Variable Score (GVS) (an index that measures single gait variable deviation), for nine key relevant kinematic variables. The GPS is presented with the nine GVS in a bar chart, creating a Movement Analysis Profile (MAP). The MAP describes the magnitude of deviation of the nine individual variables averaged over the gait cycle, thus providing insight into which variables are contributing to an elevated GPS.

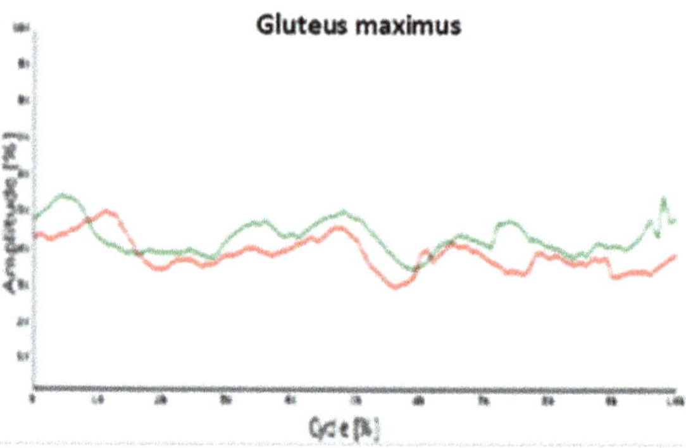

Fig. 10.11 An example of electromyography of *m. gluteus maximus* in qualified dancer's gait. The root mean square (RMS) value of the myoelectric signal within 50 ms mobile time window (green for the right side, red for the left side)

Schwartz and Rozumalski published an article introducing the Gait Deviation Index (GDI) in 2008 [32]. This is a score that provides a numerical value that expresses overall gait pathology (ranging from 0 to 100, where 100 indicates the absence of gait pathology).

The same sample of six dancers was subjected to EMG measurements in another study [33]. Electrodes were positioned according to SENIAM project on ten muscles, bilaterally: *rectus femoris, tibialis anterior, gluteus maximus, gastrocnemius, and soleus.* EMG amplifiers were attached on electrodes and blocked on the legs of dancers with tape to avoid any kind of "noise". The signals have been filtered and normalized on MVC. Results show that muscle activity is somewhat different from normal walking. The *rectus femoris* has two peaks (≈10–60%), while according to Perry [7], it only manifests one peak at 60% of the gait cycle. The *tibialis anterior* muscle has a peak at the beginning, another at the 60%, and then it is active during the remaining of the gait cycle. The *gastrocnemius* is a little different because in the dancers has a peak at 40–50% of the gait cycle, but it is active at the beginning of the gait cycle too. With regard to the *soleus*, there is a little difference at the beginning of the gait cycle because there is a little peak of activation, and regarding the rest its trend is similar to other studies. The largest difference applies to the *gluteus maximus.* This muscle is expected to be active at the beginning of the gait cycle (≈0–25%) and at the end of it (≈95–100%), but in this study it has shown some activity for the entire cycle [33] (Fig. 10.11).

The main findings are that the *gluteus maximus* activity is persistent during all the cycle, on the contrary to normal gait findigs such as those in [7]. We can see a first burst from 0 to 20% an second one around 90–100%. One can just hypothesize that dancers use more the hip extensors to move forward with respect to the normal population.

Figure 10.12 is a 4-channel sEMG record of qualified dancer's gait

BTS MOTION ANALYSIS LAB

Analisi Elettromiografica

1. Tibiale Anteriore dx

2. Gastrocnemio Laterale dx

3. Retto Femorale dx

4. Semitendinoso dx

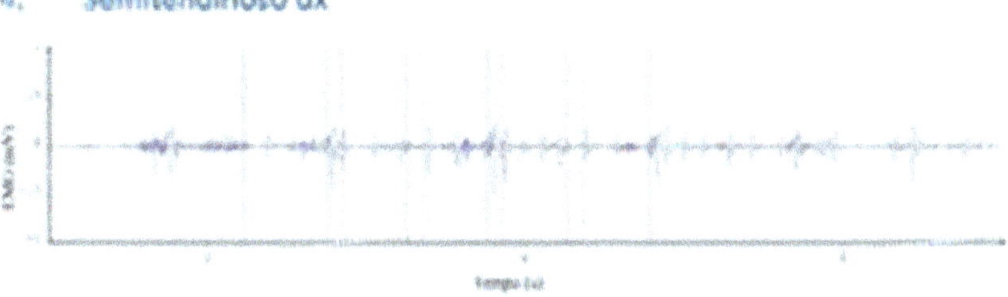

Fig. 10.12 4-channel sEMG record of qualified dancer's gait („dx“ abbreviation for the right side). EMG signals are filtered (20 Hz high pass −450 Hz low pass). The vertical lines in the graphs indicate the heel strike events of the contralateral limb. The vertical dotted lines represent the toe-off events (green for the right limb, red for the left limb)

Example 3: Measurement of Gait in Below-knee Amputees (Zagreb) Gait in below-knee amputees was the subject of Ph.D. research by Kovač [34]. The sample was drawn from the pool of war trauma related amputees that were treated in the Institute for Rehabilitation and Orthopedic Devices, University Hospital Center Zagreb; they were rehabilitated and prosthetically equipped [35]. The study population consisted of twelve males with right trans-tibial traumatic amputation, mean age 40.25 + 6 years [31–52] that volunteered to participate in this study. They were all war victims, mostly injured by means of land mines, in the period 1991–1995. All patients had completed a prosthetic training program. They all were excellent walkers who used their prosthesis on a regular basis and were leading a normal active life. They were not suffering from any severe concurrent illness. The prosthetic alignment was similar for all patients. The time lapse between the date of amputation and the time of testing ranged from 8 to 12 years (mean time 10.08 + 1.5 years). They were in good health and not specially trained in sport or another physical activity.

Measurement equipment and procedure were the same as in [18] (Example 1). In addition, surface electromyography has also been provided.

Kinematic and kinetic findings are presented and discussed in [36–38]. The study has shown useful, albeit not substantially contributing to clinical praxis. One among reasons has been a complex and demanding procedure of laboratory measurement for each patient, that took place at a distant location, in another institution, „out of hand“ to the clinical environment (Figs. 10.13 and 10.14).

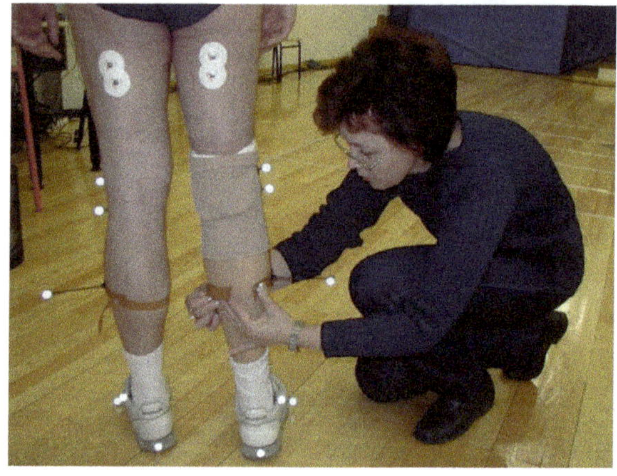

Fig. 10.13 Attaching body markers and sEMG electrodes to the subject by Ida Kovač, a Ph.D. candidate at the time

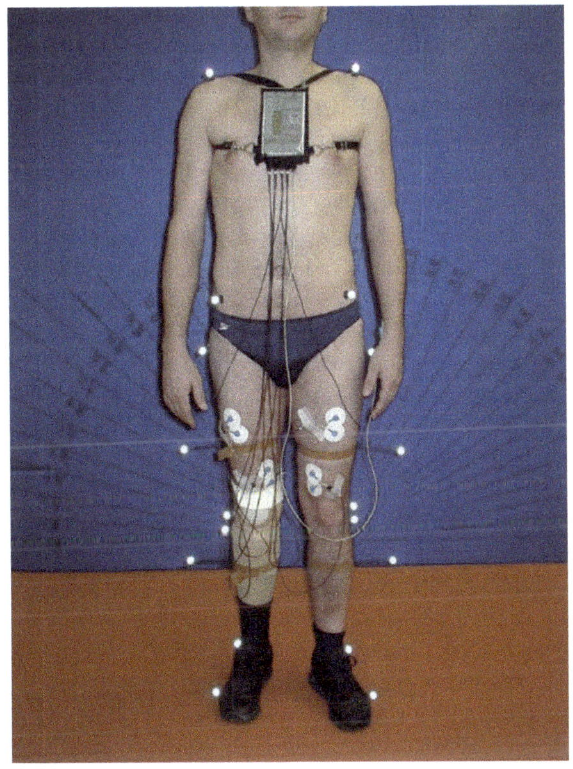

Fig. 10.14 Subject, a below-knee amputee, equipped with prosthesis and with body markers and sEMG electrodes attached. Electrodes are connected via cables to a portable unit of telemetric EMG apparatus

Example 4: Biomechanical Analysis of gait During Pregnancy (Salerno) The study was conducted on a 26-year-old pregnant woman (height 170 cm, mass 60 kg). In the 23rd, 28th and 33th week of pregnancy, the woman underwent anthropometric measurements and gait data acquisition. The anthropometric variations that have been most observed refer to the mass, as at the 23th week a mass of 62 kg was recorded on the force platform, at the 28th week 62.8 kg, while at the 33th week a mass of 66 kg was recorded. Anthropometric measurements were recorded for the reconstruction of the biomechanical model based on the Davis gait analysis protocol. Fifteen gait acquisitions were performed for all three sessions at a speed considered optimal by the subject. The results obtained show that with the progress of pregnancy there is an increase in the width of the step and a decrease in the length, a greater anteversion of the pelvis which involves a whole series of compensations at the level of the lower limbs. Some of these modifications are adopted by pregnant women but also occurs in elderly subjects, to avoid a loss of balance and fall. Stride length shows a linear downward trend as the pregnancy progresses, this variation agrees with research by Gilleard et al. [39]. The stride length may have decreased, as during pregnancy there is an increase in muscle tension resulting from an increase in the anteversion of the pelvis and a more pronounced lumbar lordosis which affects pelvic and thoracolumbar rotations. Furthermore, the reduction in stride length could be caused by incompleteness of the phases of the gait cycle, because during the third trimester of pregnancy there is an increase in flexion of the hip and knee during the intermediate and terminal swing phase which cause early swing and do not guarantee adequate initial contact. There was an increase in anteversion of the pelvis, in hip flexion during the terminal swing phase, in knee flexion during the intermediate swing phase, greater in the left limb than in the right, a factor that could be determined by the position of the baby in the mother's womb. There is also an increase in hip adduction during the intermediate support phase for the left limb, while for the right the adduction is reduced. On the other hand, the maximum extension of the hip is reduced during the terminal stance phase. These changes occur due to anteversion of the pelvis and displacement of the COM due to pregnancy. These mechanisms allow to bring the center of gravity as close as possible to the line of progression, thus avoiding situations of imbalance and the energy cost, a fundamental mechanism for making movement efficient and effective. On the other hand, there are no changes in the kinetic variables. The alterations of the kinematic parameters detected would seem to be compensations able to guarantee a walk as normal as possible despite the postural changes. The changes that arise, therefore, are mechanisms that are put in place to guarantee a walking pattern that, even if it is characterized by a greater articular excursion of the hip, avoids the occurrence of imbalance situations, allows the pregnant woman to respond adequately to any external perturbations and which allows, however, the lowest energy expenditure by keeping the COM as close as possible to the line of progression and within the support base. The study has resulted with a student's thesis [40].

10.4 On the Interpretation of Results

Interpretation of measurement results plays a key role in gait analysis. In literature, elaboration of the notion of normality of gait in a population can be found [41, 42]. The first step to evaluate the gait patterns of a patient is comparing these data with norms, but it depends from the purpose of the analysis because it could be also a comparison in the same patient before and after surgery, for example. It is important to consider the variability of interpretation of data. This because measurement record could be red by many, and everyone could interpret the data based on his/her experience and his/her knowledge, so the element of subjectivity is present [43].

Winter presented an elaborated strategy of the clinical application of gait analysis by listing observed kinematic abnormalities, their possible causes at the motor level and biomechanical and neuromuscular diagnostic evidence, forming Gait Diagnostic Chart [44] (Table 10.1).

Winter has based his Chart on the ground of a lot of practical experience researching human locomotion, in particular the biomechanical features of gait. The Chart, which is a kind of diagnostic checklist or a decision tree, can be of help to a

Table 10.1 An elaborated strategy of the clinical application of gait analysis (from Winter [44]. With permission)

Observed abnormality	Possible causes	Biomechanical and neuromuscular diagnostic evidence
Foot slap at heel contact	Below normal dorsiflexor activity at heel contact	Below normal tibialis anterior EMG or dorsiflexor moment at heel contact
	(a) Hyperactive plantar-flexor activity in late swing	(a) Above normal plantar-flexor EMG in late swing
	(b) Structural limitation in ankle range	(b) Decreased dorsiflexion range of motion
	(c) Short step-length	(c) See (a), (b), (c), and (d) immediately below
Short step-length	(a) Weak push-off prior to swing	(a) Below normal plantar-flexor moment or power generation or EMG during push-off
	(b) Weak hip flexors at toe-off and early swing	(b) Below normal hip flexor moment or power or EMG during late push-off and early swing
	(c) Excessive deceleration of leg in late swing	(c) Above normal hamstring EMG or knee flexor moment power absorption late in swing
	(d) Above normal contralateral hip extensor activity during contralateral stance	(d) Hyperactivity in EMG of contralateral hip extensors
Stiff-legged weight bearing	(a) Above normal extensor activity at ankle, knee, or hip early in stance	(a) Above normal EMG activity or moments in hip extensors, knee extensor, or plantar-flexors early in stance
Stance phase with flexed, but rigid knee	(a) Above normal extensor activity in early and middle stance at the ankle and hip, but with reduced knee extensor activity	(a) Above normal EMG activity or moments in hip extensors and plantar-flexors in early and middle stance
Weak push-off accompanied by observable pull-off	(a) Weak plantar-flexor activity at push-off; normal or above normal hip flexor activity during late push-off and early swing	(a) Below normal plantar-flexor EMG, moment or power during push off; normal or above normal hip flexor EMG or moment or power during late push-off and early swing

(continued)

Table 10.1 (continued)

Observed abnormality	Possible causes	Biomechanical and neuromuscular diagnostic evidence
Hip hiking in swing (with or without circumduction of lower limb)	(a) Weak hip, knee, or ankle dorsiflexor activity during swing	(a) Below normal tibialis anterior EMG or hip or knee flexors during swing
	(b) Overactive extensor synergy during swing	(b) Above normal hip or knee extensor EMG or moment during swing
Trendelenburg gait	(a) Weak hip adductors	(a) Below normal EMG in hip abductors: gluteus medius and minimus, tensor fasciae latae
	(b) Overactive hip abductors	(b) Above normal EMG in hip adductors, adductor longus, magnus and brevis, and gracilis

clinical gait researcher when interpreting measurement findings, aiming to reach a clinical benefit.

Gage and Stout [45] give a thorough elaboration of the subject of gait analysis as a result of rich clinical experience, treating various pathologies, first without the aid of gait analysis and later including gait analysis so that objective quantification of treated locomotions helped in improving overall care. Quantitative evaluations of gait of a „pre-aft“ type were made possible. Of special importance is the potential contribution of gait analysis to separate out „coping responses“ in a particular patient. It concerns individuals with abnormal cerebral control, muscle contractures, and/or lever arm dysfunction that are forced to introduce other abnormalities into their gait to compensate or „cope“ with the problems imposed on them by their condition.

Interpretation of motion analysis record should not be done in isolation. It always requires consideration in conjunction with the patient's medical history as well as other measures of patient assessment. A careful history, which elucidates previous treatment such as orthopedic surgery or something else may well provide the explanation for an otherwise obscure gait deviation. Confounding factors such as pain, emotional stress, and/or medications also can impose significant changes on an individual gait over a very short interval of time, which may lead to invalid interpretation of the gait data. Interpretation of motion analysis data should incorporate information regarding the speed of walking, presence or absence of orthoses, and/or balance aids, all which may significantly alter gait [45].

Baker, based on rich experience in clinical gait analysis, advocates a four-stage process to arrive at a better diagnostic report [4, pp. 164–176]. The first stage is *orientation*: it concerns to understand the background of the patient for first, visual interpretation of the gait and a check of the obtained data before detailed interpretation. This stage is important to avoid mistakes during results interpretation starting from the beginning of acquisitions. The second stage is *mark-up*: using some symbols, one can read the report easier. Symbols have to be intuitive and marked with a letter so it is easier to fit every peculiarity into the table. The third stage is *grouping*: in this stage, one has to group all associated data, for example all data about the pelvis, hip, knee or foot. These may be, for instance, suplementary data provided by the Thomas Test (TT) (Sect. 10.2.2.2). The last stage is *reporting*. It requires documentation and interpretation of the results. The report has to give information about the patient (personal data, anatomical data, anamnesis) and then it has to show all the results. In his argumentation [4], Baker uses an example of cerebral palsy (CP), the most elaborated clinical entity relying on gait analysis.

10.5 Additional Instrumentation

In addition to instruments and procedures described above, two other devices make desirable addendum for gait analysis.

Pedobarograph enables detecting pressure distribution between the foot and the ground, therefore it may be used in addition to conventional force measuring device, a force plate(s), where deemed appropriate. Chap. 8 has presented a number of clinical findings where pedobarography has been used in diagnosing and quantifying various locomotor pathologies. In principle, it is also possible to combine this measurement possibility with improved kinematic modeling of the foot (positioning a number of markers on the foot surface) and in this way adding to the inverse dynamics standardly provided. While technically complicated, simultaneous kinematic measurement of the body as a whole and of the foot is possible, albeit technically rather complex.

The other device is a portable oxygen consumption measurement apparatus. So, besides biomechanical estimation of energy expenditure (power), physiologically-based estimation is also possible. Gage and Stout give a succinct explanation of the use of this kind of measure primarily to CP patients [45].

10.6 Introduction of Subject-Specific Neuro-Musculo-Skeletal Modeling

Conventional gait analysis is focused on angles, moments and powers as the main result of the measurements carried out. However, there are alternative approaches to the problem. The direct dynamics approach has been mentioned previously (Sect. 10.2.2.3) and it can, in principle, be is used in biomechanics when it is of interest to learn the effect of muscle forces on joint kinematics [46–49]. Usually, direct dynamics is used in an effort to obtain information regarding the strategies used by the central nervous system in controlling movement. Patterns of selected muscle forces are generated and after having imposed the various initial positions and speeds, they are applied to a model of the musculoskeletal system of the human body to observe its movement. All degrees of freedom of the joints should be included in the model and the external reaction forces must be calculated [23].

Other methods use biomechanical models to estimate, for example, the length of the muscles or the moment of force experienced by a given muscle. The length of the muscle–tendon unit expressed as the distance between the point of origin and of insertion is generally modeled as a straight line joining the two points of origin and insertion, however in the case of some muscles that pass around bony structures the model foresees a series of points corresponding to these structures and the final length will be the result of the lines joining the different points present between origin and insertion. Among the first software to offer the opportunity to have a model of muscle length is SIMM (software for interactive musculoskeletal modeling) [46, see Fig. 2.19 in Chap. 2; 47]. It is to be emphasized that these modeling techniques, offering a simplification of reality, can be really useful only for describing the length of some muscles, in particular the biarticular muscles, since for the mono-articular graphs the curves relative to the length of the muscle and to the joint angle are very similar and therefore showing them both would be of little use. As for the biarticular muscles, such as *biceps femoris*, *rectus femoris*, *gastrocnemius*, the use of modeling techniques to estimate muscular length can provide additional information, since the simple graph of the joint angle is not sufficient to describe the length state of these muscles. A similar modeling technique is used to estimate the moment of force exerted by a muscle on a given joint, however these techniques have a high degree of approximation, which makes them hardly usable in a clinical setting. This approximation is mainly due to the nature of the concept of biomechanical model, which is a simplified representation of the complex reality of the human body. Furthermore, these models exploit measures obtained from studies on cadavers that cannot be representative of each individual characterized by a particular and unique anatomical structure. Again it is necessary to highlight the fact that errors in the positioning of the markers or their movements due to the artefacts of the soft tissue, provide the modeling software with an incorrect measurement of the length of a bone segment, increasing the error. Anyhow, the SIMM software provides an excellent opportunity for both researchers and clinicians to model and simulate complex musculoskeletal system's action.

10.7 Clinical Benefit of Gait Analysis

Not pretending to give a systematic overview, one may list a number of benefits of gait analysis in current clinical practice. Various conditions can interfere with normal gait, the most important are: cerebral palsy, head trauma, amputation, multiple sclerosis, muscular dystrophy, myelodysplasia, osteoarthritis, Parkinson, rheumatoid arthritis, spinal nerve lesions and stroke [50]. However, not all of these diseases can benefit from gait analysis. The first condition, historically, that showed benefits from gait analysis was cerebral palsy [51–53], particularly spastic diplegia and spastic hemiplegia seem to be the conditions under which gait analysis is more promising. Gait analysis has also been used for the diagnosis of pathologies, for example distinguishing between pathological walking on toes, due to spasticity of the *triceps surae*, and normal walking [54]. It has also been used to evaluate the potential benefit of an osteotomy in patients with osteoarthritis. Also in the prosthetic prescription the analysis proved to be useful for optimizing the prosthesis-limb alignment through an objective assessment of gait. The gait analysis proved to be promising also in the evaluation of the results deriving from the different clinical treatments. In muscle–tendon transfers, EMG may be of great help [55]. Among principal domains of clinical application of gait analysis is also traumatic brain injury [5]. The clinical psychiatric assessment also is related to gait features [56].

A short comment on the legislative aspect of clinical gait laboratories follows. It is worth noting that in order to secure clinical credibility of laboratory-obtained findings, laboratories must adhere to accreditation procedures. In more general terms, as instrumentation used has to comply with regulations covering medical equipment, an ISO certificate of a laboratory is to be aimed at [4] (http://www.cmlainc.org/Portal.html). In the USA, by the year 2020, all clinical motion laboratories should have achieved full accreditation from CMLA (The Commission for Motion Laboratory Accreditation), this being is a nonprofit organization established and operated to enhance the clinical care of persons with disorders of human movement by evaluating and accrediting clinical motion laboratories by a set of evaluative criteria. It was incorporated in Delaware in 1997, including several member societies: American Academy of Orthopaedic Surgeons (AAOS), American Academy of Physical Medicine and Rehabilitation (AAPM&R), American Physical Therapy Association (APTA) and Gait and Clinical Movement Analysis Society (GCMAS).

The reader seeking for in-depth knowledge on clinical gait analysis is referred to excellent references [4, 6, 7], as well as to the technically and methodologically oriented publications by Cappozzo group, referred to in Sect. 5.2.2.

10.8 Non-linear Analysis

In recent years, several authors have proposed a different approach to the movement analysis based on the study of the variability present in human movement. The notion of variability refers to the set of typical variations present in a gesture that can be observed between multiple repetitions of the same action. This phenomenon has been described as "repetition without repetition" by Bernstein in reference to the problem of redundancy [57]. The variability, although inherent in human movement, has always been considered with a negative connotation such as noise within the movement control system. It has also always been measured using traditional statistical tools such as the standard deviation, which in general quantifies the deviation from a central value, considered the norm, given by the average. In this perspective, the more a value deviates from the average and therefore from normalcy, the more it is considered non-optimal. The most recent research in many scientific disciplines has shown instead that many apparently "noisy" or "random" phenomena actually follow precise patterns. In order to be able to observe these schemes and measure variability in a different way, it is necessary to resort to non-linear mathematical analysis techniques. These non-linear analysis techniques can also be applied to complex biological systems and the study of human movement.

The study of variability in gait analysis, in this perspective, seems promising for the identification and evaluation of adaptations of the neuromuscular system following traumas or pathologies. Specifically, for the gait analysis, in the non-linear perspective, the variability of the time series of the step and joint angles is evaluated, this analysis seems to provide information on the way in which subjects control movement. Furthermore, once an optimal level of variability is obtained from a population of healthy subjects, it is possible to make inferences about the effects of a pathology. In this sense, this type of analysis seems to be an excellent tool for diagnostic and prognostic evaluation.

The analysis of variability was applied in different pathological conditions of walking, such as amputation of the lower limb, stroke and peripheral arterial disease, obtaining interesting results and providing useful information not obtainable with traditional analysis.

In the amputee subject equipped with a prosthesis, for example, the neuromuscular system must learn new strategies in order to integrate the prosthesis and perform the task of locomotion. Researching the variability of the time series at the level of the joint angles of the ankle, knee and hip a subject with different prostheses was evaluated. Using the "maximum Lyapunov exponent" (LyE) as an indicator of variability, the researchers found a strong correlation between this index and the patient's preference for the prosthesis [58], in fact, LyE showed that the subject had a lower variability of the step with the appropriate prosthesis compared to the less appropriate one.

Researching gait following a stroke it emerged that traditional variability measures such as standard deviation were able to show the difference between healthy and pathological subject, but were unable to show the difference between the healthy limb and the pathological limb in patients with stroke, on the contrary, measures such as sample entropy have made it possible to identify these differences.

In general, to resume, variability in certain temporal gait features must be carefully distinguished from the "gait stability," i.e., greater variability does not necessarily mean less stability. Here lies the importance of nonlinear analysis techniques from which the measures for analyzing the dynamic characteristics of a quasi-periodic process, like human gait, can be applied for gait analysis.

10.9 Pervasive Gait Analysis Using Wearable Sensors

The recent development of inertial-type wearable kinematic sensors (Chap. 5), in-shoe pressure sensors for pedobarography (Chap. 7), sEMG detection using wireless amplifiers (Chap. 9), as well as interfacing possibility with smart devices have enabled the proliferation of these technologies into measurements of walking, bringing simplicity and freedom of movement. Chen et al. [59] named it pervasive gait analysis. Classical gait analysis, namely, has grown in an environment of laboratory-type 3D kinematic measurement systems, marker-based, and taking advantage of close range photogrammetry. In these conditions, and with force plate(s) as a standard sensor implied, appropriate measurement methodology could have been developed and standardized (see papers by Cappozzo and Woltring referred to in Chaps. 5 and 6). But, this approach is rather time consuming and space limited and unable to secure the acquisition of long series of signals (the use of treadmill not considered at present). Therefore, although such instrumentation and methodology have galvanized gait analysis research in the past, they are not pervasive enough among clinics for gait analysis to realize its full potential. Many studies continue to use goniometers for measuring joint angles, tools also being cumbersome to use and able to only provide limited types of information [59; already recognized previously in Sect. 10.2.2.1].

In order for gait analysis to gain popularity in clinics, Chen et al. claim to be essential to replace the current gait analysis systems with easier to use, more economical, and portable measurement setups. They provide a systematic review of current techniques for quantitative gait analysis and propose key metrics for evaluating both existing and emerging methods for qualifying the gait features extracted from wearable sensors. By detailed evidence and comparisons of used measurement tools and obtained findings as reported in the literature, in the context of clinical problems researched and/or diagnosed, they put forward a claim of pervasive approach being, in principle, more suitable than the standard one. To highlight the advantages of wearable sensors over the current laboratory systems, Chen et al. have tabulated comparison of laboratory gait analysis tools with their wearable counterparts. In their reasoning, besides classical approach to gait analysis, they also consider the non-linear approach (discussed in previous sections) which, they claim, is also more easily implementable by emerging pervasive methods. They give some examples in support of their reasoning.

A number of nonlinear analysis techniques center around one simple but important and informative presentation of gait signals—phase portrait. A phase portrait is a geometric representation of the trajectories of a dynamical system in the phase plane (Jose and Saletan, citation after 49). In this representation, the position information is often plotted against its first time derivative. Certain gait measures can then be extracted by quantifying this geometric shape, including for example, gait regularity, gait mechanical energy, gait complexity, and gait stability. Besides, it is a valuable visualization tool for data presentation and clinical interface. Visualization tools, such as phase portrait, can play a vital role in promoting pervasive gait analysis. First, current gait analysis results can only be understood by gait experts by reading critical gait measures from a lengthy report (as claimed when presenting Heimer's Example 1 in 10.3). Since humans interpret images better than data, visualization of gait can provide a vivid and memorable impression of the severity of gait abnormalities for clinical staff. Second, these phase portraits can even be quantified to provide a more sensitive and precise characterization of gait patterns [59].

Another example concerns quantitative data captured by inertial sensors treated by non-linear analysis techniques, so that gait stability is characterized by the Lyapunov exponent (LyE) (see previous Sect. 10.8) which describes how a quasi-periodic dynamic system (e.g., human gait function) responds to "very small perturbations continuously in real time" (Dingwell and Kang, citation after 56).

Particular applications of this kind advocate use of wearable-type sensors in gait analysis and proliferation of similar solutions will no doubt continue. But, at present, there is no alternative for a classical laboratory-type approach regarding quantitative precision and comprehensiveness of a biomechanical evaluation.

10.10 Evolving Methodology of Clinical Gait Analysis

As a practical outgrow of the field of biomechanics of human movement and locomotion, modern gait analysis is evolving in a synergy of biological and engineering knowledge and achievements, being shaped by requirements of medical clinical use. The idea of human locomotion as a mechanical phenomenon that has resulted with a kinematic model of gait as introduced by the Berkeley Group [60] has been well accepted by biomechanics community and this model is lasting, with small modifications, until today, as commented in Sect. 2.1. Gait determinants are recognized as a fundamental frame for defining kinematics of walking [14, 60–62].

Recent literature reports critically evaluate the present situation in clinical gait analysis searching for future solutions [63, 64]. Tesio and Rota [63] develop an argument in favor of introducing into the gait analysis routine additional information on the kinematics of the center of mass (COM). Namely, presently this information is not included and analysis report is based essentially on segmental kinematics and changes of joint angles as principal information carriers of diagnostic information. Added to this is information of physiological character, obtained separately via oxygen consumption measurement. But, taking a broader view and having as a background in mind the evolutive uprising of humans on lower legs (see Chap. 3), bipedal locomotion has been defined as essentially a vehicle carrying the body weight through space, via translation of its COM. Precisely through considering gait determinants, knowledge on what should be a normal, healthy trajectory of COM during walking has been established and a quasi-sinusoidal amplitude of this trajectory is appreciated as a mark of normal gait. Tesio and Rota consider COM to be a summary index of both balance and the neural maturation of walking. They suggest the inclusion of 3D trajectory of COM into routine gait report, which is possible in practice by means of developing corresponding algorithms and joining them to existing standard software being a part of automated 3D kinematic measurement systems. Besides using standard laboratory 3D kinematic measurement equipment, COM trajectory is also obtainable in approximate means by using an inertial-type sensor positioned adequately on the human body, which should be applicable in the clinical environment.

At the engineering side one witneses both standardization of "classical" laboratory equipment for gait analysis [4, 6, 14, references 19–22 in Chap. 5] and—as mentioned in previous sections—simultaneous emerging of novel "pervasive" measurement tools that offer, at the moment, only partial solutions, however.

Current state of art in gait analysis methodology is characterized by the creation of techniques to capture data rapidly, accurately, and efficiently, and to interpret such data by an assortment of modeling, statistical and wave interpretation techniques, as well as by avant-garde methodologies like AI, as was forcasted in [26], while the importance of the field of mathematics in this realm has already been accentuated by Nigg (as mentioned in Chap. 2, Fig. 2.13). The complexity of human body anatomy as well as of neurophysiological control of locomotion have resulted with many a research endeavor pursuing novel approaches to data analysis (besides inverse dynamics and standard signal analysis) including data mining, neural networks, AI as mentioned, expert systems, etc. Gait analysis is a challenging field, and observation by some leading research groups is that today it is focused dominantly on peripheral body biomechanics, while one should seek in addition for quantitative evidence of brain control of movement (brain scanning, EEG).

To conclude the subject of gait analysis, a reminiscence on the interplay of art and science in approaching the phenomenon of human locomotion is worthwhile. We refer to Rodin's sculpture from the beginning of the twentieth century: The Walking Man (L'homme qui marche) (Fig. 10.15).

Auguste Rodin French (https://www.metmuseum.org/art/collection/search/198565).

https://www.metmuseum.org/about-the-met/policies-and-documents/open-access.

An artistic view on the phenomenon is nicely expressed as: „*The Walking Man displays not only Rodin's fascination with partial figures, reminiscent of antique sculptural fragments, but also his interest in the sculptural representation of the human body in sequential motion. By showing both feet planted firmly on the ground, the sculptor attempted to record not a realistic depiction of a man walking, but instead the movements at the beginning and at the end of his step, producing the impression of a movement which, in fact, takes several moments to accomplish*" (citation taken from https://www.metmuseum.org/art/collection/search/198565). Thus, through the sculpture, the artist has successfully captured the dynamics of the walking phenomenon.

Fig. 10.15 The Walking Man (L'homme qui marche) modeled before 1900, cast before 1914

References

1. Nigg BM (1994) Selected historical highlights. In: Nigg B Herzog W (eds) Biomechanics of the musculo-skeletal system. New York, Wiley, pp 3–35
2. Baker R (2006) Gait analysis methods in rehabilitation. J NeuroEng Rehab 3(1):4
3. Gerber L, Rasch EK (2013) In memoriam: a tribute to Jacqueline Perry. Arch Phys Med Rehabil 94(2036):c1-e4
4. Baker R (2013) Measuring walking: a handbook of clinical gait analysis. Mac Keith Press, London
5. Baker R, Esquenazi A, Benedetti MG, Desloovere K (2016) Gait analysis: clinical facts. Eur J Phys Rehab Med 52(4):560–574
6. Gage JR, Schwartz MH, Koop SE, Novacheck TF (2009) The identification and treatment of gait problems in cerebral palsy, 2nd edn. McKeith Press, London
7. Perry J (1992) Gait analysis: normal and pathological function, Thoroafare, NJ, SLACK
8. Kaufman KR, Sutherland DH (2006) Kinematics of normal human walking (Chapter 3). In: Rose J, Gamble JG (eds) Human walking. 3rd edn. Lippincott Williams & Wilkins, Philadelphia, Pa, pp 33–51
9. Perry J, Antonelli DJ, Bontrager EL (1976) VA-Rancho gait analyzer, final project report, Rancho Los Amigos Hospital, Downey, California
10. Hannah R, Cousins S, Foort J (1978) The CARS-UBC electrogoniometer—a clinically viable tool, 7th Canadian Medical & Biological Engineering Conference

11. Medved V (1980) Initial evaluation of CARS-UBC electrogoniometer system. Biomechanics Laboratory, Division of Applied Sciences, Harvard University, technical note
12. Koopman B, Grootenboer HJ, de Jongh HJ (1995) An inverse dynamic model for the analysis, reconstruction and prediction of bipedal walking. J Biomech 28(11):1369–1376
13. Forner-Cordero A, Koopman HJFM, van der Helm FCT (2006) Inverse dynamics calculations during gait with restricted ground reaction force information from pressure insoles. Gait Posture 23(2):189–199
14. Chambers HG, Sutherland DH (2002) A practical guide to gait analysis. J Am Acad Orthop Surg 10(3):222–231
15. Kiseljak D, Bolčević F, Ćurko D, Pažin K, Ujaković F, Gruić I, Medved V (2017) The modified Thomas Test realized by optoelectronic kinematic measurement. Functional Neurology, Rehabilitation, and Ergonomics. 7(3):39–44 (and as abstract in: Book of Abstracts of the 2017 Conference on Movement: Brain, Body, Cognition (Leisman, Gerry (ur.)—Oxford University, Oxford, July 9–11, 2017. International association of functional neurology and rehabilitation, Oxford, UK, p 95)
16. Davis RB III, Ounpuu S, Tyburski D, Gage JR (1991) A gait analysis data collection and reduction technique. Hum Mov Sci 10(5):575–587
17. Kadaba MP, Ramakrishnan HK, Wootten ME, Gainey J, Gorton G, Cochran GVB (1989) Repeatability of kinematic, kinetic, and electromyographic data in normal adult gait. J Orthop Res 7(6):849–860
18. Heimer, Ž. (2005) Automatizirano kliničko mjerenje biomehanike i kineziologije hoda (Automated clinical measurement of biomechanics and kinesiology of gait). [master's thesis]. [Zagreb (Croatia)]: University of Zagreb
19. Cappozzo A, Catani F, Della Croce U, Leardini A (1995) Position and orientation in space of bones during movement: anatomical frame definition and determination. Clin Biomech 10(4):171–178
20. Rabuffetti M, Crenna P (2004) A modular protocol for the analysis of movement in children. Gait Posture 20:S77–S78
21. Leardini A, Sawacha Z, Paolini G, Ingrosso S, Nativo R, Benedetti MG (2007) A new anatomically-based protocol for gait analysis in children. Gait Posture 26(4):560–571
22. Frigo C, Rabuffetti M, Kerrigan DC, Deming LC, Pedotti A (1998) Functionally oriented and clinically feasible quantitative gait analysis method. Med Biol Eng Comput 36(2):179–185
23. Di Prampero PE, Cappello A, Cappozzo A (Eds.) (2003) Bioingegneria della postura e del movimento. Patron
24. Manal K, Stanhope SJ (2004) A novel method for displaying gait and clinical movement analysis data. Gait Posture 20(2):222–226
25. Kasović M, Heimer Ž, Medved V (2011) Coloured signal coding in computerized human gait analysis. Proceedings 6th international scientific conference on kinesiology „Integrative power of kinesiology", Opatija, Croatia, pp 155–157
26. Simon RS (2004) Quantification of human motion: gait analysis—benefits and limitations to its application to clinical problems. J Biomech 37:1869–1880
27. Wu G, Cavanaugh PR (2001) ISB recommendations for standardization in the reporting of kinematic data. J Biomech 28(10):1257–1261 (Appendix A)
28. Wu G, Siegler S, Allard P, Kirtley C, Leardini A, Rosenbaum D, Whittle M, D'Lima DD, Cristofolini L, Whittle H, Schmidt O, Stokes I (2002) Letter to the editor—ISB recommendation on definitions of joint coordinate system of various joints for the reporting of human joint motion—part I: ankle, hip, and spine. J Biomech 35(4):543–548 (Appendix B)
29. BTS GAITLAB—Analysis protocols davis protocol version 1.0.0 (2016) BTS S.p.A
30. Astone S, Mignone A, Albano D (2019) Comparison of gait kinematics and kinetics between qualified dancers and non-dancers. J Phys Educ Sport 19:2289–2291
31. Baker R, McGinley JL, Schwartz MH, Beynon S, Rozumalski A, Graham HK et al (2009) The gait profile score and movement analysis profile. Gait Posture 30:265–269
32. Schwartz MH, Rozumalski A (2008) The gait deviation index: a new comprehensive index of gait pathology. Gait Posture 28(3):351–357

33. Astone S, Mignone A, Coppola S (2020) Comparison of EMG profile during gait between qualified dancers and non dancers. RRISS 2nd edn, vol II, pp 80–84
34. Kovač I (2007) Biomehanička analiza hoda osoba s potkoljenom amputacijom opskrbljenih protezama s osloncem na ligament patele (Biomechanical analysis of walking in lower extremity amputees' equipped with Patella Tendon Bearing (PTB) prostheses). [dissertation]. [Zagreb (Croatia)]: University of Zagreb
35. Jelić M, Kovač I (2002) Epidemiological analysis of war amputees. In: Book of abstracts—third ISPO Central and Eastern European Conference, Dubrovnik
36. Kovač I, Medved V, Ostojić Lj (2009) Ground reaction force analysis in traumatic transtibial amputees' gait. Coll Antropol 33(Suppl. 2):107–114
37. Kovač I, Medved V, Ostojić Lj (2010) Spatial, temporal nad kinematic characteristics of traumatic transtibial amputees' gait (2010) Coll Antropol 34(Suppl. 1):205–213
38. Kovač I, Medved V, Kasović M, Heimer Ž, Lužar-Stiffler V, Pećina M (2010) Instrumented joint mobility analysis in traumatic transtibial amputee patients. Period Biol 112(1):25–31
39. Gilleard WL, Crosbie J, Smith R (2002) Static trunk posture in sitting and standing during pregnancy and early postpartum. Arch Phys Med Rehabil 83(12):1739–1744
40. Cilio A (2019) Analisi biomeccanica del cammino in gravidanza (Biomechanical analysis of gait during pregnancy). [dissertation]. [Salerno (Italy)], University of Salerno
41. Perry J, Davids JR (1992) Gait analysis: normal and pathological function. J Pediatr Orthopaed 12(6):815
42. Whittle MW, Levine D, Richards J (2012) Whittle's gait analysis. Churchill Livingstone, Elsevier
43. Skaggs DL, Rethlefsen SA, Kay RM, Dennis SW, Reynolds RAK, Tolo VT (2000) Variability in gait analysis interpretation. J Pediatr Orthopaed 20(6):759–764
44. Winter DA (1991) An elaborated strategy of clinical application of gait analysis (Gait Diagnostic Chart). In: Boenick U, Nader M, Mainka (eds) Gangbildanalyse-Stand der Messtechnik und Bedeutung für die Orthopadie-Technik. Duderstadt, Mecke Druck, pp 266–277
45. Gage JR, Stout JL (2009) Gait analysis: kinematics, kinetics, electromyography, oxygen consumption and pedobarography. In: Gage JR, Schwartz MH, Koop SE, Novacheck TF (2009) The identification and treatment of gait problems in cerebral palsy. 2nd edn. McKeith Press London, pp 260–284
46. Delp SL, Loan JP, Hoy MG, Zajac FE, Topp EL, Rosen JM (1990) An interactive graphics-based model of the lower extremity to study orthopaedic surgical procedures. IEEE Trans BME 37(8):757–766
47. Delp SL, Loan JP (1995) A graphics-based software system to develop and analyze models of musculoskeletal structures. Comput Biol Med 25(1):21–34
48. Seth A, Hicks JL, Uchida TK, Habib A, Dembia CL, Dunne JJ et al (2018) OpenSim: simulating musculoskeletal dynamics and neuromuscular control to study human and animal movement. PLoS Comput Biol 14(7):e1006223. https://doi.org/10.1371/journal.pcbi.1006223
49. Hicks JL, Schwartz MH, Delp SL (2009) Modeling and simulation of normal and pathological gait. In: Gage JR, Schwartz MH, Koop SE, Novacheck TF (2009) The identification and treatment of gait problems in cerebral palsy. 2nd edn. McKeith Press London, pp 285–305
50. Whittle MW (1993) Gait analysis. In: The soft tissues. Butterworth-Heinemann, pp 187–199
51. Perry J, Hoffer MM (1977) Preoperative and postoperative dynamic electromyography as an aid in planning tendon transfers in children with cerebral palsy. J Bone Joint Surg Am 59(4):531–537
52. Sutherland DH (1978) Gait analysis in cerebral palsy. Develop Med Child Neurol 20(6):807–813
53. Gage JR (1983) Gait analysis for decision-making in cerebral palsy. Bull Hosp Joint Dis Orthop Inst 43(2):147–163
54. Hicks R, Durinick N, Gage JR (1988) Differentiation of idiopathic toe-walking and cerebral palsy. J Pediatr Orthop 8(2):160–163
55. Gage JR, Perry J, Hicks RR, Koop S, Werntz JR (1987) Rectus femoris transfer to improve knee function of children with cerebral palsy. Develop Med Child Neurol 29(2):159–166

56. Sanders RD, Gillig PM (2010) Gait and its assessment in psychiatry. Psychiatry (Edgmont) 7(7):38–43
57. Bernstein, N. (1966). The co-ordination and regulation of movements. Pergamon-Press
58. Wurdeman SR, Myers SA, Jacobsen AL, Stergiou N (2014) Adaptation and prosthesis effects on stride-to-stride fluctuations in amputee gait. PloS one 9(6):e100125
59. Chen S, Lach J, Lo B, Yang G-Z (2016) Toward pervasive gait analysis with wearable sensors: a systematic review. IEEE J Biomed Health Inform 20(6):1521–1537
60. Rose J, Gamble JG (eds) (2006) Human walking. 3rd edn. Lippincott Williams & Wilkins, Philadelphia, Pa
61. McMahon TA (1984) Muscles, reflexes, and locomotion. Princeton University Press, Princeton, NJ
62. Medved V (2001) Measurement of human locomotion. CRC Press Inc., Boca Raton, Fl
63. Tesio L, Rota V (2019) The motion of body center of mass during walking: a review oriented to clinical applications. Fron Neurol 10:999. https://doi.org/10.3389/fneur.2019.00999
64. Lu TW, Chang CF (2012) Biomechanics of human movement and its clinical applications. Kaohsiung J Med Sci 28:513–525

Chapter 11
Concerns of a Modern Orthopedic Traumatologist

Nikica Daraboš

Abstract Modern orthopedics and traumatology are nowadays impacted ever more by a field of bioengineering. In this chapter the reflections of an orthopedic traumatologist on the area of human locomotion are given. An orthopedic traumatologist has a goal to treat the injuries or diseases of patient's locomotor system with the highest respect to its biomechanical features and relationships. It is very important for him/her to understand the individual anatomical characteristics and biomechanical performances of each patient's joints. Therefore, empirically investigated values are described of locomotion capacity and mobility of individual regions of the locomotor apparatus of the lower extremities of an average patient, known so far. Practical examples of treatment of patients with locomotor system injury are presented through surgical treatment techniques commonly used by the orthopedic traumatologist, including the use of computer-aided surgery (CAS) technology when appropriate. In this way, the surgeon tries to bring the end result of treatment as close as possible to the imagined performance of the patient's locomotor system up to the moment of injury. Demonstrated are some practical applications of such treatment options through individual cases of treated patients. The chapter covers critical diagnostics and treatment aspects of hip, knee and ankle pathology through presented examples of individual cases.

11.1 Introduction

Biomechanics of bone and joint movements of human body is one of the premier interests in the fields of orthopedics and traumatology. This is especially pronounced when treating patients with injuries or diseases of the locomotor system. Namely, due to injuries or diseases of the locomotor system, performances of locomotion that affect and are affected by the outcome of treatment change.

The aim of modern treatment in orthopedics and traumatology is to achieve as high as possible biomechanical features and relationships within the patients' locomotor

N. Daraboš (✉)
School of Medicine, University of Split, Šoltanska 2, 21000 Split, Croatia

V. Medved (ed.), *Measurement and Analysis of Human Locomotion*,
Series in Biomedical Engineering, https://doi.org/10.1007/978-3-030-79685-3_11

system. Given individual anatomical characteristics and biomechanical performances of each patient, for the best outcome of comprehensive treatment it would be ideal to be able to look into the measures and analysis of his locomotion capacity and mobility before and after the injury or illness of the locomotor system. Of course, these values are variable in each individual from the moment of birth until the onset of the pathological condition, and it is impossible to identify and process them objectively for each patient individually. This is especially important in emergency orthopedics or traumatology, where orthopedic traumatologist must treat emergency pathological conditions, most commonly in patients with bone and joint injuries.

At the same time, the decision of the orthopedic traumatologist on the type of fracture treatment is also influenced by the clinical status of the muscles, tendons, ligaments and other structures of the patient's locomotor system. His clinical status is beyond doubt a focus of the orthopedic traumatologist's interest, whether or not it be pathologically altered, whether it is the sole source of pathological symptoms, or just an integral part of the pathology of injuries and/or diseases of multiple affected parts of the locomotor system. In these cases, when approaching the treatment choice, orthopedic traumatologist usually is not able to have an objective insight into the condition of the injured region before the actual injury. For this reason, when selecting a treatment of particular pathological conditions of a locomotor apparatus, measures and analysis of the values of locomotion capacity and mobility of an average patient within a population class (empirically obtained on the basis of research) must be referred to.

Therefore, in continuation of this chapter we shall describe empirically investigated values of locomotion capacity and mobility of individual regions of the locomotor apparatus of the lower extremities of the average patient, known so far. The possibilities of their practical application in the treatment of patients with locomotor system injury will be presented through surgical treatment techniques used by the orthopedic traumatologist during surgery. In this way, the surgeon tries to bring the end result of treatment as close as possible to the imagined performance of the locomotor system of the patient up to the the moment of injury.

We will demonstrate some practical applications of such treatment options in orthopedics and traumatology by presenting examples of individual cases of treated patients. The clinical cases presented and discussed in this Chapter arose from the practice of the author of the Chapter while having been affiliated with the General Hospital Varaždin, in Varaždin, Sestre milosrdnice University Hospital Center, in Zagreb, and the University Clinical Hospital Center Zagreb, in Zagreb.

11.2 Hip

Biomechanical specificities of fracture resolution in the hip joint region will be presented.

Bone transforms depending on the function it performs, that is, depending on the mechanical stress to which it is exposed. Human bone undergoes continuous processes of degradation and formation, and in the parts of bone that are loaded, the formation dominates, while in unloaded parts the bone deteriorates [1].

The same is expressed in all lower extremity bones, especially the femur, in its proximal part, which together with the acetabulum of the pelvis, forms the articular bodies of the hip joint. The biomechanics of femur in the hip joint is exerted by pressure forces due to the load, but stretching forces also occur that affect the morphology of the proximal femur as well. The femur withstands a high level of forces, but parts of the proximal femur that are not under the influence of compressive forces have lower bone density and firmness [2].

Risk factors for femoral fracture in the hip joint region are related to the aging process—osteoporosis, femoral neck variation; then the weakening of the psychophysical abilities of the patient—lack of neuromuscular coordination, fear of activity and accompanying diseases that give a high probability of hip fracture in old age. Furthermore, the biomechanical reasons that cause a high incidence of this injury and hinder or impede treatment are: close contact and possible injury to the blood vessels supplying the femoral head with sharp dislocated neck fragments, intra-articular fracture localization with consequent fissure fracture flushing with synovial fluid, increased intra-capsular pressure (impaired circulation), and the presence of angiogenic inhibitory factors in the absence of periosteum in the femoral neck with consequent endostal healing [3, 4]. The risk of fracture of the proximal femur has been found to increase with increasing neck length and greater colo-diaphysal angle, while the intertrochanteric width, thickness of the cortex of the femur body is lower in patients with fracture.

In addition, there are negative (unfavorable) biomechanical relationships. The hip is involved in all major body movements, with the transfer of body weight to the legs over the horizontal part of the femoral neck and the action of the force of movement over the shorter arm of the lever (1:3 ratio) [5]. The risk of fracture of the proximal part of the femur depends primarily on the bone density of the proximal femur and on the microarchitecture of the bone. Adaptation of the trabecular bone structure to the load is a very important factor when it comes to surgery for osteosynthesis of bone fracture by alenthesis or by implantation of a prosthetic hip joint prosthesis. The reason for this is the change in natural stretching and the so-called "stress shielding" phenomenon implying that due to the greater strength of the implant than the bone, force shifts towards the implant, which results in less loading of certain parts of the bone. Garden's analysis of the hip function confirmed that the femoral neck is angled to promote the transmission of the loads arising from weight-bearing [6] (Fig. 11.1).

The role of bone stretching is even more significant in osteoporotic bone, where these forces are distributed unevenly [7]. Intraoperative implant's positioning during the per trochanteric fracture surgery should be performed with the respect that secondary trabecular groups of hip region cross at acute angles, and not at right angles, to resist the stresses induced by the complex combination of forces the femur is subjected to. Irreverence of these rules lead to postoperative complications and this explains the occurrence of recurrent and/or peri-prosthetic fractures (Figs. 11.2, 11.3, 11.4, 11.5, 11.6 and 11.7).

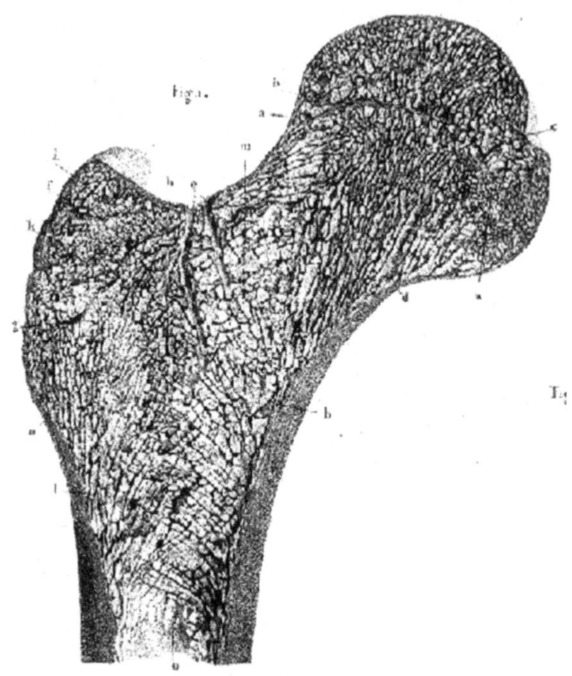

Fig. 11.1 Garden's analysis of the hip's trabecular bone microarchitecture

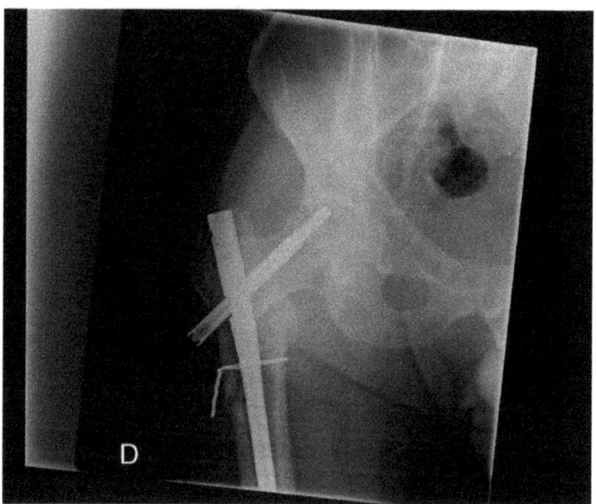

Fig. 11.2 Case of pertrochanteric refracture—False implant's position—X ray anteroposterior view

Femoral fractures in the hip joint in the older population usually occur during stumbling and falling. In younger people, they are caused by a large force such as a fall from a high altitude or in a car accident and often include fractures of the acetabulum. The biomechanical mechanism of injury involves: the action of direct force of lateral impact in the area of a large trochanter or an excessive leg abduction

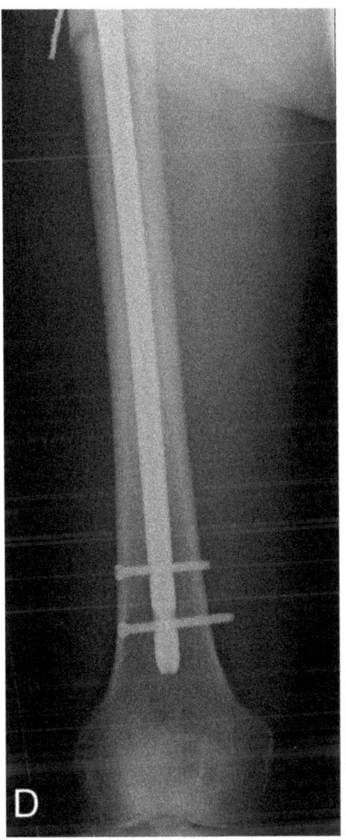

Fig. 11.3 Case of pertrochanteric refracture—False implant's position—X ray laterolateral view

(abduction fractures). Fractures of this kind are usually stable and, with an increased collo-diaphysal angle, do not represent significant dislocations of bone fragments, but the action of the abductor muscles results in the impaction of fragments into each other. In contrast, indirect force action along the longitudinal extremity axis (fall to the extended leg or knee) or excessive foot adduction (adjunctive fractures) results in unstable fractures, which with reduced collo-diaphysal angle and due to the action of the adductor muscle results with fragmentation and shortening of leg. Furthermore, fractures of the upper femur can be anatomically localized in the area inside or outside the hip joint capsule. Intra-capsular fractures are divided into those below or within the femoral head, and at the level of the middle or base of the femoral neck as the weakest part of the femur. Garden's classification is based according to the risk of a) bone death due to damage to the blood circulation of the femoral head and b) impaired fracture healing of these fractures [8]. Pauwels's classification considers the risk due to increased shear stresses according to the angle of the vertical line in the fracture region. Fractures of the proximal femur outside the articular capsule have a better prognosis of healing and are divided into those between the large and

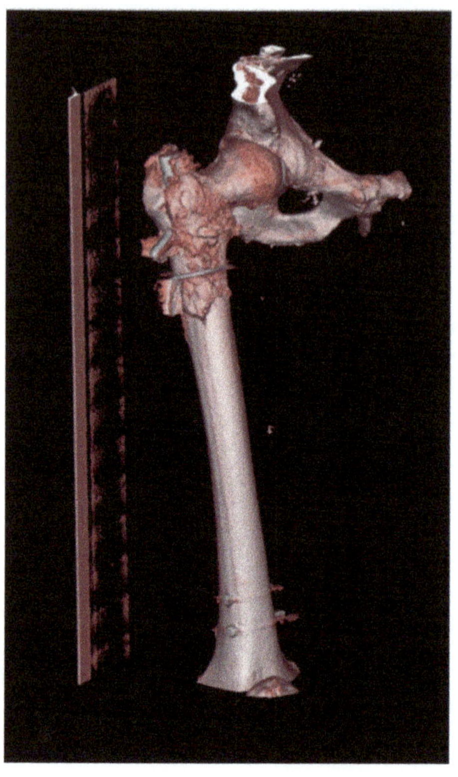

Fig. 11.4 Case of pertrochanteric refracture—False implant's position, anteroposterior view—CT 3D image

small trochanter (intertrochanteric region) and those in the subtrochanteric region [9].

Technological challenges in resolving fractures in the proximal femur region are based on the demanding biomechanics of the hip joint. This is especially important during preoperative planning, intraoperative repositioning of fracture fragments, and osteosynthesis of fractures or implantation of the artificial hip joint. The consequences of not adhering to the biomechanical principles of joint function in the hip joint adversely affect post-operative rehabilitation and the result of overall treatment. In order to prevent this, in the age of modern traumatology and orthopedics we are trying to make more use of the principles of computer-aided surgery (CAS) technology. CAS enables us to locate the position of surgical instruments in three dimensions and to present them to the surgeon in real time. In this domain, CAS technologies present a dual challenge (clinical and technological) for surgeons and engineers. Since the beginning of the 1990s, CAS technology has gained popularity because of the potential to enhance the precision of orthopedic traumatologist's procedures and its application has been extended to clinical applications in spine, knee and hip surgery (this applies therefore also to 11.3) [10].

Although hip surgery is significantly associated with prosthetic hip surgery, total arthroplasty is not the only clinical application in hip surgery and there are numerous

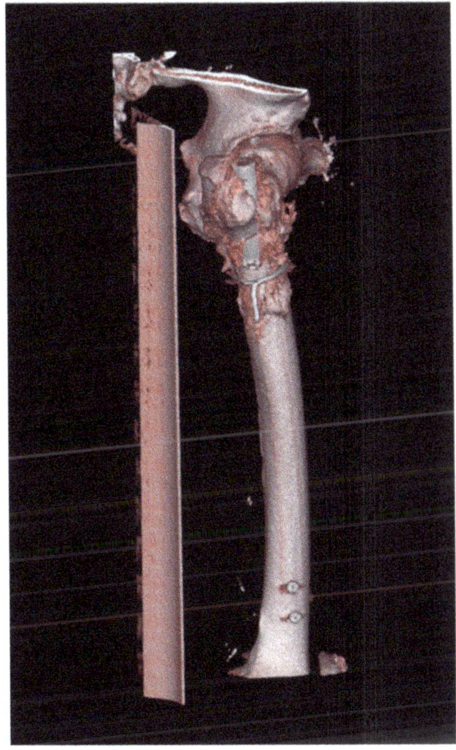

Fig. 11.5 Case of pertrochanteric refracture—False implant's position, latero-lateral view—CT 3D image

other clinical applications of CAS technology including fractures of the femur and pelvic acetabulum in the hip region (Figs. 11.8 and 11.9).

CAS technology uses a combination of various hardware and software technology components.

The seven major technologies used for computer-assisted hip surgery are:

- The preoperative CT based navigation system has a key element based on the system of the registration of preoperative CT scans with intraoperative surgical space. Although this technology provides complete 3D representations of bone anatomy and surgical tools, the use of CT-based navigation in orthopedics is limited because the registration process between preoperative CT scans and surgical space is error-prone and time consuming [11].
- An alternative to using CT scans is with an intraoperative non-CT, non-RTG navigation system. It is the "Image Free" system, based on cinematic (point-based) and/or anatomic (surface-based) markers that have been collected intra-operatively. This technique considers tool calibration, dynamic characteristics of joints, and anatomical points digitized by the surgeon (Bone Morphing). These systems are accurate and reliable and do not require preoperative imaging. Currently, in clinical practice, more than 90% of the systems used are image-free [12].

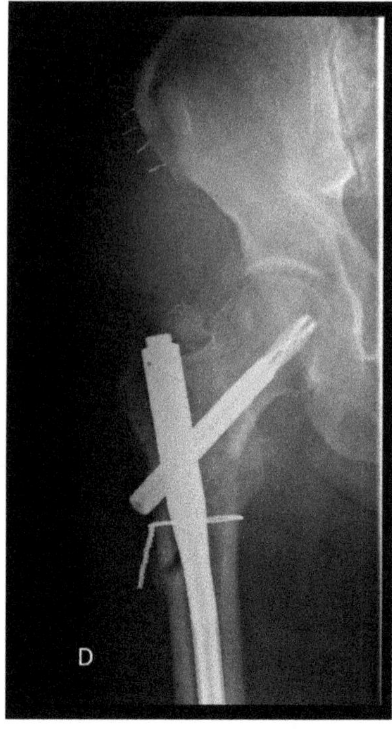

Fig. 11.6 Case of pertrochanteric refracture—Correct re-fracture's refixation with new implant and bone grafting—X ray anteroposterior view

- Virtual fluoroscopy is an intraoperative navigation system without the use of CT. The principle is to navigate calibrated fluoroscopic images. After collecting the RTG images, the surgeon can locate the position of the instruments in real time on the pre-registered recordings. This technology is closest to conventional fluoroscopy. No registration process is required, but views are always in two dimensions. Virtual fluoroscopy is suitable for hip navigation where CT or non-CT imaging is deficient (as in revision total hip arthroplasty) and in cases of previous hip fusion (where routine CT registration or surface-based method is not possible) [13].
- Surgical procedures using custom made femoral components have been developed for primary total hip arthroplasty and or for revision procedures. Computer-assisted system using CAD/CAM technology enables producing custom made femoral components. Custom made implants are a well-accepted tool in orthopedic surgery. The only missing information is the exact relationship between bone, surgical planning and tools. This system is also called "semi-active" in contrast to "active" systems (robots) [14].
- Rapid prototype and template fabrication is similar to the previous technique: patient CT images are used for simulation and computer preoperative planning of osteotomy incisions. This technique uses rapid prototyping technology to show the shape of the bone in the area of the planned incision and to produce drilling guidelines. The templates are affixed intra-operatively to the appropriate position

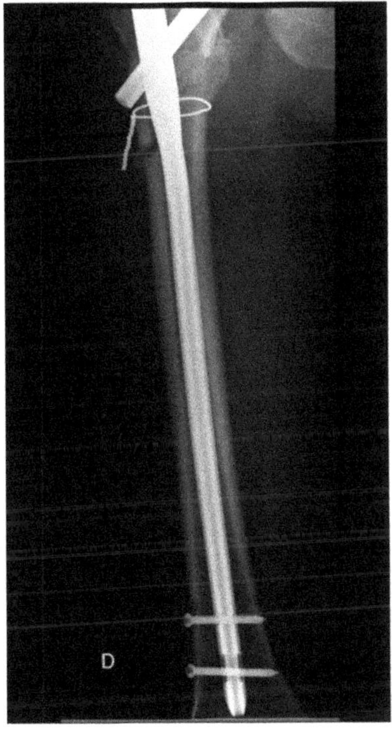

Fig. 11.7 Case of pertrochanteric refracture—Correct re-fracture's refixation with new implant and bone grafting—X ray laterolateral view

on the bone, thanks to a precise representation of the bone surface. Each osteotomy is performed using a template as planned. This technology does not require the intraoperative need for a specific navigation device [15].

- A recent addition to orthopedic intraoperative imaging methods is CT-like imaging using an isocentric 3D fluoroscope. New type of fluoroscope is a motorized C-arc unit that captures up to 100 2D fluoroscopic sequences using a 180° C-arc rotation around the iron region. This system provides intraoperative CT-like images without the need for a registration process [16].
- Fracture repositioning is considered a key step in the treatment of bone injuries and it requires widespread use of fluoroscopy. One solution to navigating during repositioning is to insert a tracking device into one bone fragment and to monitor another fragment using a bone tracking device to reposition the fragments [17].

In parallel, 2D virtual fluoroscopy can be used in the following clinical situations:

- Fixation of some simple and non dislocated acetabular fractures. This technique can assist the surgeon in minimizing the surgical dissection associated with the formal approach. A classic example is the placement of the anterior column screw with the transverse component of the acetabular fracture with a posterior approach.
- Removal of pelvic fixation screws is indicated in some cases. Fluoroscopically based navigation can be used to guide the surgeon toward the implant, thereby reducing the risk of soft tissue damage and reducing pelvic radiation.

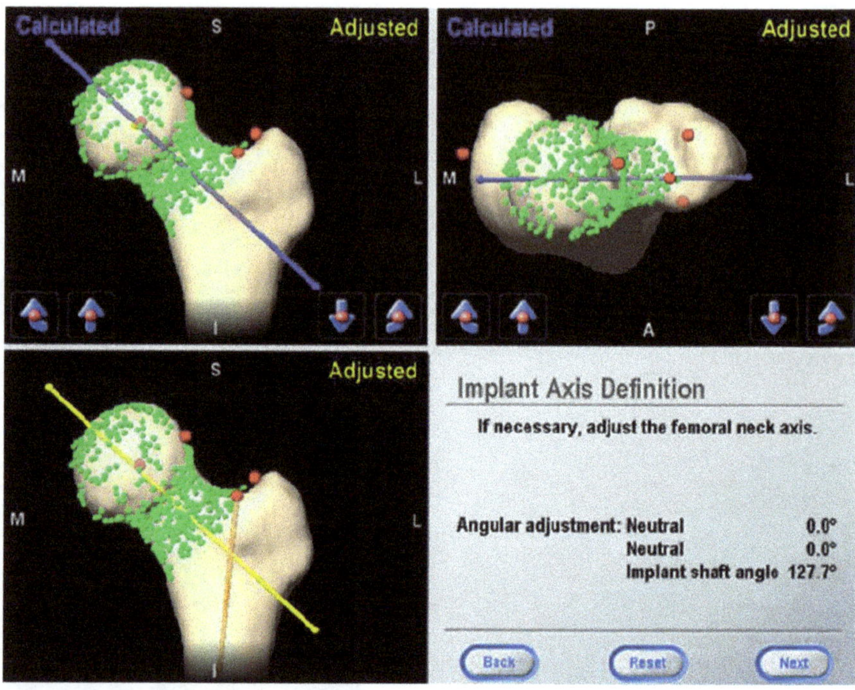

Fig. 11.8 CAS technology of the proximal femoral fracture (from Pitto RP, Malak S, Anderson IA (2009) Accuracy of a computer-assisted navigation system in resurfacing hip arthroplasty. International Orthopaedics (SICOT) (2009) 33:391–395 DOI https://doi.org/10.1007/s00264-008-0644-6. Open access. With authors' permission)

- Hip fracture fixation is generally performed using cannulated screws with conventional fluoroscopy. Certain bolt expansion and configurations are preferred to prevent complications such as collapse. Navigation can be used to achieve the desired spread. A study comparing 2D virtual fluoroscopy and conventional fluoroscopy has shown superior screw positioning in terms of parallelism and propagation with fewer overall complications. Paired with navigation 3D virtual fluoroscopy provides axial and spatial views without the need for registration. Percutaneous fixation of non-dislocated and temporarily reduced acetabulum fractures (anterior column, acetabulum roof) can be done using these devices. Intraoperative assessment of screw position is also possible. The current limitations of navigated 3D fluoroscopy are narrow viewing area, radiation dose, and price [18, 19].

At the same time as CAS technology in surgery in the hip joint region has been significantly less successful, over the last decade, computer-assisted total knee arthroplasty has become a well-accepted procedure in orthopedic surgery. This can be explained in many ways: time spent; the added value in relation to conventional neck/acetabulum methods, dislocation rate, and faster functional recovery remains

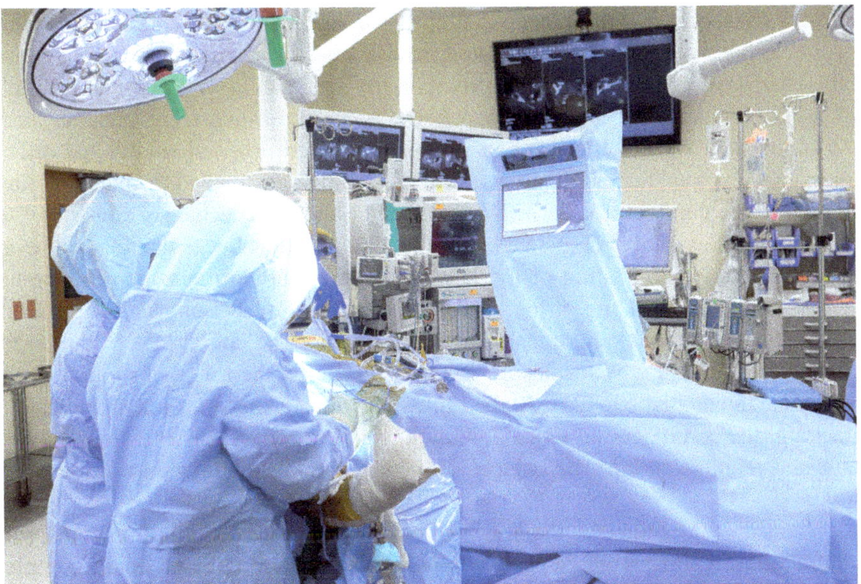

Fig. 11.9 CAS technology usage in the operation room. University of Florida Health surgeon performs first-of-its-kind computer-assisted surgery in U.S. using new technology (Published: Jun 15, 2017 By: Bill Levesque https://m.ufhealth.org/news/2017/uf-health-surgeon-performs-first-its-kind-computer-assisted-surgery-us-using-new. Photo credit to University of Florida Health. With permission)

to be proven; mechanical guides are used by orthopedic surgeons despite their relative imprecision.

In the future, it will be possible to further optimize CAS-total hip arthroplasty collision analysis, by diaphysis ante-torsion measurement, further incorporating diaphysis navigation and implant data. Navigation, however, should only be used as an aid to orientation and not to replace surgeons' lack of knowledge and experience, especially with total hip revision arthroplasty [20].

Finally, it is worth noting that although the time period of rehabilitation after surgery for femur osteosynthesis and/or implantation of the hip joint prosthesis is relatively short, its quality and extensiveness of performance significantly affect the overall treatment outcome, which seeks to restore all previous biomechanical activity values in the hip joint.

11.3 Knee

Knee can be characterized as a complex set of asymmetrical moving parts acting together as a living biologic transmission. The purpose of this system is to accept, transfer, and dissipate loads generated at the ends of the long mechanical lever arms of the femur and tibia (and patella). Knee joint with harmonious relationship of its elements of the femur, tibia and patella, and the tendon-ligamentous apparatus, cartilage and meniscus, with its mobility and stability is a perfect example of joint biomechanics in the human body. Knee movements are complex with different patterns, which can be attributed to different internal (individual differences) and external factors (ligamentous joint stability, tendon and ligamentous elasticity, muscle contractility, etc.). Normal human knee biomechanics consists of an internal rotation of tibia associated with the femur that enhances flexion, following a large posterior translation of the lateral femoral condyle more than the medial femoral condyle. Loading of the knee by the force of quadriceps and of posterior lobe reduces rotational and translational movements. The lateral lobe of the posterior lobe of thigh's muscles has a greater influence on the range of motion of the knee than the medial one [21]. At the same time, during knee movement and patella tracking, the patella and trochlea articulation is biomechanically individually variable and requires an intricate balance of the soft tissue structures that surround the joint.

When there is injury in the knee region, the load on the joint apparatus with contractions of surrounding muscle groups, and joint stability based on the complex ligamentous knee apparatus are important segments in planning overall surgical treatment of a knee joint injury. An important clinical question is how to surgically treat a knee injury and how it affects specific anatomical variation and geometric features, while still allowing the surplus kinematic freedom to adapt the knee joint to external influences. When planning for surgical treatment of bone injury in the region of knee joint, while aiming at postoperative knee functionality, it is inevitable to consider possible complexity of the injury, which, in addition to bone destruction, may include collateral ligaments, tendons, joint capsule, and intra-articular structures of the cruciate ligaments, meniscus and cartilage.

In the case of bone fractures in the knee joint region, clinical and radiological CT diagnostics and 3D reconstruction of individual fragments of the femur, tibia and patella provide information on the internal or extra-articular position and complexity of bone tissue destruction. While knee MRI presents with frequent soft injuries, clinical examination provides insight into the dynamic stability and elasticity of the same structures.

When selecting the type of bone injury treatment in the knee joint region, the most appropriate surgical technique, the most appropriate approach to the bone fracture fragments, the possibility of repositioning, and the type of implant used in the surgical treatment should be determined based on the diagnostics. The goal is to enable achieving the best postoperative and post-rehabilitation outcome of our treatment while meeting the criteria for biomechanical ability to move the bones and joints closest to those before the injury. Today, we use modern surgical tools, a traction table for the dosed traction of the patient's legs, instruments for minimally invasive surgical techniques with radiological guidance, and implants made of modern materials and design, maximally adapted to biomechanical needs. Furthermore, it is equally important to meet the criteria for biological ability of post-operative and post-rehabilitation healing of bone and other tissues, which is not the subject of this article.

It is very important to correctly identify and understand the fracture pattern. Distal femur fractures according to the AO (germ. Arbeitsgemeinschaft für Osteosynthesefragen, eng. Association for the Study of Internal Fixation) classification are divided into those of the extra-articular so-called supracondylar—A type, partially intra-articular—B type and fully intra-articular—C type [22]. Such complex injuries, given the necessity of adhering to biomechanical principles of motility of the adjacent joints of the knee and hip, and the consequent influence on the proper biological course of fracture healing, require selective approach and right indication for certain implant based on clinical and radiological evaluation.

The aforementioned new 3D imaging software enables better planning of surgery. Based on radiological diagnostics and empirically established postulates of various surgical techniques, modern computerized assistance provides an insight into the pre-operative position of bone fracture fragments, repositioning techniques and desired intra-operative post-reposition of fragments, and the most favorable implant position and operative technique.

In realms of minimally invasive plate osteosynthesis (MIPO) technique in surgical treatment, we use locking plates for one or both condyles of femur, depending on the type of fracture, or the retrograde intramedullar nail. Proximal tibial or tibial plateau fractures are challenging to treat from biomechanical point of view because they result in joint incongruity, axial mal-alignment, instability and overload of the knee joint. They are typically caused by a high-energy mechanism where strong deforming forces are involved with high incidence of substantial soft-tissue injury.

According to the AO classification of proximal tibial fractures, Type A are extra-articular, type B are partially articular, and type C are articular with a metaphyseal component [22]. A "Three-collumn Classification" is often pointed out as a new tibial plateau classification system, and it describes fractures involving the medial, lateral or posterior tibial column. It differs from the standard classification systems because it is based upon CT imaging, identifying the fracture in a three-dimensional view, enabling a better representation of the fracture site, unlike the radiograph including only antero-posterior and lateral views [23, 24].

Restoration of the mechanical axis, articular congruity and knee motion without soft-tissue complications is the goal of operative treatment. Both IM nails and plates are appropriate for treating type A and simple intra-articular type C1 fracture patterns. Treatment with a plate or multiple plates fixation could be a treatment of choice for all proximal tibial fractures with anterior surgical approach. When a plate is used, the surgical approach and technique should minimize soft-tissue damage and account for future surgical procedures that may be needed. At the same time, with respect to the staged treatment, an arthroscopically assisted ligamentar, chondral and menisceal reconstruction is recommended. Fractures with intra-articular involvement and/or comminution of the medial metaphyseal region are appropriately treated with dual plating, while extra-articular fractures without major medial comminution may be treated with a locked lateral plate [25]. Posterior surgical approach for multi-fragmentar tibial plateau fractures involving the posterior column enables proper plate fixation and potentially better biomechanical stability of the knee joint because of the common avulsed joint capsule and posterior cruciate ligament stabilisation (Figs. 11.10, 11.11, 11.12, 11.13, 11.14, 11.15, 11.16, 11.17 and 11.18).

Combined injuries to multiple knee joint tissues are common, but complete dislocations of the knee joint are rare and may include other neurovascular comorbidities. As we have introduced bone, joint capsule and ligaments congruity as factors of biomechanically proper knee joint stability it is highly important to recognise accompanied cruciate and/or collateral ligaments' injuries. These injuries could lead to the rare knee dislocation associated with increased morbidity. It could be more extensive

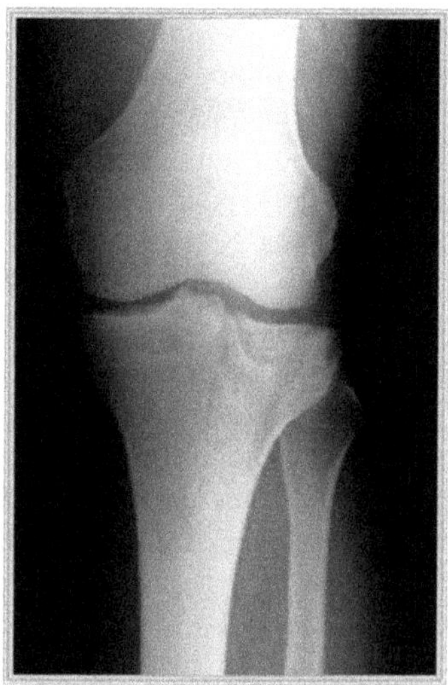

Fig. 11.10 Case of lateral tibial condyle fracture—X ray anteroposterior view

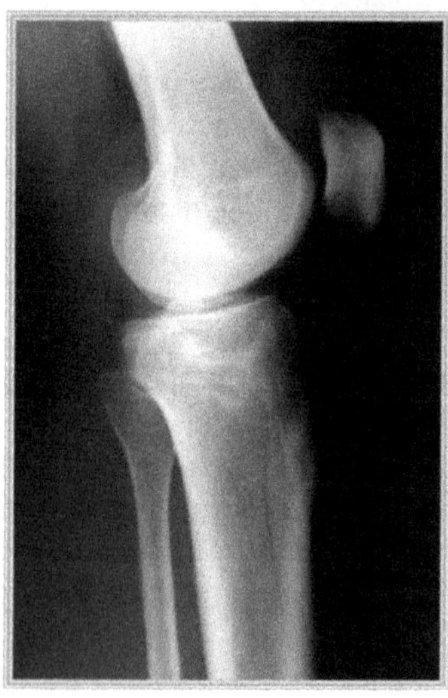

Fig. 11.11 Case of lateral tibial condyle fracture—X ray laterolateral view

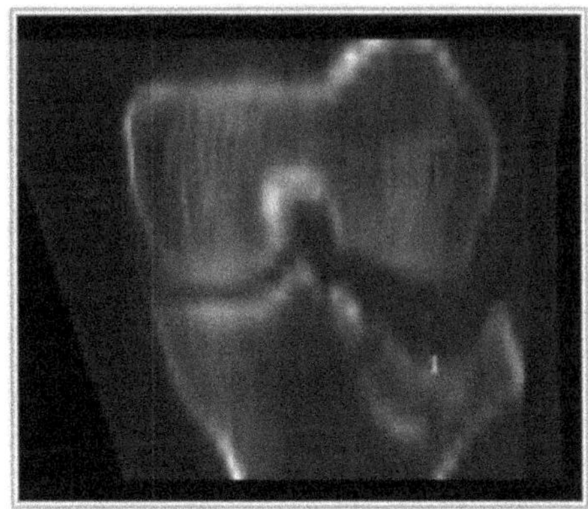

Fig. 11.12 Case of lateral tibial condyle fracture—CT anteroposterior view

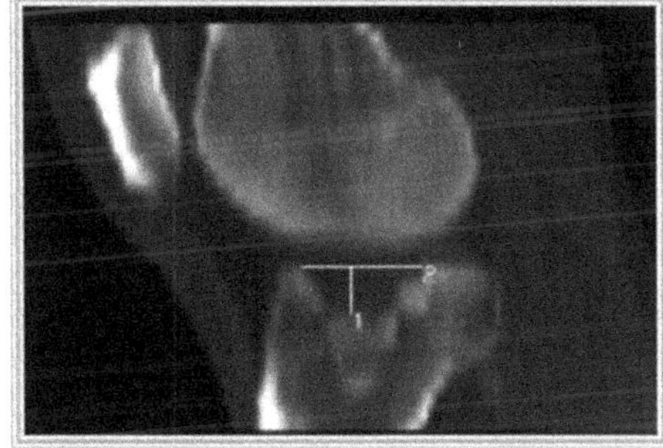

Fig. 11.13 Case of lateral tibial condyle fracture—CT laterolateral view

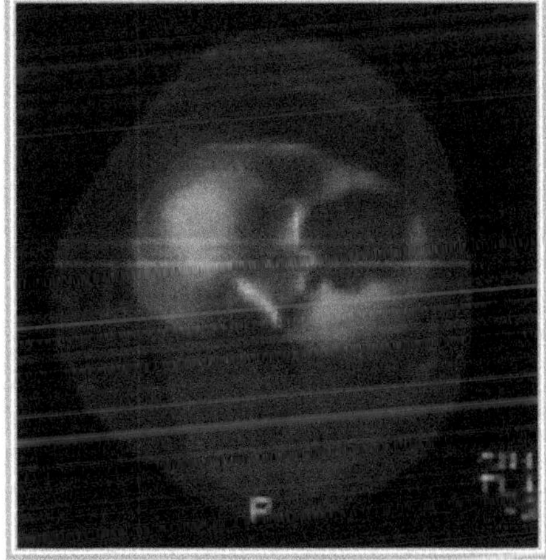

Fig. 11.14 Case of lateral tibial condyle fracture—CT axial view

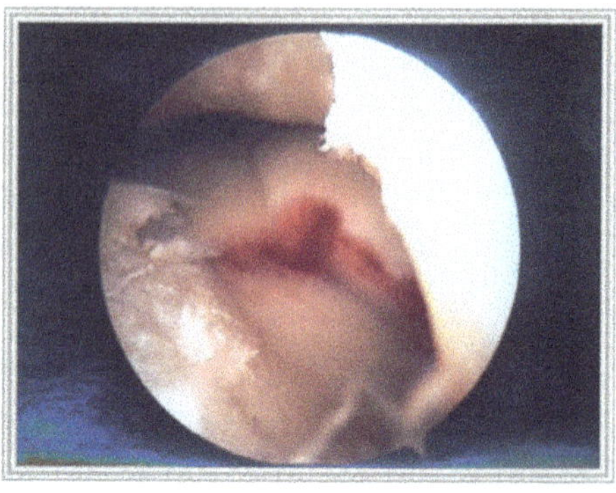

Fig. 11.15 Case of lateral tibial condyle fracture—intraoperative arthroscopic intraarticular view

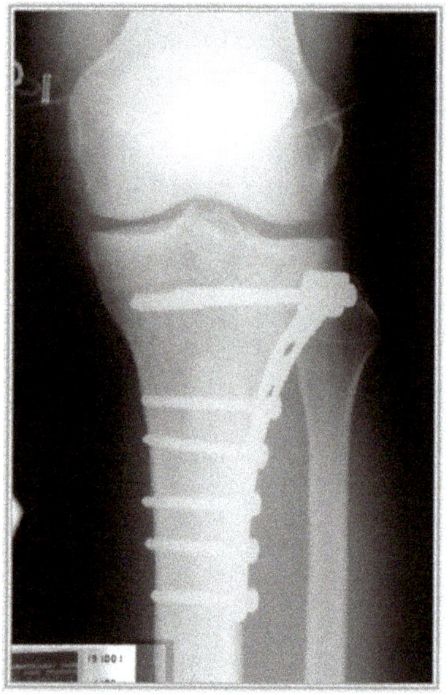

Fig. 11.16 Operative fixation of lateral tibial condyle fracture—X ray anteroposterior view

if it involves bone, ligament, vascular and neural tissue damage. There are several principles in the management of that kind of injury but the repair of bone and parallel or postponed ligaments' injuries is the imperative for full biomechanical recovery of knee joint. Major goal of surgical reconstruction is the best possible anatomical restoration of joint biomechanics that can only be achieved by a complete restoration of the primary and secondary knee stabilizers. To avoid further compromise to soft tissue and perfusion, sometimes a temporary joint and fracture stabilization is required [26].

For fractures with minimal intra-articular extension, fracture fixation with an intramedullary nail provides better axial load sharing than a plate [27]. Using the semi-extended technique, choosing the correct starting portal, incorporating blocking screws or stability screws into the fixation construct, and using mini-open reduction

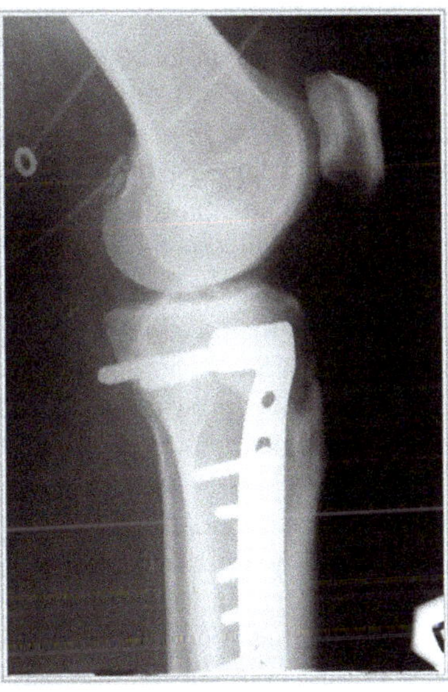

Fig. 11.17 Operative fixation of lateral tibial condyle fracture—X ray laterolateral view

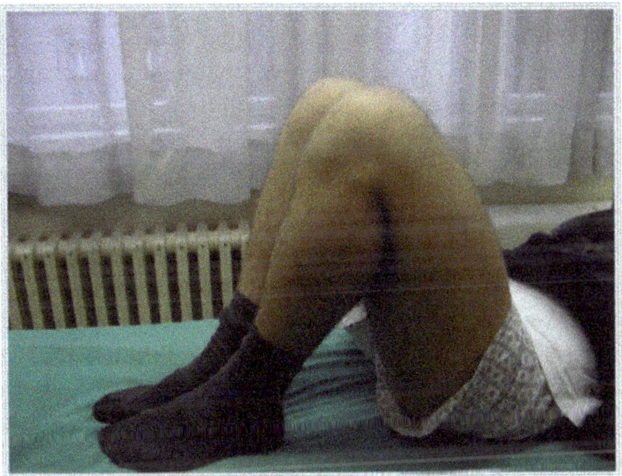

Fig. 11.18 Postoperative correct functional status

and internal fixation of the fracture will help achieve the goals of fracture fixation with an intramedullary (IM) nail.

Treatment of extra-articular proximal tibial fractures is challenging, and it is associated with higher rates of complications when compared with treatment of diaphyseal tibial fractures. Most of these fractures result from high energy, direct trauma that causes extensive injury to the soft tissue and bone; therefore, it is desirable to limit the extent of direct open exposure. A high incidence of wound breakdown, infection, and fixation failure has been observed in association with conventional plating of the proximal tibia. To prevent these complications, irrespective of whether an IM nail or a locking plate is used, employment of minimally invasive techniques has been advocated. Comparing the biomechanical properties of the IM nail versus the locking plate for treatment of a comminute fracture of the proximal tibia, with respect to the concepts of minimally invasive surgery and enhanced fracture healing, IM nail is an ideal load sharing implant, and its bending stiffness is superior to that of the locking plate. Improvements in oblique interlocking screws used in treatment of fractures of the proximal metaphysis leaded to greater stability of modern tibial nails compared to that of old nails with two transverse interlocking screws. From the biomechanical viewpoint, IM may be the ideal option for treating extra-articular proximal tibial fractures. However, a higher incidence of malreduction with malalignment resulting in non-union has been reported in treatment of fractures with a short proximal end segment, compared with locked plating. Soft tissue attachment of tendons or muscles could cause malreduction, also [28].

Early motion and weight-bearing are helpful for rehabilitation; therefore, maintaining the reduction of proximal tibial fractures is important. However, due to its lesser ability to tolerate axial load, loss of reduction after LP is not uncommon during the healing period. Many surgeons prefer to delay weight-bearing for a few months in patients treated with LP.

There are multiple loading forces on the knee joint, including bending, torsional, axial, and shear load. The axial load is the most important and it causes fixation failure in a clinical situation, since main forces during gait are applied in the axial direction [29, 30]. The Dual Plating (DP) construct with a medial locked plate may be the most favorable for use in treatment of fractures of the proximal tibia that need plating. During treatment of bi-condylar fractures of the proximal tibia, no differences were observed in the stiffness of the dual plating (DP) construct when compared to that of the lateral fixed angle plate. However, DP was found to be the least soft tissue-friendly method when used during an open approach and therefore it is not favorable to use it in clinical situations. Also, minimally invasive plate osteosynthesis (MIPO) on the medial side is also gaining in popularity. From the biomechanical point of view, the IM nail was found to be the most stable implant for use in treatment of comminute extra-articular fractures of the proximal tibia. However, when it is difficult to achieve satisfactory reduction by nailing, dual locked plates may be a stronger implant for use in MIPO when compared to lateral locked plating [31, 32].

Patella fracture, and soft tissue-quadriceps and patellar tendon injuries could be inadequately treated, with the biomechanical dysfunction as the most common final consequence. For the most effective treatment of patellar pathology, it is imperative to understand that individually variable patella-trochlear (femoral) transfer requires an intricate balance of the soft tissue structures that surround the joint. Biomechanically unequal pull from one set of structures can cause increased force distribution between the patella and femur leading to the new pathology. During a femoro-tibial motion, patellar tracking is dependent upon the extensibility of the connective tissue about the patella, the active contraction of the *quadriceps* and the geometry of the trochlear groove and patella. As a gliding joint, the patella has motions in multiple planes: superior/inferior glide, medial and lateral glide, medial and lateral tilt, and medial and lateral rotation. During open chain knee motion, besides the lateral-medial–lateral motions the patella glides inferiorly with knee flexion and superiorly with knee extension, following the path of the tibia due to the distal insertion of the patellar tendon to the tibial tubercle. The overall pattern of patellar contact area increases with increasing knee flexion, which serves to distribute joint forces over a greater surface area. In those with normally aligned patello-femoral joints, this distribution of force allows the knee to resist the deleterious effects that could occur from routine exposure to high compressive forces [33–35].

In closed kinetic chain movements, patella is relatively tethered within the quadriceps tendon so as the femur rotates in the transverse plane, it is the femoral surface that glides behind the patella.

High patella-femoral joint stress may be harmful to the patellar fracture healing and recovery. It is the consequence of patella-femoral resultant compression joint reaction force which is dependent on knee joint angle and muscle tension [36].

Operative treatment of patella fractures is frequently associated with implant failure and secondary dislocation which can be attributed to the employed hardware. Biomechanically interesting is a treatment of operative stabilization of transverse fractures of the patella. The treatment techniques used are: (a) standard modified tension band (AO technique); (b) two parallel 4.5-mm inter-fragmentar lag screws; or (c) four-millimetre cannulated lag screws with a tension band wired through the screws. Results of mechanical testing of two different fixation techniques of transverse patella fractures showed combined inter-fragmentar screw fixation with the tension band principle provides improved stability over the standard modified tension band or screws alone [37] (Figs. 11.19 and 11.20).

Lag screws with tension wiring showed significant fracture displacement after the initial cycle during the biomechanical repetitive testing over 100 cycles of the simulated knee motion from 90° flexion to full extension. The bilateral fixed-angle plate showed no significant fracture gap widening In contrast to cannulated lag screws with anterior tension wiring. It stabilizes transverse patella fractures securely and sustainably [38].

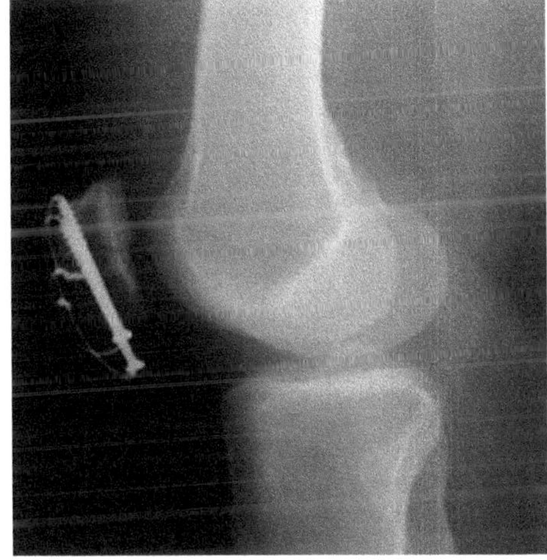

Fig. 11.19 Combined inter-fragmentar screw fixation with the tension band principle—X ray laterolateral view

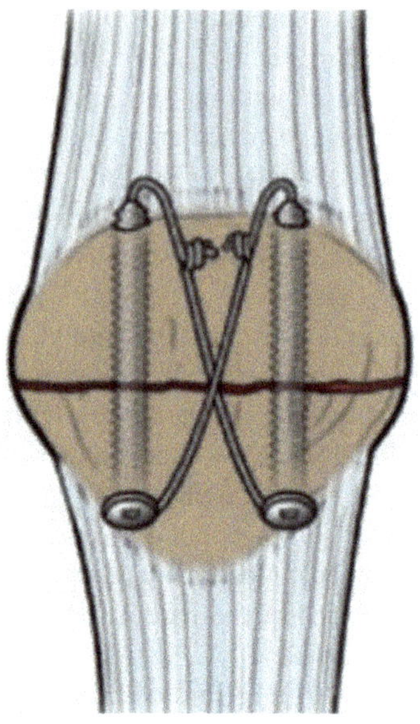

Fig. 11.20 Combined inter-fragmentar screw fixation with the tension band principle—scheme anteroposterior view

11.4 Ankle

Ankle is a transmitter of forces in both directions between knee and hip on the pad in different positions of the body. The normal amplitude of ankle movements of 20° dorsal flexion and 25–35° of plantar flexion is achieved by movements of the tibio-talar and fibulo-talar joints. The distal tibiofibular joint contributes with movements of several degrees (in plantar flexion). Ankle joint kinematics results from morphology and a biometric organization of the joint surfaces and a very specific multi-axial ligament system. During normal body oscillations in its static posture, the ankle bones, then the tibiofibular ligaments and the interosseous membrane take over the body weight, while the lower leg muscles and the Achilles tendon actively participate in their balance. The same anatomical structures participate in gait, by the influence of main compressive forces through the ankle by contractions of the m. gastrocnemius and m. soleus transmitted through the Achilles tendon. In the dynamics of the ankle, m. tibialis anterior produces only 20% of force to exert weight on the body at an early stage of support [1]. During gait cycle, at the beginning of standing phase, at heel strike, the foot slows down and with the initial shock wave the heel touches the ground. Magnitude of load varies with speed of gait. Initial vertical load is about 80% of body weight. The next peak takes place at mid-cycle, exceeding body weight by approximately 10%. Slow walking reduces the load, while a fast walking increases it [39].

In the injured ankle, maintaining balance and the gait cycle are disrupted as all forces occur earlier than in healthy walking. In case of mechanical stress, anatomical or other deviations, overload etc. in the ankle, its function will change very quickly. In addition to possible ligamentous lesions and fracture of the talus bone, the greatest concern for the orthopedic traumatologist are fractures of the lower parts of tibia and fibula. Therefore, insight into the status of all these structures is an essential imperative for clinical and radiological (RTG, CT, MR) diagnostics in the planning of surgical treatment of ankle injury.

Distal tibial fracture in ankle region is a devastating injury and its treatment options and outcomes depend on the proper planning and timing of surgical treatment. Usually patients suffer from that kind of injury upon the axial loading [40] and direct high energy trauma or rotation with lower energy. Involvement of an articular surface with severe comminution and soft tissue injuries is not rare. According to the AO classification beside A nad B types, the most complicate for surgical treatment are type C, a so called Pilon or Plafond fractures: C1—articular simple, metaphyseal simple, C2—articular simple, metaphyseal complex, C3—articular and metaphyseal complex [41].

Surgical techniques are aimed to re-establish the predispositions of the normal anatomical shape and the relationship of the ankle joint, which will allow normal statics of the ankle and the dynamics of gait cycle. The degree of destruction of bone and surrounding ligamentous-tendon apparatus of the ankle will dictate the success of the results of the surgical treatment and the subsequent instability of the ankle joint. Of course, a lack of post-traumatic functional ankle stability may be based on a lack of proprioception and muscular and postural (valgus, varus etc.) instability. Poor functional results upon the multiple reoperations are common in the treatment of pilon fractures and even foot amputation or arthrodesis are its possible final result. Timing of surgery is determined by the condition of the severely injured soft tissues that usually dictates the choice of procedure: single or multiple stage procedure. Single stage procedure for minimally displaced pilon fractures with closed soft tissue injury include four classical principles: reconstruction of the fibula, reconstruction of the tibial joint surface, autogenous cancellous or corticocancellous bone grafting (from iliac crest), and a fixation with implants (Figs. 11.21, 11.22, 11.23, 11.24, 11.25, 11.26, 11.27, 11.28, 11.29 and 11.30).

Multiple stage procedures are reserved for grossly displaced fracture and/or fracture with severe closed tissue injury. Closed reduction and joint bridging external fixator present Stage 1 procedure, while a timing for a definitive reconstruction of fracture with the same principles as the single stage procedure is in the period of 7 to 21 days until the soft tissues recover and presents second stage of procedure. Different techniques and implants for the fracture's fixation are used in the surgical treatment of pilon fractures: (a) ExFix—conventional, articulated, fine wire Ilizarov or Hybrid—Ilizarov distally, conventional fixator proximally (b) ORIF with the minimally invasive articular reconstruction, (c) combination of both techniques. Imperative for the most successfull rehabilitation process upon the pilon's surgery should respect soft tissues and minimize complications. Early ROM with the staged

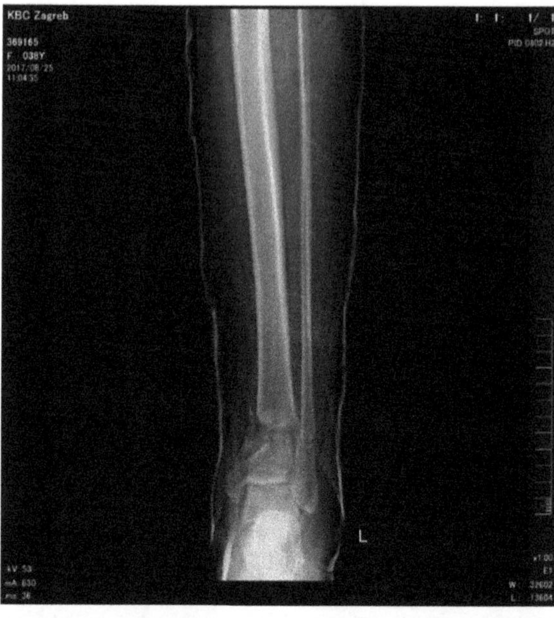

Fig. 11.21 Pilon fracture of distal tibia—X ray anteroposterior view

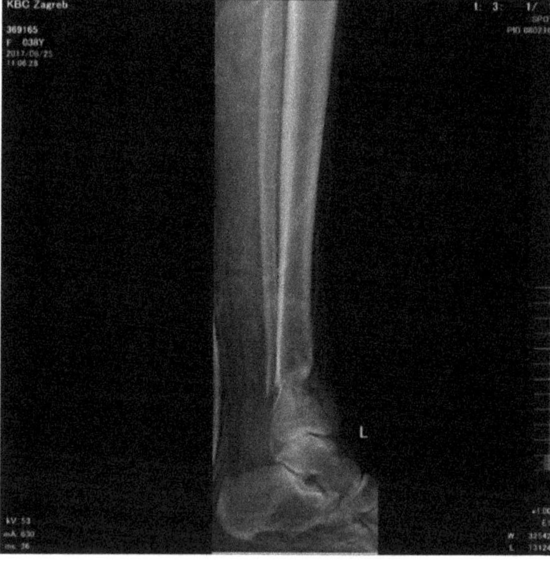

Fig. 11.22 Pilon fracture of distal tibia—X ray laterolateral view

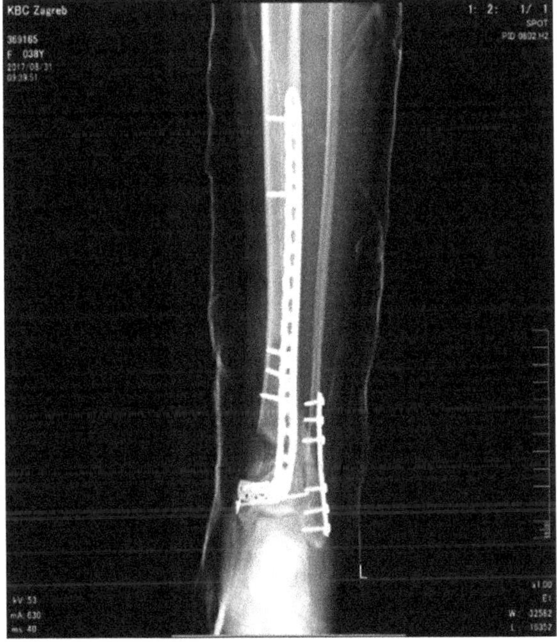

Fig. 11.23 Postoperative X ray anteroposterior view after fixation of the pilon fracture with the minimally invasive articular reconstruction during the single stage procedure

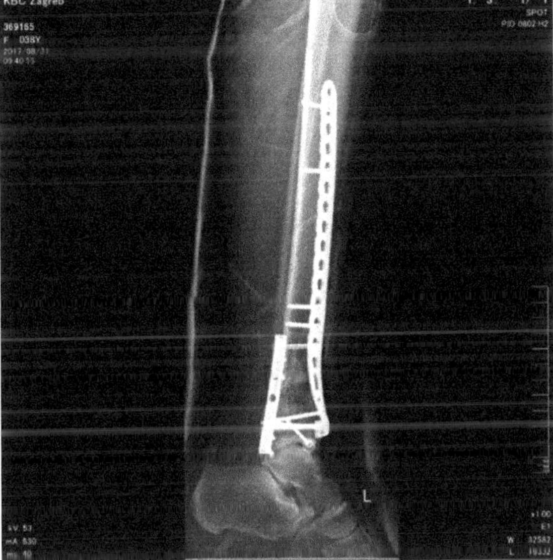

Fig. 11.24 Postoperative X ray laterolateral view after fixation with the minimally invasive articular reconstruction during the single stage procedure

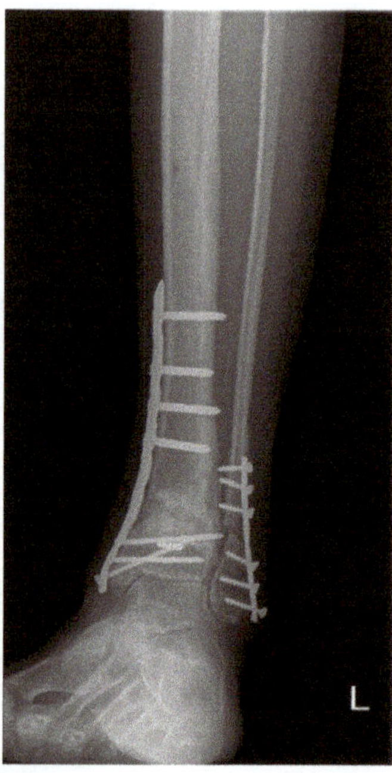

Fig. 11.25 Non union of fibular fracture—X ray anteroposterior view

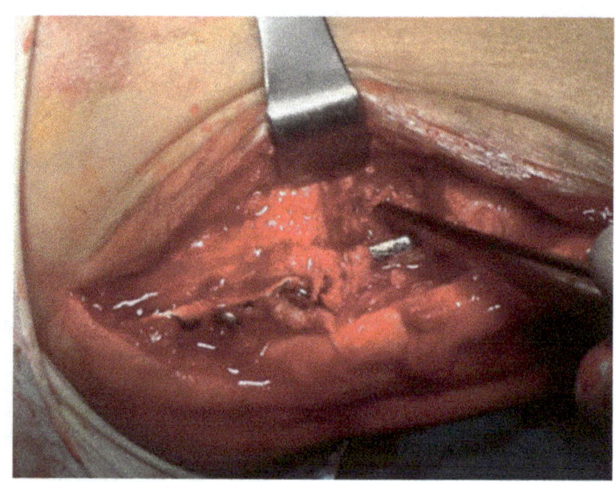

Fig. 11.26 Bone defect upon the resection of fibular non union part

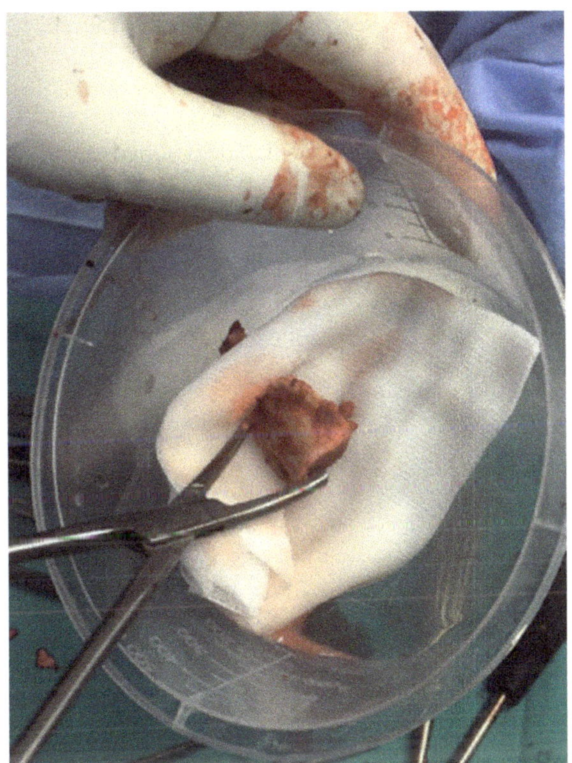

Fig. 11.27 Autogenous corticocancellous bone graft from iliac crest

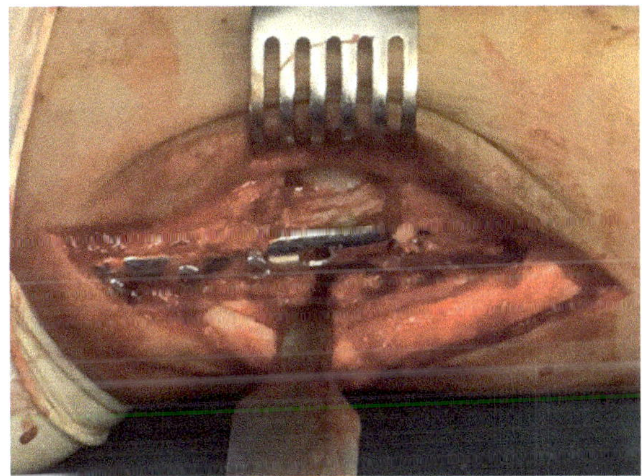

Fig. 11.28 Fibular refixation upon the bone grafting

protocol is highly recommended to restore the ankle 's biomechanics and qualitative walking cycle.

If the anatomical and functional criteria for a favourable postoperative course of rehabilitation cannot be met with surgery as a result of treatment, the instability or contracture of the ankle are left behind, and other surgical and conservative methods are needed to correct it.

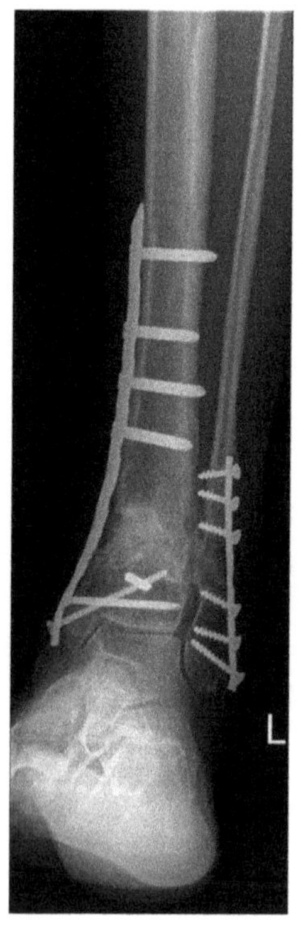

Fig. 11.29 X ray anteroposterior view, 1 year after the bone grafting and fibular refixation, final correct fracture healing

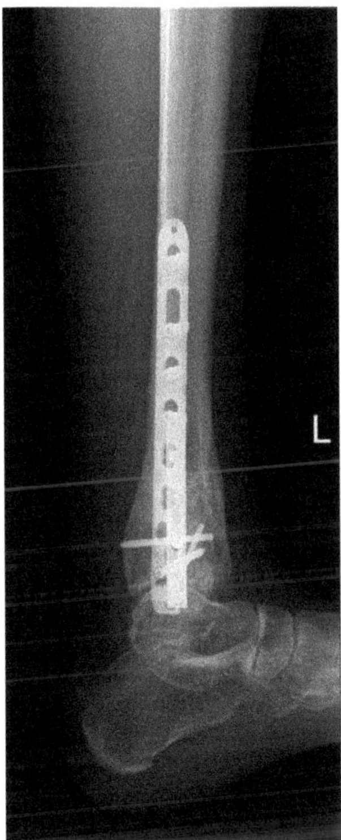

Fig. 11.30 X ray laterolateral view, 1 year after the bone grafting and fibular refixation, final correct fracture healing

11.5 Conclusion

Application of biomechanics in orthopedics and traumatology is impossible to avoid and we find it in every scientific analysis or evaluation of achievements in the prevention and treatment of injuries or diseases of the locomotor system. Only with an interdisciplinary approach can individual medical problems of a respective pathology be properly addressed.

Knowledge of individual biomechanics of the hip, knee, ankle and associated bones, muscles, tendons and ligaments, as well as joint action of these anatomical structures during the movement of the leg give the orthopedic traumatologist the necessary knowledge for planning and deciding on various details of surgery and its performance.

Not complying with biomechanical principles in hip joint function may adversely affect post-operative rehabilitation and the result of overall treatment. Therefore, nowadays, the computer-aided surgery (CAS) technology is used ever more. The most demanding trauma in the knee region is multifragmental proximal tibial fracture combined with ligamentar, chondral and meniscal injury. From the biomechanical point of view the IM nail was found to be the most stable implant for fixation of proximal extraarticular tibial fracture, while the dual locked plates may be a stronger implant for use in MIPO of intraarticular bicondylar tibial fracture with arthroscopically assisted ligamentar, chondral and menisceal reconstruction during the staged treatment.

Although the functional results upon the multiple reoperations of ankle pylon fracture are often poor, surgeons should choose the best kind of treatment: single or multiple stage procedure determined by the condition of the severely injured soft tissues.

It is important that each orthopedic traumatologist, in his/her education, acquires knowledge of the biomechanics of the locomotor system so that he/she can approach each case objectively from a biomechanical point of view. In this way many mistakes and non-optimal solutions in the treatment of injuries or diseases of the locomotor system can be avoided.

References

1. Wolff J (1986) The law of bone remodeling. Springer, Berlin
2. Huston RL (2008) Principles of biomechanics. CRC Press Taylor and Francis Group, Boca Raton
3. Hankenson KD, Dishowitz M, Gray C, Schenker M (2011) Angiogenesis in bone regeneration. Injury 42(6):556–561
4. Dwek JR (2010) The periosteum: what is it, where is it, and what mimics it in its absence? Skeletal Radiol 39(4):319–323
5. Byrne DP, Mulhall KJ, Baker JF (2010) Anatomy & biomechanics of the hip. Open Sports Med J 4:51–57
6. Skedros J, Baucom S (2007) Mathematical analysis of trabecular 'trajectories' in apparent trajectorial structures: the unfortunate historical emphasis on the human proximal femur. J Theor Biol 244:15–45
7. Ridzwan MIZ et al (2007) Problem of stress shielding and improvement to the hip implant designs: a review. J Med Sci 7(3):460–546
8. Kazley JM (2018) Classifications in brief: Garden classification of femoral neck fractures. Clin Orthop Relat Res 476(2):441–445
9. Caviglia HA, Osorio PQ, Comando D (2002) Classification and diagnosis of intracapsular fractures of the proximal femur. Clin Orthop Relat Res 399:17–27
10. Cartiaux O et al (2010) Computer assisted and robot assisted technologies to improve bone cutting accuracy when integrated with a freehand process using an osscilating. J Bone Joint Surg Am 92(11):2076–2082
11. Wood M, Mannion RJ (2011) A comparison of CT-based navigation techniques for minimally invasive lumbar pedicle screw placement. Spinal Disord 24(1):E1-5
12. Merloz H (2008) Computer assisted hip surgery by using modular femoral neck component: principles, technique, advantages and limits. Interact Surg 3:139–147
13. Murphy SB (2008) Comparison of experiences with CT and fluoroscopy-based surgical navigation for total hip arthroplasty. In: Langlotz F, Davies BL, Schlenzka D (eds) Computer assisted orthopaedic surgery. Pro Business. Berlin, pp 334–335
14. Muirhead-Allwood S, Sandiford NA, Skinner JA, Hua J, Muirhead W, Kabir C, Walker PS (2010) Uncemented computer-assisted design-computer-assisted manufacture femoral components in revision total hip replacement: a minimum follow-up of ten years. J Bone Joint Surg Br 92(10):1370–1375
15. Cartiaux O et al (2010) Computer assisted and robot assisted technologies to improve bone cutting accuracy when integrated with a freehand process using an osscilating. J Bone Joint Surg Am 92(11):2076–2082
16. Fleute M, Lavallée S, Julliard R (1999) Incorporating a statistically based shape model into a system for computer-assisted anterior cruciate ligament surgery. Med Image Anal 3(3):209–222

17. Weil YA, Liebergall M, Mosheiff R, Khoury A (2018) Fluoroscopic based navigation in orthopaedic trauma—a review of a large center's experience. Harefuah Mar 157(3):145–148 Hebrew
18. Crowl AC, Kahler DM (2002) Closed reduction and percutaneous fixation of anterior column acetabular fractures. Comput Aided Surg 7(3):169–178
19. Liebergall M, Ben-David D, Weil Y, Peyser A, Mosheiff R (2006) Computerized navigation for the internal fixation of femoral neck fractures. J Bone Joint Surg 88(8):1748–1754
20. Kahler DM (2017) Computer-assisted surgery: the use of stored intraoperative images for accurate restoration of femoral length and rotational alignment after fracture. Injury 48(Suppl 1):S35–S40
21. Victor J (2010) A comparative study on the biomechanics of the native human knee joint and total knee arthroplasty. Katholieke Univesiteit Leuven
22. Wagner M, Frigg R (2011) AO manual of fracture management: internal fixators: concepts and cases using LCP/LISS (AO Manual of Fracture Management Series) Thieme/AO. 2011
23. Prat-Fabregat S, Camacho-Carrasco P (2016) Treatment strategy for tibial plateau fractures: an update. EFORT Open Rev 1:225–232
24. Luo CF, Sun H, Zhang B, Zeng BF (2010) Three-column fixation for complex tibial plateau fractures. J Orthop Trauma 24(11):683–692
25. Kurylo JC, Tornetta P 3rd (2013) Extra-articular proximal tibial fractures: nail or plate? Instr Course Lect 62:61–77
26. Darabos N, Gusic N, Vlahovic T, Darabos A, Popovic I, Vlahovic I (2013) Staged management of knee dislocation in polytrauma injured patients. Injury 44(3):40–45
27. Perren SM (2002) Evolution of the internal fixation of long bone fractures. The scientific basis of biological internal fixation: choosing a new balance between stability and biology. J Bone Joint Surg Br 84(8):1093–110
28. Lee SM et al (2014) Biomechanical analysis of operative methods in the treatment of extra-articular fracture of the proximal tibia. Clin Orthop Surg 6(3):312–317
29. Mueller KL, Karunakar MA, Frankenburg EP, Scott DS (2003) Bicondylar tibial plateau fractures: a biomechanical study. Clin Orthop Relat Res 412:189–195
30. D'Lima DD, Patil S, Steklov N, Slamin JE, Colwell CW Jr (2006) Tibial forces measured in vivo after total knee arthroplasty. J Arthroplasty 21(2):255–262
31. Mueller CA, Eingartner C, Schreitmueller E et al (2005) Primary stability of various forms of osteosynthesis in the treatment of fractures of the proximal tibia. J Bone Joint Surg Br 87(3):426–432
32. Oh CW, Oh JK, Kyung HS et al (2006) Double plating of unstable proximal tibial fractures using minimally invasive percutaneous osteosynthesis technique. Acta Orthop 77(3):524–530
33. Insall J, Salvati E (1971) Patella position in the normal knee joint. Radiology 101(1):101–104
34. Powers CM, Lilley JC, Lee TQ (1998) The effects of axial and multi-plane loading of the extensor mechanism on the patellofemoral joint. Clin Biomech Bristol Avon 13(8):616–624
35. Feller JA, Amis AA, Andrish JT, Arendt EA, Erasmus PJ, Powers CM (2007) Surgical biomechanics of the patellofemoral joint. Arthroscopy 23(5):542–553
36. Powers CM, Ward SR, Fredericson M, Guillet M, Shellock FG (2003) Patellofemoral kinematics during weight-bearing and non-weight-bearing knee extension in persons with lateral subluxation of the patella: A preliminary study. J Orthop Sports Phys Ther 33(11):677–685
37. Carpenter JE, Kasman RA, Patel N, Lee ML, Goldstein SA (1997) Biomechanical evaluation of current patella fracture fixation techniques. J Orthop Trauma 11(5):351–356
38. Thelen S, Schneppendahl J, Jopen E, Eichler C, Koebke J, Schönau E, Hakimi M, Windolf J, Wild M (2012) Biomechanical cadaver testing of a fixed-angle plate in comparison to tension wiring and screw fixation in transverse patella fractures. Injury 43(8):1290–1295
39. Brockett CL, Graham J (2016) Chapman biomechanics of the ankle. Orthop Trauma 30(3):232–238
40. Saad BN, Yingling JM, Liporace FA, Yoon RS (2019) Pilon fractures: challenges and solutions. Orthop Res Rev 11:149–157
41. Mak KH, Chan KM, Leung PC (1985) Ankle fracture treated with the AO principle—an experience with 116 cases. Injury 16(4):265–272

Chapter 12
Studying Sportive Movement Patterns: Selected Examples

Igor Gruić

Abstract Cross section of biomechanical, predominantly kinematics-oriented, research in the 'Zagreb kinesiology circle' is often categorized within division related to: monostructural, polystructural, complex, and aesthetically-conventional kinesiological activities. Movement patterns in sporting activities are represented through reductionist-constructivist dualism. Reductionist sports, colloquially called 'gram-meter-second (GMS) sports', are essentially reduced to the goal of achieving results through international standards of measurement and measures, e.g. the SI system. Movement analysis is mostly performed in controlled space (either closed, indoors, such as in a motion analysis laboratory, or outdoors) and with technologies for 'extended' sensory-motor circuits, and motion control in general. Examples of javelin throw and sprinting discipline are presented. Constructivist sports, in addition to the physical manifestation of the individual movement, include additional aesthetic component, cooperation, opposition, or some achievement of an externally ruled goal. Movement analysis should imply both GMS dimension of movement and psychosocial and systemic elements, such as tasks, tactics, strategy, momentum, etc. Examples of tennis serve and team handball jump shot are presented.

12.1 Introduction

The evolution of human needs—from involuntary compulsive inherent instincts to civilization, culture and art—has influenced the formation of human body morphology, functional-motor patterns, psychosocial adaptations of movement in space and time. Nomadic and tribal patterns of movement and dormancy have been replaced by *urban jungle* frames, edges, algorithms and alternative energy and means of movement. Existential hunting and fighting skills have been replaced by activities aimed at symbolic conquest, construction or destruction/dissolvement—sport and occupational practices, in narrowest meaning. Methods for treatment of injuries after hunting have been improved and standardized to medical extremes.

I. Gruić (✉)
Faculty of Kinesiology, University of Zagreb, Horvaćanski zavoj 15, 10000 Zagreb, Croatia
e-mail: igor.gruic@kif.hr

V. Medved (ed.), *Measurement and Analysis of Human Locomotion*,
Series in Biomedical Engineering, https://doi.org/10.1007/978-3-030-79685-3_12

In order to be able to quantify this transition, it is necessary to understand *form, content and function* of movement. Narrow and exact example of specific movement (opposite to broad perspective), understandable to practitioner in sportive and medical occupations, would more or less relate to definition of *a skill*. Numerous sources of variability operate intertwined with exact, individual performance—beyond (epi-) genetics, compulsivity of thirst and hunger, psychosocial and economic goals, treathening or challenging surroundings. All possible variations operate within physical laws of movement. However, cascades of exertion of power, mechanical momentum and skill affect (and has effect within) different criteria and impact. All variations and properties affected e.g. by size are described with numerous concepts—"dimensional analysis" (limitations), "relative growth" (i.e. allometry), on the other hand e.g. with media for gross body movement, furthermore with relations to animal world (swimming, running and flying speed and length in animals) and so on (many synthetic and plastic introductions presented in [1]). Within all concepts, rooted by mathematical, physical, biochemical etc. rules and laws, it is obvious that evolution of human locomotion transits and supports for (most) activities performed in the *open* natural space to "flew" into closed *controlled*, *automated* patterns and spaces—that aim to allow directing and expression of the physical instinctive human nature into conditions which are *safe* both for humans and the environment. From a medical perspective, this refers more to the *compensation of disadvantages*, and from a sportive perspective, to the attainment of *higher dimensions* of the relationship between the physical and mental capacities of human nature.

Depending on the structure of need, medical or sportive, or both at the same time, the rough structure of activities aimed at achieving different goals categorizes all movements mainly into domains of: (1) recreation, (2) kinesitherapy, (3) sports, and (4) physical and health culture in and out of the framework of regular education during formal schooling.

Basic muscular, mechanical analysis of a skill (presented in e.g. [2]), i.e. application of mechanical laws and principles to a skill, should be introduced trough cross section with application of kinesiology to basic performance patterns. Furthermore, above biotic patterns (i.e. abilities and skills important for survival and everyday activities), there is a distinction between and conjugation of motor performance and motor learning ('The Chicken and the Egg' of the movement and motor skill [3]). Moreover, if performance&learning dualism is portrayed with respect to laws and alghoritms of biological apparatus (i.e. body), and than superimposed on technological system (e.g. humanoid robot)—vast field of potential improvements and upgrades opens to researcher, technologist, consumer, even society and ecology of automated and random processes. Further developments in understanding totality of intertwinded layers of human locomotion are related to (ibid. [3])—movement production, movement programs, motor control, then—discrete, serial, continuous skills, dualities such as capabilities and abilities within morphology variable, limitations in information-processing capacity, etc.

More or less, all ideas, definitions, forms and explanations converge into mechanisms of *control*, internal or external, more or less automated.

Previous chapters have predominantly covered the *general* health and medical aspects of gross body movement and locomotion analysis (measurement, diagnostics, prevention and/or rehabilitation of injuries, orthopedics, physical therapy, etc.), but in the present Chapter, the framework has been narrowed primarily to sports, sports science and supporting technologies. Sports can be divided and categorized by several criteria. For the purpose of producing a sort of cross section of research within the 'Zagreb kinesiology circle', the clearest picture may be obtained through the division into: (1) monostructural, (2) polystructural, (3) complex, and (4) aesthetically-conventional kinesiological activities (presented in [4]). To make it even easier for reader, introduction of the 'anatomy' of movement patterns in sporting activities, all sports (regardless of Olympic or other status) can be represented through *reductionist-constructivist dualism*, commonly used for the purposes of structuring science (e.g. in [5]).

In *constructivist* sport, with respect to reductive background/basis, advantage gained by 'winner' may be explained trough *cascade*: (1) ability of instantaneous situational (even vicariant) recognition of desired skill (when/where best performance should be 'immerged' into, and will be a result of, convergence of both all possible developments/extrapolations of situations in sorroundings and *catalogue* of abilities to overcome problem/challenge, (2) exertion of (muscular) power through 'open' and flexible engrams (neural patterns/matrices)—(3) production of maximal/optimal (rotatory/translatory) momentum —(4) control and transfer of the momentum into desired skill, and finally—(5) transfer of exerted momentum/skill into most efficacious tactical solution, i.e. 'fine tunning' of the performance (often includes skills for providing mechanical impetus to objects/projectiles).

Fine tunning in *reductive* sport reduces athlete towards SI (or equivalent) measurement system of recognizing Human within Nature—mostly directed towards '*automation*' of one (often complex) pattern within the discipline. On the other side, fine tunning in constructivist sport positions Human within Nature as a measure of Nature within the rules of (selected) Game—mostly directed towards '*diversification*' of the human nature.

12.2 Reductionist Sports

Colloquially called '*gram-meter-second (GMS) sports*', those are essentially reduced to the goal of achieving results through international standards of measurement and measures, e.g. the SI system. This approach greatly simplifies measurement and final motion analysis because it 'boils down' (through rather sophisticated cascade) to classic Newtonian mechanics. Therefore, the very location of the competition through mostly individual activities allows one to experience the same as a kind of laboratory mission in which measurements and treatments can be changed and supplemented during training, therapy, competition, etc. Movement analysis in this context can then be performed in closed and controlled space (whether indoors or outdoors) and with technologies for 'extended' sensory-motor circuits and motion control in general (examples of javelin throw and 100 m sprint to be presented in 12.4).

Within the analysis of movements, in predominantly monostructural cyclic and acyclic activities (athletics of running and throwing, swimming, etc.), and partly within polystructural activities, the spatial displacements of the body (and possibly missiles), angles, angular velocities and accelerations and comparative analyzes relative to the geometrical, Cartesian coordinatesystem system as a frame of reference, are predominantly measured. Kinematic parameters are often further correlated with latent and manifest features of functional-motor status, morphology, physiological and psychosocial traits, and motor knowledge in general.

12.3 Constructivist Sports

Constructivist sports are activities that, in addition to the 'GMS' physical manifestation of the individual movement, include additional aesthetic component, cooperation, opposition, or some achievement of an externally ruled goal. Within these activities, the analysis of movement should imply both an approach from a reductionist perspective (GMS dimension of movement), together with the understanding and registration of different psychosocial and systemic elements, such as tasks, tactics, strategy, mechanical momentum, etc. Without the former, it is not possible to have the power to understand, analyze and interpret real activities in their relation to success. The level of utilization of the potentials measured by Newtonian mechanics is not a sufficient guarantee of success. This is precisely the reason for the different, often smaller, interest in classical biomechanical analysis of the technique of performance of particular specific-situational elements.

In constructivist activities, with polystructural (martial arts, etc.), aesthetically-conventional (gymnastics, rhythmic gymnastics, dance, etc.) and complex (handball, football, volleyball, basketball, water polo) elements, analysis of spatial movements, angles, angular velocities and accelerations, are complemented by notational analysis as well with all the accompanying comparative analyzes which characterize analysis of reductionist activities. Examples of tennis serve and handball jump shot are presented in following text.

12.4 Examples: Reductive and Constructive

(1) For the purposes of historical overview of methodology (without regard to findings of the research) kinematic analysis of the javelin ejection phase is presented (citation from [6]): "The recording was done with two VHS cameras, at a rate of 60/s, set so that the footage allowed transformation into a three-dimensional space. Among the 12 successful attempts, the best throw (75.5 m) was further analyzed. The resulting recordings were digitized using APAS (Ariel Performance Analysis System; it is a 3-D high speed video motion analysis system developed for conducting biomechanical measurement). The 18-point coordinates that define the 14 segments of the human body, along with the 3 points that define the spear, were digitized for each frame. Transformation into 3D space was performed using DLT (Direct Linear Transformation) algorithm

(Reference [9] in Chap. 5, editor comment). The digitized 3D coordinates of the segments of the body and spear parts were smoothed out by a natural Cube spline (cubic spline, editor comment) function. The smoothed coordinates were used to calculate the various kinematic parameters needed to characterize the best spear throwers at the 1992 Barcelona Olympics.". "Since the length of the shot is primarily conditioned by the speed and angle of the spear and the height of the grip during the ejection phase, these parameters were further analyzed".

(2) Further, Čoh et al. [7] investigated maximal sprint velocity of the World's fastest 100 m sprinter, Usain Bolt. Two high-speed video cameras recorded kinematics from 60 to 90 m during the men 100 m final at the IAAF World Challenge Zagreb 2011, Croatia (IAAF goes for International Association of Athletics Federations; in 2019 the name has changed to World Athletics). Despite a relatively slow reaction time (194 ms), Bolt won in 9.85 s (mean velocity: 10.15 m/s). His fastest 20-m section velocity was 12.14 m/s, reached between 70 and 90 m, by 2.70-m long strides and 4.36 strides/s frequency. At the maximal velocity, his contact and flight times were 86 and 145 ms, respectively, and vertical ground reaction force generated equalled 4.2 times his body weight (3932 N). The braking and propulsion phases represented 37 and 63% of ground contact, respectively, with body's center of mass (CoM) exhibiting minor reductions in horizontal velocity (2.7%) and minimal vertical displacement (4.9 cm). Bolt's maximal sprint velocity and international predominance emerged from coordinated motor abilities, power generation capacities, and effective technique. This study confirms that the athlete's maximal velocity was achieved by means of relatively long strides, minimal braking phase, high vertical ground reaction force, and minimal vertical displacement of CoM. According to authors, this study is the first in-depth biomechanical analysis of Bolt's maximal sprinting velocity with the segmental reconstruction.

(3) From a constructivist perspective, kinematic analysis of tennis serve of 205 km/h is presented (citation from [8]): "In the research, a serve with a speed of 205 km/h was analyzed, from which a direct point was scored. The serve performer was Goran Ivanišević, one of the best serve performers in the World". "The video on the basis of which the kinematic analysis was performed was recorded on the ATP Tour in Zagreb in 1997 (ATP goes for Association of Tennis Professionals). The recording was done with two video cameras so that they closed the angle of 100° with each other. The shooting speed was 60 frames per second, and due to the high speed of the recorded movement pattern, the camera shutters were set to 1/1000 s. Before recording the serve, for the purpose of later transformation of two-dimensional camera recordings into a real three-dimensional space at the site where the serve is performed, a reference frame measuring 180 × 180 x 180 cm was recorded. Video processing: digitization, transformation and data filtering, were performed by APAS. The mathematical model of the body and the racket was specially constructed for the purposes of this analysis. It is described with 18 reference points defining the body of the subject, 1 point for the ball and 5 points for the racket. Based on the extracted data, all relevant spatio-temporal parameters of the analyzed

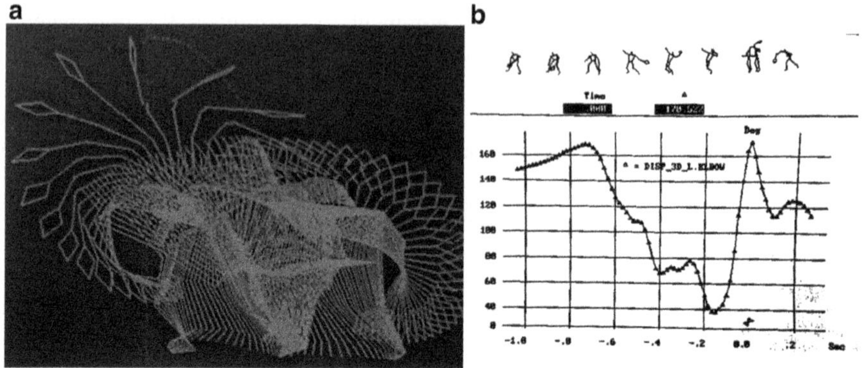

Fig. 12.1 **a** Contourograph of the analyzed tennis serve. A serve with a speed 205 km/h was analyzed, from which a direct point was scored. The serve performer was Goran Ivanišević, one of the best serve performers in the World (from Zmajić and Hraski [8], with permission); **b** In a methodological contribution, among other relevant moments, displacement of left elbow while serving is presented (Goran's dominant serving hand was left) - it varied in final critical phase from less than 40 deg to approximately 170 deg (almost maximal amplitude) within approximately 0.2 s (from Hraski and Mejovšek [9], with permission)

serve (linear displacements, velocities, accelerations, angles, angular velocities and angular accelerations) were then calculated (Fig. 12.1).

(4) Ohnjec et al. [10] compared kinematic parameters of jump shot performance by female handball players of different age. Citation from [10]: "The basic aim of this research was to analyze kinematic parameters when performing a jump shot. The sample of entities consisted of four female handball players, potential candidates for the Croatian national teams in their respective age categories (a junior, an under eighteen, an under-fifteen and a girl). The kinematic variables sample set was made up from the parameters related to the specific phases of a jump shot and they referred to: movement of the body's center of gravity (CoG) in a horizontal and vertical plane, velocity of the body's CoG in a horizontal and vertical plane, maximum linear velocity of some particular body segments and their actuation in time. From the series of seven attempts, the throws with the highest ball flight velocities were chosen for each of the subjects and they were explained in detail by kinematic variables observed. It is possible to use registered kinematic parameters to explore the performance of the jump shot by the subjects being measured, with the purpose to detect the constituent characteristics of the jump shot. These features, then, might be used to improve and correct the performances of the players within the process of their technical development (correcting mistakes), i.e. they might be considered as indicators relevant for directing the future trainingKinematic process in general" (Fig. 12.2).

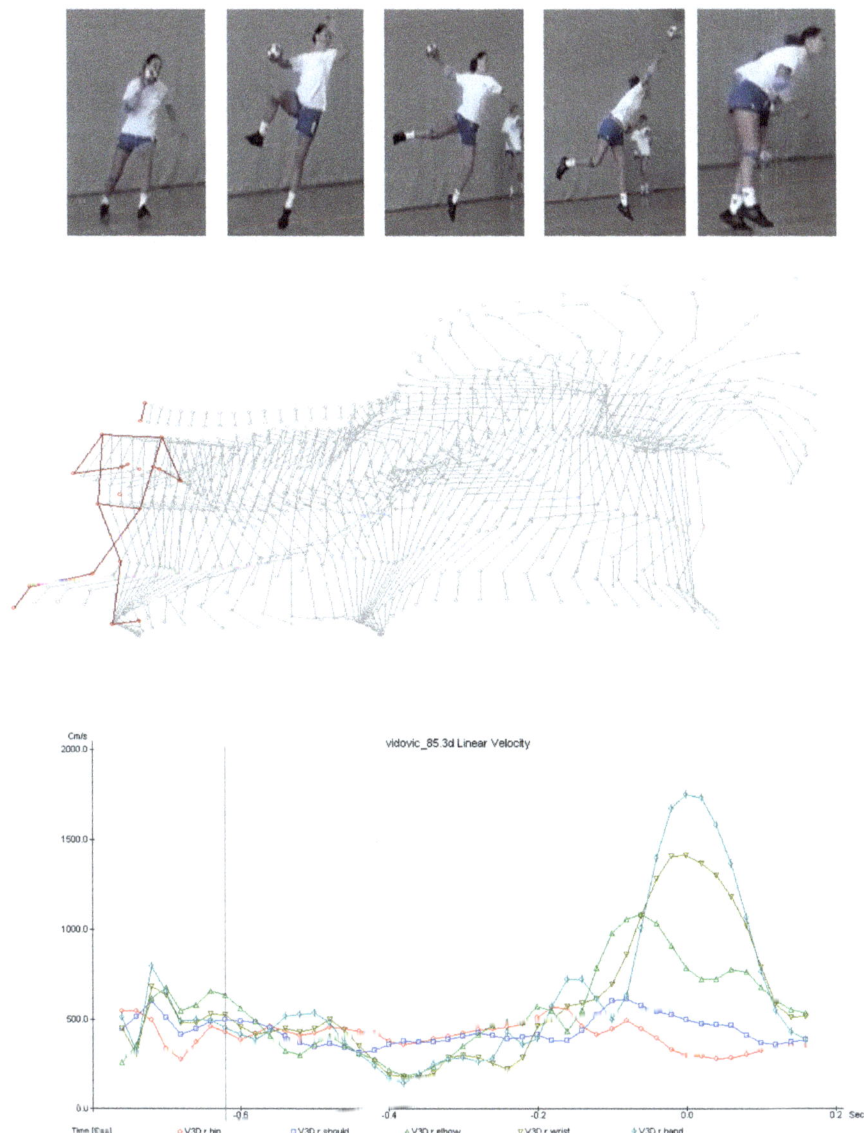

Fig. 12.2 Contourograph of one analyzed jump shot by a female right-handed left-wing handball player, junior (body height 182.0 cm, body mass 63.0 kg), a potential candidate for Croatian national team, participant in the study. Linear velocities of actuated right-sided body segments are presented with peak values in kinetic chain characterized by sequential activation of joints (hip-shoulder-elbow-wrist-hand, plotted in colors, see legend below abscissa). One hip acts as a 'stabilizer' and other later as 'initiator' within the jump shooting pattern. Right hand reveals dual function with two velocity peaks—one swinging phase (potentiator of the vertical one leg explosive jumping component) and other one while shooting ('force speed transition') (modified from Ohnjec et al. [10]. With permission)

12.5 Discussion

Reductionist-constructivist dualism of movement implies a cross-sectional area of research and methodology of biomechanical analysis of movement and effects on final and ultimate performance. The causes of additional entropy are also found in various experimental designs, diagnostic power of instrumentation, set goals and research hypotheses. Supporting visualizations of gymnastics/acrobatics, karate, volleyball and handball techniques (such as in [10], for instance) may be found in many publication sources of that time, characterizing the approach within the *'Zagreb kinesiology circle'*.

Faculty of Kinesiology fosters tradition in organizing international scientific symposia in the field of kinesiology since 1997 (usually at a triennial basis), covering a number of its branches including biomechanics, as well as scientific/professional symposia on sports diagnostics since 1997, co-organizer being Zagreb Sports Association, and later on physical conditioning of athletes since 2003 (annually), co-organizer being Croatian Physical Conditioning Association. At all occasions research of a kind briefly illustrated in this Chapter is being reported. Biomechanics sessions, including both movement analysis and motor control aspects, characteristically attract interesting contributions, and over the years have reflected evolving engineering technology and new instrumental solutions to address measurement and signal/data processing tasks (Fig. 12.3: Session chairing team at an international scientific-professional symposium on physical conditioning of athletes).

Among other sources, the contribution of [10, 11, 12 - these later two related to computer-supported 3D kinematic patterns capture and analysis in handball: see Figs. 5.6 and 5.7 in Chap. 5], could serve the purpose of this Chapter due to its cross-sectional character. Another kinematic measurement instrument has been used in this later—more recent—research, i.e. the automated 3D kinematic measurement system ELITE by BTS [11, 12], which, however, brought both advantages and disadvantages compared to APAS.

Further approaches and developments in understanding deeper roots of movement would direct reader into biochemistry of muscular function, social, statistical methodology (e.g. PCA—principal component analysis ensures approach to higher levels of movement inspection and control, in [10]) or 'reversed' biomechanical methodology (kinetics extracted and approximated from kinematics, ibid. [11]) or other. Main goal would be recognizing patterns. After that—extrapolating possibilities. At last—aquiring and conquesting needed ones (patterns).

Even further, kinetic analysis that may support previous lines of methodology may be found in [13]. Conclusions based on analysis of handball players uncovered the complexity and importance of this approach (citation form [13]): "Differences between [handball players'] left and right foot in foot rotation degree, percentage of contact time with heel, and time to change heel to forefoot were probably normal in general population, but by extensive and intensive unilateral loads in regular handball training and competition, as well. Decision mechanisms in terms of motor control probably alter automatic and somatic mechanisms. It can be assumed that specific

Fig. 12.3 Mario Cifrek of the Faculty of Electrical Engineering and Computing, and Vladimir Medved and Željko Hraski of the Faculty of Kinesiology, prior to co-chairing a session at the International scientific-professional symposium "Kondicijska priprema sportaša" (Physical Conditioning of Athletes) in realms of the 12th Zagreb Sports and Nautical Fair (February 21–22 2003)

footwear already influences neuromuscular and motor control(learning) 'decision mechanisms'. However, partialization offers deeper insight into the phenomena on the homogeneity/heterogeneity verge—between genders, age, longitudinal skeletal dimensionality, barefoot and footwear gait, complexity of kinesiological activity etc."

Context of recent biomechanical research at the Faculty of Kinesiology, University of Zagreb, was presented in ([11], citation follows): "Among additional measurement and sensing possibilities at disposal in our institution we would like to mention the 8-channel telemetric sEMG system TELEMG by BTS and pedobarograph pressure measuring device (Zebris Medical, Gmbh) that can be included according to modified measurement protocols". In addition to the equipment existing in the Biomechanics Laboratory, there are also other systems that might be used such as the Microsoft Kinect Sensor, property of Kinanthropometry Laboratory, acquired through the project [14].

There is a tradition of collaboration between the Biomechanics Laboratory group within Department of General and Applied Kinesiology, Faculty of Kinesiology, UniZg, Croatia, and Institute for Sport Science, Department for Biomechanics, Kinesiology, and Computer Sciences in Sport, UniVie, Austria. A bilateral project entitled: "Computer aided neuro-muscular biomechanical analysis and diagnostics of complex movements" has been realized in 2006 and 2007. Recently, within the partnership program with this institution (they house the VICON kinematic measurement system in the Laboratory. VICON goes for VIdeo CONvertor for biomechanics) we strive to quantify elementary dynamics and contact between two sport teams through the synthesis of large-scale standard biomechanics with a small-scale sport game analysis. The framework of this research project is entitled "System dynamics and contact between teams in handball", and it has been announced through lecture entitled: "Kinesiology of sports: From biomechanics to sociodynamics", given by Igor Gruić (at the Department for Biomechanics, Kinesiology, and Computer Sciences in Sport, UniVie, Austria) in Vienna, January 10, 2018.

12.6 Instead of Conclusion

Simplification of research to reductive-constructive dualism and distinction was heavily influenced by reductive-constructive orientation of thoughts and writings by Thomas A. McMahon: 'On size and life' [1], 'Muscles, reflexes and locomotion' [15]), and constructive-reductive orientation of thoughts and writings by Schmidt and Wisberg [3], as well by contemporary writings and thoughts of pleiad of (sometimes understood as marginal) research in biomechanics, motor control, sociocybernetics, behavioral psychology, statistics etc. (e.g. [16]). Previous statistical racours of Milas, Mraković, Momirović and Štalec enabled interdisciplinarity and convergence of most relevant ideas into final contributions for the purposes of this chapter within whole context of the book.

To embrace all of the world's 'schools' of research approaches to analyzing human movement in just one chapter of the book, is much broader mission of shortening hypotheses, postulates and protocols to text less burdened with technical and systemic descriptions, and more prone to emanate messages and guidelines for understanding history, current status and trends in which this area develops from the perspectives of researcher, clinician, practitioner. Although any phenomenon or hypothesis could be interpreted through any research 'circuit', in this Chapter it was done through the 'Zagreb kinesiology circle', whose presence is reflected in the sports science and medical and health aspects of kinesiology as components of the kinesiology curriculum, curriculum at some other university institutions, laboratories, scientific projects, cooperation with industry, etc.

It is fair to say that, in a way, a bernsteinean, body centered and neurophysiologically based approach from the 1930es, and reassessed later [17], is transformed and extrapolated today to also embrace sports game and societal levels of studying human movement [18]. Biotechnological, psychosocial, economic-political and other consequences of automation, urbanization, informatization, robotization and computerization (even immersive 'virtualization') are changing the modern man in ways that he can or choose to articulate with his genetic catalog—above all through the growing dualism of diversification and equalization.

Living space, work environment, recreational and serious sporting activities, medical procedures and programs contribute designing movement protocols in ways that an individual chooses to manage with his/her own body, or oppositely—are left to certified and/or automated (public and commercial) agents who offer a 'guarantee' of achieving, returning, promoting or maintaining Health (with a large 'H').

References

1. McMahon TA, Bonner JT (1983) On size and life. Scientific American Books, New York
2. Jensen CR, Schultz GW (1977) Applied kinesiology. McGraw-Hill Book Company. New York, pp 189, 271
3. Schmidt R, Wisberg A (2000) Motor learning and performance (2nd ed.) Human Kinetics. Champaign, Il
4. Mraković M (1971) Kineziologija. Kinesiology 1(1):2–5
5. Milas G (2009) Istraživačke metode u psihologiji i drugim društvenim znanostima (Research methods in psychology and other social sciences). Slap. Zagreb
6. Hraski Ž, Milanović D, Mejovšek M (1999) Kinematička analiza faze izbačaja koplja. In: Hraski Ž, Matković B (eds) Znanstveno-stručno savjetovanje Trener i suvremena dijagnostika. Zbornik radova. 8. zagrebački sajam sporta, 24–28. veljače,1999, Zagreb: Fakultet za fizičku kulturu, Sveučilište u Zagrebu, Hrvatski Olimpijski Odbor, Zagrebački Sportski Savez, Zagrebački velesajam. pp 83–89
7. Čoh M, Hébert-Losier K, Štuhec S, Babić V, Supej M (2018) Kinematics of Usain Bolt's maximal sprint velocity. Kinesiology 50(2):172–180
8. Zmajić H, Hraski Ž (1999) Kinematička analiza tenis servisa od 205 km/h. In: Hraski Ž, Matković B (eds) Znanstveno-stručno savjetovanje Trener i suvremena dijagnostika. Zbornik radova. 8. zagrebački sajam sporta, 24–28. veljače,1999, Zagreb: Fakultet za fizičku kulturu, Sveučilište u Zagrebu, Hrvatski Olimpijski Odbor, Zagrebački Sportski Savez, Zagrebački velesajam. pp 91–97
9. Hraski Ž, Mejovšek M (1999) Primjena sustava za kinematičku analizu sportskih tehnika. In: Hraski Ž, Matković B (eds) Znanstveno-stručno savjetovanje Trener i suvremena dijagnostika. Zbornik radova. 8. zagrebački sajam sporta, 24–28. veljače,1999, Zagreb: Fakultet za fizičku kulturu, Sveučilište u Zagrebu, Hrvatski Olimpijski Odbor, Zagrebački Sportski Savez, Zagrebački velesajam. pp 17–28
10. Ohnjec K, Antekolović LJ, Gruić I (2010) Comparison of kinematic parameters of jump shot performance by female handball players of different ages. Acta Kinesiologica 4(2):33–40
11. Gruić I, Medved V (2018) Computer supported 3D kinematic patterns capture and analysis in handball. In Andročec V (ed) Jubilee annual 2017–2018 of the Croatian Acedemy of Engineering, Croatian Academy of Engineering, Zagreb, pp 217–222
12. Pažin K, Bolčević F, Gruić I (2016) Kinematička analiza tehnike u rukometu. In: Jukić I, Gregov C, Šalaj S, Milanović L, Wertheimer V, Knjaz D (eds) 14. međunarodna konferencija Kondicijska priprema sportaša, 26. i 27. veljače 2016. Kineziološki fakultet Sveučilišta u Zagrebu, Udruga kondicijskih trenera Hrvatske, pp 63–67
13. Gruić I (2017) Pedobarographic assesment of male handball players gait. In: Proceedings of 8th international scientific conference on kinesiology—20th Anniversary, Opatija, Kineziološki fakultet, Hrvatska, pp 149–154
14. Gruić I, Katović D, Bušić A, Bronzin T, Medved V, Mišigoj-Duraković M (2019) Construction and validation of protocol for digital measurement of human body. In: Cabri J et al (eds) icSPORTS 2016/2017, CCIS 975, Springer Nature Switzerland AG 2019, pp 1–17 (https://doi.org/10.1007/978-3-030-14526-2 1)

15. McMahon TA (1984) Muscles, reflexes, and locomotion. Princeton University Press, Princeton, NJ
16. Federolf P (2014) Studying the coordination patterns in human motion. icSPORTS 2014, 14–16 Oct, Rome, Italy (invited speaker) https://player.vimeo.com/video/110567628?title=0&portra
17. Whiting HTA (1984) Human motor actions: Bernstein reassessed. Elsevier
18. Special Symposia/Session on 'Kinesiology in Sport and Medicine: from Biomechanics to Sociodynamics K-BioS' within icSPORTS 2018&2019. (https://icsports.scitevents.org/K-BioS.aspx?y=2018; https://icsports.scitevents.org/K-BioS.aspx?y=2019)

Appendices

The following five articles are reprinted here by kind permission of Elsevier:

Appendix A: Wu G, Cavanagh PR (1995) ISB recommendations for standardization in the reporting of kinematic data. J Biomech 28(10):1257–1261 (Copyright (1995), with permission from Elsevier.)
Appendix B: Wu G, Siegler S, Allard P, Kirtley C, Leardini A, Rosenbaum D, Whittle M, D'Lima DD, Cristofolini L, Witte H, Schmid O, Stokes I (2002) ISB recommendation on definitions of joint coordinate system of various joints for the reporting of human joint motion-part I: ankle, hip, and spine. J Biomech 35(4):543–548 (Copyright (2002), with permission from Elsevier.)
Appendix C: Wu G, van der Helm FCT, Veeger HEJ (DirkJan), Makhsous M, Van Roy P, Anglin C, Nagels J, Karduna AR, McQuade K, Wang X, Werner FW, Buchholz B (2005) ISB recommendation on definitions of joint coordinate systems of various joints for the reporting of human joint motion-Part II: shoulder, elbow, wrist and hand. J Biomech 38(5):981–992 (Copyright (2005), with permission from Elsevier.)
Appendix D: Derrick TR, van den Bogert AJ, Cereatti A, Dumas R, Fantozzi S, Leardini A (2020) ISB recommendations on the reporting of intersegmental forces and moments during human motion analysis. J Biomech 99:109533 (Copyright (2019), with permission from Elsevier.)
Appendix E: Standards for reporting EMG data (2014) JEMG Kinesiol 24(2):Pages I–II (Copyright (2019), with permission from Elsevier.)

V. Medved (ed.), *Measurement and Analysis of Human Locomotion*,
Series in Biomedical Engineering, https://doi.org/10.1007/978-3-030-79685-3

Appendix A
ISB Recommendations for Standardization in the Reporting of Kinematic Data

Ge Wu and Peter R. Cavanagh

Since 1990, the Standardization and Terminology Committee of the International Society of Biomechanics has been working towards a recommendation for standardization in the reporting of kinematic data. The paper, which is a result of those efforts (including broad input from members of the Society), is intended as a guide to the presentation of kinematic data in refereed publications and other materials. It is hoped that some uniformity in presentation will make publications easier to read and allow for the more straightforward comparison of data sets from different investigators. It is not intended to restrict individual investigators in the manner in which they collect or process their data, Rather, it could be viewed as a "output filter" applied to a variety of data formats to provide uniformity in the final product.

The ISB is cognizant of the various attempts at standardization that are being pursued by other organizations—such as CAMARC in Europe, the Clinical Gait Laboratory Group in the U.S.A., and the efforts of individual professional groups such as the Scoliosis Research Society. Where possible, we have sought unanimity with these groups, but on issues where the members of our society expressed strong opinions, we have—at times stated a contrary view. One example in point is the use of center of mass-based segmental reference frames. Since such reference frames are needed for conventional dynamic analysis, we make the recommendation that such frames should be routinely used. We anticipate that extension to the present document in the future will include recommendations for joint coordinate systems and the definition of anatomical landmarks for locating other segmental reference frames.

The committee recognizes that standardization of the description of movement at individual joints is best left to those who are intimately involved in the study of those joints, and we have therefore appointed subcommittees for various joints to provide recommendations. Groups are currently active for the ankle, hand, shoulder, spine, temporomandibular joints, whole body and wrist; members with interests

G. Wu (✉) · P. R. Cavanagh
The Center for Locomotion Studies, Penn State University, University Park, PA 16802, USA

V. Medved (ed.), *Measurement and Analysis of Human Locomotion*,
Series in Biomedical Engineering, https://doi.org/10.1007/978-3-030-79685-3

and expertise in other joints are being actively sought. The initial recommendations of some of these groups have already been published in the ISB Newsletter, and once these recommendation have been discussed by the membership, a subsequent document on joint coordinate systems will be published.

The present recommendations are presented as a framework on which future progress can be based. We are grateful to former members of the Standardization and Terminology Committee (notably Professors John Paul, David Winter, and Don Grieve) and to the many ISB members who have commented on earlier drafts of this recommendation. The present paper owes much to the work of Sommer [3].

Part l: Definition of a global reference frame

Need: A Global Reference Frame with the direction of the global axes being consistent, no matter which activities or subjects are being studied, or which investigator is conducting the experiment.

Notation: X, Y, Z

Recommendation: A right-handed orthogonal triad fixed in the ground with the + Y axis upward and parallel with the field of gravity, X and Z axes in a plane that is perpendicular to the Y axis.

Notes: (a) Where there is clear direction of travel or work (as is the case for level gait), the $+X$ axis is defined as the direction of travel or work (see Fig. A1).

(b) In case of locomotion on inclined planes, the Y axis will remain vertical and the X and Z axes will be in the same horizontal plane.

(c) Where there is no clear direction of travel or work (as is the case for insect flight), the $+X$ axis should be defined by the investigator.

(d) In tasks such as exercise in zero gravity, the $+X$ axis should be defined according to some arbitrary but visible surface in the environment and in the direction that is meaningful to the task.

(e) We acknowledge that there may be situations where non-Cartesian axes are more appropriate to the task being studied (for example, cylindrical coordinates are useful for the study of asymmetric manual exertion). Since the majority of studies use a Cartesian approach, it will be left to individual investigators to devise systems for the reporting of more specialist situations.

(f) The directions of the X, Y and Z axes have been chosen so that for those conducting two-dimensional studies, X, Y will lie in a sagittal plane. This will be consistent with the three-dimensional convention.

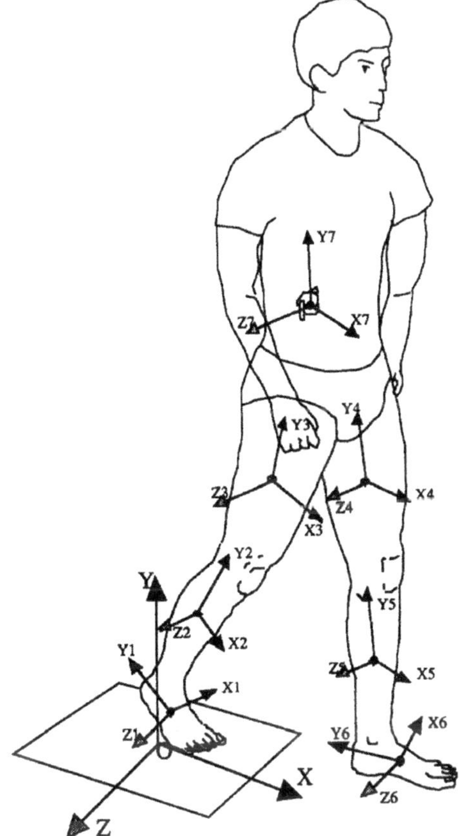

Fig. A1 Conventions for global reference frame and segmental local center of mass reference frame

Part 2: Definition of segmental local center of mass reference frames

Need: A coherent frame to describe segment pose (position and orientation) with respect to the global frame.

Notation: X_i, Y_i, Z_i.

Recommendation: A series of right-handed orthogonal triads fixed at the segmental centers of mass with the $+X_i$ axis anterior, $+Y_i$ axis proximal, and $+Z_i$ being defined by a right-hand rule.

Notes: (a) In general, the anterior—posterior, proximal—distal, and medial—lateral directions are defined in relation to the standard anatomical position.

(b) The use of right-hand reference frames for both left- and right-body segments implies that for the segments on the right side of the body the $+Z_i$ is pointing laterally, and for the segments on the left side of the body the $+Z_i$ is pointing medially (Fig. A1). As a result, the positive movements about the X and Y_i axes of a segment on the side of the body will have

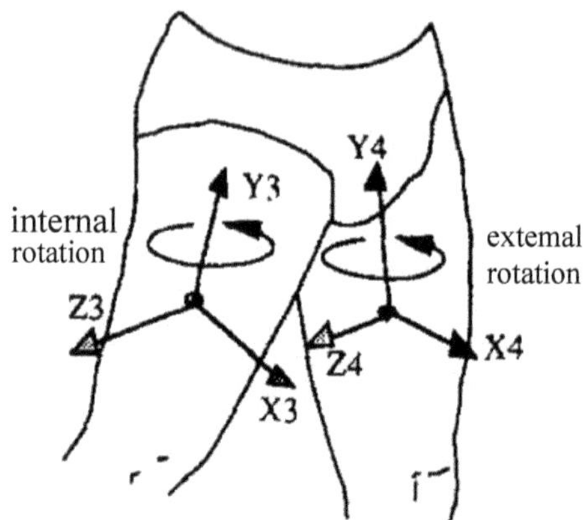

Fig. A2 The same rotations about segmental local center of mass reference frames produce anatomically different motions on the left and right sides of the body

opposite effects of movements of similar sign on the right side of the body (Fig. A2). This difference should be accounted for by describing the movements in their anatomical terms in any presentation of the data.

Part 3: Global displacement

Need: Specification of displacement of a segment with respect to the Global Reference Frame.

Notation: x_i, y_i, z_i

Recommendation: Report the coordinates of the origins of the Segmental Local Center of Mass Reference Frames with respect to the global origin in meters. The position of the local origin will be represented by the first column of the 4 x 4 matrix in the Local to Global transformation matrix $[T_{\text{lg}}]$ (see below).

Part 4: Global orientation

Need: Specification of orientation of a segment with respect to the Global Reference Frame.

Notation: $\alpha_i, \beta_i, \gamma_i$

Recommendation: A standard *ZYX* decomposition [3] of the 3 x 3 rotation submatrix of the 4 x 4 matrix will be used to define the Local to Global transformation matrix $[T_{\text{lg}}]$:

$$[T_{\text{lg}}] = \begin{bmatrix} 1 & 0 & 0 & 0 \\ x_i & C_{11i} & C_{12i} & C_{13i} \\ y_i & C_{21i} & C_{22i} & C_{23i} \\ z_i & C_{31i} & C_{32i} & C_{33i} \end{bmatrix},$$

where C_{11i}, C_{21i}, C_{31i} are the direction cosines of the local X_i axis with respect to the Global X, Y and Z axes, respectively.

Note: If $\alpha_i, \beta_i, \gamma_i$ is an ordered series of rotations about Z_i, Y_i, and X_i axes, respectively, then

$$[T_{1g}] = \begin{bmatrix} 1 & 0 & 0 & 0 \\ x_i & c\alpha_i c\beta_i & c\alpha_i s\beta_i s\gamma_i - s\alpha_i c\gamma_i & c\alpha_i s\beta_i c\gamma_i + s\alpha_i s\gamma_i \\ y_i & s\alpha_i c\beta_i & s\alpha_i s\beta_i s\gamma_i + c\alpha_i c\gamma_i & s\alpha_i s\beta_i c\gamma_i - c\alpha_i s\gamma_i \\ z_i & -s\beta_i & c\beta_i s\gamma_i & c\beta_i c\gamma_i \end{bmatrix},$$

where $s\alpha_i = \sin(\alpha_i)$ and $c\alpha_i = \cos(\alpha_i)$.

The individual angles can be found as follows:

$$\begin{aligned} \beta_i &= -\arcsin(C_{31i}), \\ \alpha_i &= \arcsin(C_{21i}/\cos(\beta_i)), \\ \alpha_i &= \arccos(C_{11i}/\cos(\beta_i)), \\ \gamma_i &= \arcsin(C_{32i}/\cos(\beta_i)), \\ \gamma_i &= \arccos(C_{33i}/\cos(\beta_i)). \end{aligned}$$

Part 5: Relative orientation

Need: A frame (or system) to express the relative orientation of the body segments with respect to each other.

Notation: α: rotation about one axis of the proximal segment's Local Reference Frame.

γ: rotation about one axis of the distal segment's Local Reference Frame.

β: rotation about the floating axis.

Recommendation: A joint coordinate system (which might better be called a joint rotation convention) is defined for each joint individually. This system allows rotations about axes which can be anatomically meaningful at the sacrifice of establishing a reference frame with non-orthogonal axes. As long as forces and moments are not resolved along these non-orthogonal axes, this does not present a problem. This approach allows the preservation of an important linkage with clinical medicine where the use of independent paired rotations (ab/ad, internal/external, etc.) is common usage.

The most well-known examples of such systems are those developed for the knee by Chao [1] and Grood and Suntay [2] (Fig. A3). Two body fixed axes are established relative to anatomical landmarks, one in each body on opposing sides of the joint. The third axis, called the floating axis, is defined as being perpendicular to each of the two body fixed axes.

Notes: (a) The orientation of the axes in the Local Reference Frames of the proximal and distal segments must be clearly specified.

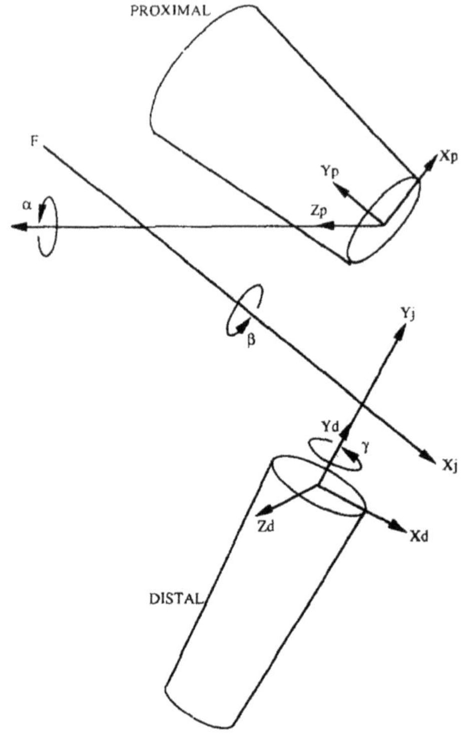

Fig. A3 A joint coordinate system for the knee joint

(b) The choice of the location of the origins drastically affects the distraction displacement terms.
(c) The Euler angle set in Part 4 (Global orientation) should match the angle decomposition for the joints as closely as possible.

References

1. Chao EYS (1986) Biomechanics of human gait. In: Schmid-Schonbein GW, Woo SL-Y, Zweifach BW (eds) Frontiers in biomechanics. Springer, New York
2. Grood ES, Suntay WJ (1983) A joint coordinate system for the clinical description of three-dimensional motions: application to the knee. J Biomechanical Eng 136–144
3. Sommer HJ Ill (1991) Primer on 3-D kinematics. Handout to the American Society of Biomechanics Meeting. Tempe, AZ, USA

Editorial Comment

The following recommendations on terminology for the reporting of kinematic data represent a thoughtful approach arising from a committee of the International Society of Biomechanics. The material was developed over some years, and involved a number of individuals so credited. The recommendations are published without peer review, since they arose from a committee of one of our participating organizations, and since they represent a compilation of commonly accepted practices.

R. A. Brand
Editor

Appendix B
ISB Recommendation on Definitions of Joint Coordinate System of Various Joints for the Reporting of Human Joint Motion—Part I: Ankle, Hip, and Spine

Ge Wu, Sorin Siegler, Paul Allard, Chris Kirtley, Alberto Leardini, Dieter Rosenbaum, Mike Whittle, Darryl D. D'Lima, Luca Cristofolini, Hartmut Witte, Oskar Schmid and Ian Stokes

Abstract The Standardization and Terminology Committee (STC) of the International Society of Biomechanics (ISB) proposes a general reporting standard for joint kinematics based on the Joint Coordinate System (JCS), first proposed by Grood and Suntay for the knee joint in 1983 (J. Biomech. Eng. 105 (1983) 136). There is currently a lack of standard for reporting joint motion in the field of biomechanics for human movement, and the JCS as proposed by Grood and Suntay has the advantage of reporting joint motions in clinically relevant terms. In this communication, the STC proposes definitions of JCS for the ankle, hip, and spine. Definitions for other joints (such as shoulder, elbow, hand and wrist, temporomandibular joint (TMJ), and

---

G. Wu (✉)
Department of Physical Therapy, University of Vermont, 305 Rowell Building, Burlington, VT 05405-0068, USA
e-mail: ge.wu@uvm.edu

S. Siegler
Department of Mechanical Engineering and Mechanics, Drexel University, Philadelphia, PA, USA

P. Allard
Research Center, Laboratoire D'Etude du Mouvement, Sainte-Justine Hospital, Montreal, Canada

C. Kirtley
School of Physiotherapy, Curtin University of Technology, Perth, Australia

A. Leardini
Movement Analysis Laboratory, Istituti Ortopedici Rizzoli, Bologna, Italy

D. Rosenbaum
Movement Analysis Laboratory, Department of Orthopaedics, University of Muenster, Muenster, Germany

M. Whittle
Cline Chair of Rehabilitation Technology, The University of Tennessee at Chattanooga, Chattanooga, TN, USA

D. D. D'Lima
Joint Mechanics Laboratory, Scripps Clinic, La Jolla, CA, USA

V. Medved (ed.), *Measurement and Analysis of Human Locomotion*,
Series in Biomedical Engineering, https://doi.org/10.1007/978-3-030-79685-3

whole body) will be reported in later parts of the series. The STC is publishing these recommendations so as to encourage their use, to stimulate feedback and discussion, and to facilitate further revisions. For each joint, a standard for the local axis system in each articulating bone is generated. These axes then standardize the JCS. Adopting these standards will lead to better communication among researchers and clinicians.

Article Note

Corresponding author: G. Wu

Experts in ankle joint: S. Siegler, P. Allard, C. Kirtley, A. Leardini, D. Rosenbaum, M. Whittle

Experts in hip joint: A. Leardini, D. D. D'Lima, L. Cristofolini, H. Witte, O. Schmid

Expert in spine: I. Stokes

Introduction

Since November 1993, the Standardization and Terminology Committee (STC) of the International Society of Biomechanics (ISB) has begun its journey of developing a set of standards for reporting joint motion. Headed by Drs. Peter Cavanagh and Ge Wu in 1993, an initial decision was made to adopt the Joint Coordinate System (JCS), first proposed by Grood and Suntay in [6], as the standard. This decision was publicized to the biomechanics community via Biomech-L, an electronic discussion network. With the enormous amount of support received from the Biomech-L subscribers, the STC then decided to move forward with this decision. A group of volunteers was recruited via Biomech-L who would like to participate in the effort of developing the JCS for each of the major joints in the body. To date, nine subcommittees involving a total of 25 people have been established and, so far, eight proposals have been completed. They include ankle, hip, spine, shoulder, elbow, hand and wrist, TMJ, and whole body.

L. Cristofolini
Laboratorio di Tecnologia Medica, Engineering Faculty, Istituti Ortopedici Rizzoli, University of Bologna, Bologna, Italy

H. Witte
Institut fur Spezielle Zoologie und Evolutionsbiologie, Friedrich-Schiller-Universitat, Jena, Germany

O. Schmid
Orthopadische Klinik mit Poliklinik der, Friedrich-Alexander-Universitat Erlangen-N. Urnberg, Erlangen, Germany

I. Stokes
Department of Orthopaedics and Rehabilitation, University of Vermont, Burlington, VT 05405, USA

There are two main reasons as to why these JCSs are established. First, there is a lack of standard for reporting joint motion in the field of biomechanics for human movement. This makes the comparisons among various studies difficult, if not impossible. Secondly, the use of JCS as proposed by Grood and Suntay has the advantage of reporting joint motions in clinically relevant terms. This makes the application and interpretation of biomechanical findings easier and more welcoming to clinicians.

Although all of the JCS recommendations have been published in various forms, such as in previous ISB Newsletters, and on the ISB Home Page, only a few of them have been test-used and subsequently revised. The purpose of this paper is to present these JCS definitions to the biomechanics community so as to encourage the use of these recommendations, to provide first hand feedback, and to facilitate the revisions. It is hoped that this process will help the biomechanics community to move towards the development and use of a set of widely acceptable standards for better communication among various research groups, and among biomechanists, physicians, physical therapists, and other related interest groups.

Overview of JCS

All recommendations of JCS for various joints follow the similar procedures as proposed by Grood and Suntay [6]. First, a Cartesian coordinate system (CCS) is established for each of the two adjacent body segments. The axes in these CCSs are defined based on bony landmarks that are either palpable or identifiable from X-rays, and follow the ISB general recommendations [15]. The common origin of both axis systems is the point of reference for the linear translation occurring in the joint, at its initial neutral position. Secondly, the JCS is established based on the two CCSs. Two of the JCS axes are body fixed, and one is "floating". Lastly, the joint motion, including three rotational and three translational components, is defined based on the JCS.

JCS for the Ankle Joint Complex

Introduction

According to our terminology, the ankle joint complex is composed of the talocrural and the subtalar joints. A complete standard must include a separate set for each of these individual joints and an additional standard for the entire ankle joint complex (i.e. calcaneus relative to tibia/fibula). However, a standard for the foot to shank system will address the needs of a great majority of the biomechanical community that is concerned with functional activities such as walking and running. In these studies external anatomical landmarks are being used and it is not possible to directly distinguish between talocrural and the subtalar joints. It was therefore decided to propose a standard for the ankle joint complex first, and to develop the standards for the talocrural joint and for the subtalar joint at a later time. Tibio-fibular articulation could also be addressed at a later time.

Terminology

Joint Definition

The ankle (talocrural) joint: The articulation formed between the talus and the tibia/fibula.

The subtalar (talocalcaneal) joint: The articulation between the talus and the calcaneus.

The ankle joint complex: The structure composed of the ankle and the subtalar joints.

Anatomical Landmarks Used in This Proposal

MM: Tip of the medial malleolus.
LM: Tip of the lateral malleolus.
MC: The most medial point on the border of the medial tibial condyle.
LC: The most lateral point on the border of the lateral tibial condyle.
TT: Tibial tuberosity.
IM: The inter-malleolar point located midway between MM and LM.
IC: The inter-condylar point located midway between the MC and LC.

Definition of Standard Anatomical Planes of the Tibia/fibula (Fig. B1)

Frontal plane: The plane containing points IM, MC and LC.

Torsional plane: The plane containing points IC, MM and LM.

Sagittal plane: The plane perpendicular to the frontal plane and containing the long axis of the tibia/fibula (the line connecting points IC and IM).

Transverse plane: The mutual plane perpendicular to the frontal and sagittal planes.

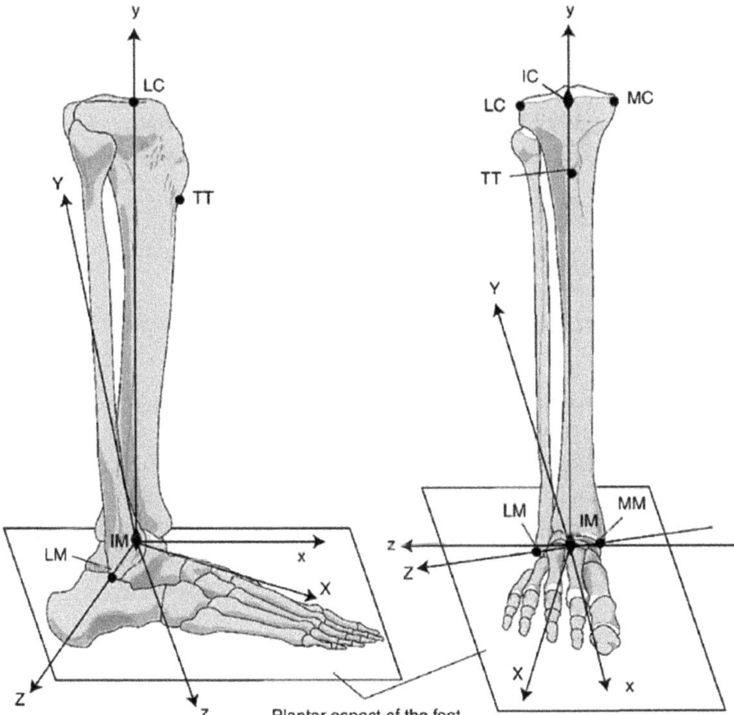

Fig. B1 Illustration of the tibia/fibula coordinate system (*XYZ*) and the calcaneus coordinate system (*xyz*) with the ankle joint complex in the neutral position

Definition of the Neutral Configuration of the Ankle Joint Complex

Neutral dorsiflexion/plantarflexion: Zero degrees between the long axis of the tibia/fibula and the line perpendicular to the plantar aspect of the foot projected onto the sagittal plane of the tibia/fibula

Neutral inversion/eversion: Zero degrees between the long axis of the tibia/fibula and the line perpendicular to the plantar aspect of the foot projected onto the frontal plane of the tibia/ fibula.

Neutral internal/external rotation: Zero degrees between the line perpendicular to the frontal plane of the tibia/fibula and the long axis of the second metatarsal, projected onto the transverse plane of the tibia/fibula.

Tibia/Fibula Coordinate System—XYZ (Fig. B1)

O: The origin coincident with IM.
Z: The line connecting MM and LM, and pointing to the right.
X: The line perpendicular to the torsional plane of the tibia/fibula, and pointing anteriorly.
Y: The common line perpendicular to X- and Z-axis.

Calcaneus Coordinate System—xyz (Fig. B1)

o: The origin coincident with that of the tibia/ fibula coordinate system (O) in the neutral configuration.
y: The line coincident with the long axis of the tibia/fibula in the neutral configuration, and pointing cranially.
x: The line perpendicular to the frontal plane of the tibia/fibula in the neutral configuration, and pointing anteriorly.
z: The common line perpendicular to *x*- and *y*-axis.

JCS and Motion for the Ankle Complex (Fig. B2)

e_1: The axis fixed to the tibia/fibula and coincident with the *Z*-axis of the tibia/fibula coordinate system.
Rotation (α): dorsiflexion (positive) or plantarflexion (negative).
Displacement (q_1): medial (negative) or lateral (positive) shift.
e_3: The axis fixed to the calcaneus and coincident with the *y*-axis of the calcaneal coordinate system.
Rotation (γ): internal rotation (positive) or external rotation (negative).
Displacement (q_3): correspond to compression (positive) or distraction (negative).
e_2: The floating axis, the common axis perpendicular to e_1 and e_3.
Rotation (β): inversion (positive) or eversion (negative).
Displacement (q_2): anterior (positive) or posterior (negative) drawer.

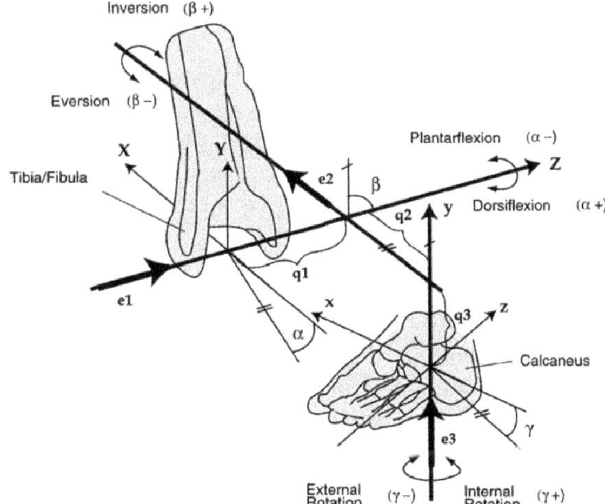

Fig. B2 Illustration of the JCS for the right ankle joint complex

JCS for the Hip Joint

Introduction

Originally, a more universally acceptable reference system was sought, to be suitable for different biomechanical investigations, including gait analysis, radiographic analysis, in vitro studies, and finite element modeling. However, it was recognized that these different fields tend to use different anatomical landmarks and different reference axes. One reason is that the most reproducible landmarks of the bones are not necessarily accessible in vivo in standard practice. The present proposal defines landmarks easily accessible in humans from external palpation or from estimation methods, therefore not necessarily optimal for in vitro investigations, when the bone is entirely accessible. For these latter applications, different reference systems have been proposed and discussed [4, 9, 16].

Definitions

Anatomical Landmarks Used

ASIS: anterior superior iliac spine (Nomina anatomica: Spina iliaca anterior superior).
PSIS: posterior superior iliac spine (Spina iliaca posterior superior).
FE: femoral epicondyle (Epicondylus femoris medialis, Epicondylus femoris lateralis).

Definition of Hip Center of Rotation

For most areas of biomechanical research, the normal human hip joint is treated as a ball and socket joint, with the center of rotation defined as the center of hip joint, even if a measurable incongruity of ball and socket does exist. The location of the hip center of rotation has been estimated using either a "functional" approach [2, 7] or a "prediction" approach [1, 5, 11]. The recommendation is to use the functional approach. This method seems appropriate when it is possible to analyze an adequate range of motion at the hip. More specific algorithms can also be examined for the optimal estimation of the center of this spherical rotation. Alternatively, when hip rotations cannot be effectively obtained, any of the prediction methods may be used. In choosing among them, a recent paper [12] has demonstrated that the large inaccuracies reported for hip joint center estimation [7] affect calculations of both angles and moments at the hip and knee joints. Hip moments showed the largest propagation error, with the flexion–extension component being particularly sensitive to errors in antero-posterior direction.

Pelvic Coordinate System—XYZ (Fig. B3)

O: The origin coincident with the right (or left) hip center of rotation.
Z: The line parallel to a line connecting the right and left ASISs, and pointing to the right.
X: The line parallel to a line lying in the plane defined by the two ASISs and the midpoint of the two PSISs, orthogonal to the *Z*-axis, and pointing anteriorly.
Y: The line perpendicular to both *X* and *Z*, pointing cranially [3].

Femoral Coordinate System—xyz (Fig. B3)

o: The origin coincident with the right (or left) hip center of rotation, coincident with that of the pelvic coordinate system (O) in the neutral configuration.
y: The line joining the midpoint between the medial and lateral FEs and the origin, and pointing cranially.
z: The line perpendicular to the y-axis, lying in the plane defined by the origin and the two FEs, pointing to the right.
x: The line perpendicular to both *y*- and *z*-axis, pointing anteriorly [3].

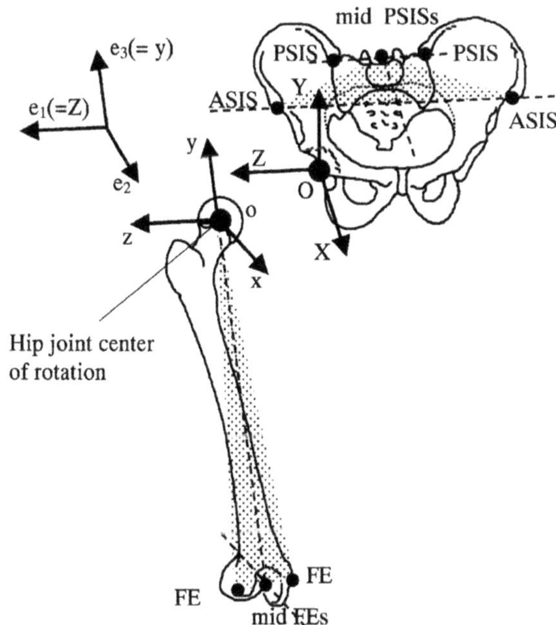

Fig. B3 Illustration of the pelvic coordinate system (*XYZ*), femoral coordinate system (*xyz*), and the JCS for the right hip joint

JCS and Motion for the Right (or Left) Hip Joint

e_1: The axis fixed to the pelvis and coincident with the *Z*-axis of the pelvic coordinate system.
Rotation (α): flexion or extension.
Displacement (q_1): mediolateral translation.

e_3: The axis fixed to the femur and coincident with the *y*-axis of the right (or left) femur coordinate system.
Rotation (γ): internal or external rotation.
Displacement (q_3): proximo-distal translation.

e_2: The floating axis, the common axis perpendicular to e_1 and e_3.
Rotation (β): adduction or abduction.
Displacement (q_2): antero-posterior translation.

JCS for the Spine

Introduction

Spinal motion occurs at intervertebral joints between adjacent vertebrae. These are the seven cervical vertebrae, twelve thoracic vertebrae and five lumbar vertebrae. Spinal motion is the summation of the intervertebral motion occurring at all mobile joints. Overall spinal motion is the motion that occurs between the head and the pelvis. Regional motion is the motion in a defined section of the spine (e.g. the lumbar spinal motion is the motion that occurs between the pelvis and the thorax).

Vertebrae articulate with their neighboring vertebrae via a flexible intervertebral disc and the two zygoapophyseal (facet) joints that are diarthroidial joints. The uppermost cervical vertebra articulates with the occiput. Thoracic vertebrae additionally articulate with the ribs [10]. The most caudal vertebra (L-5) articulates with the first of the sacral vertebrae (sacral vertebrae are fused together to form the sacrum). The sacrum articulates with the two innominate bones, also known as ilia that in turn have a flexible articulation with each other at the pubic symphysis. The ilia also include the acetabula that articulate with the femora.

This proposed standard concerns the intervertebral motion between adjacent vertebrae, but the principles can be extended to regional and overall spinal motion. The intervertebral articulations have six degrees of freedom (three translations and three rotations) each of which has a measurable stiffness. Therefore, there are six independent parameters of motion (three displacements and three rotations). The load–displacement characteristics of these joints has been described by a stiffness matrix [8]. This stiffness matrix has off-diagonal ('coupling') terms as well as diagonal terms. Therefore the pattern of motion that occurs between two vertebrae depends on the combination of forces applied, and it is only possible to define an instantaneous axis of rotation, since no fixed joint axis exists. The helical axis of motion is as alternative to the three rotations and three translations description of intervertebral motion. Using the helical axis of rotation, the motion is described by the position and direction of an axis of motion, together with a scalar translation along this axis and a scalar rotation around it.

Vertebral Coordinate System—XYZ (Proximal) and xyz (Distal) (Fig. B4)

$O(o)$: The origin is the intersection of the axes Y and y in the reference, neutral position (see Fig. B5a. The neutral position must be specified, and must be in a position where the vertebral axes Y and y are coplanar. If Y and y are parallel (do not intersect at the common origin O) the Y and y-axis are constrained to be colinear, and the origin O is the mid-point between adjacent endplates (see Fig. B5b).

$Y(y)$: The line passing through the centers of the vertebra's upper and lower endplates, and pointing cephalad.

$Z(z)$: The line parallel to a line joining similar landmarks on the bases of the right and left pedicles, and pointing to the right.

$X(x)$: The line perpendicular to the Y- and Z-axis, and pointing anteriorly.

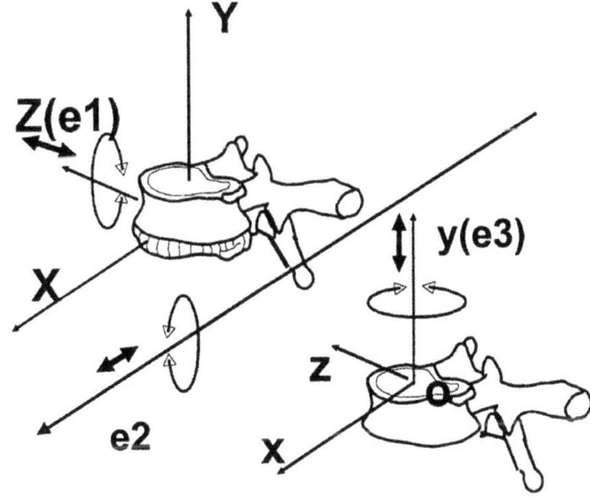

Fig. B4 Illustration of a proximal vertebral coordinate system (*XYZ*), a distal vertebral coordinate system (*xyz*), and the corresponding JCS

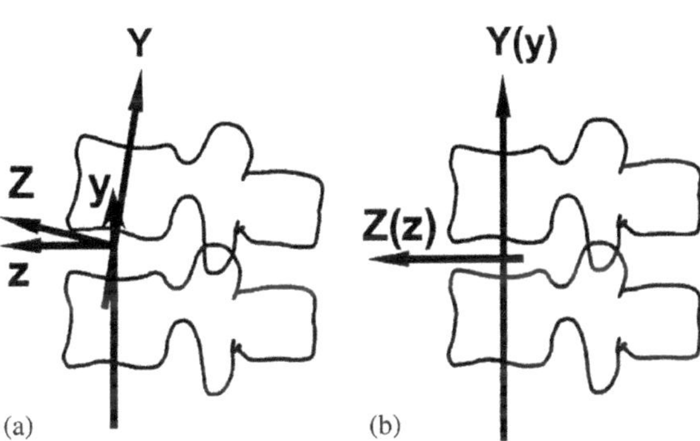

Fig. B5 Location of the common origin of axes: (a) the general case; (b) the specific case of *Y* and *y* being parallel. Note: the *Y*-and *y*-axis must be coplanar in the reference position of the two vertebrae

It should be noted that other axis conventions have been described. White and Panjabi [14] have *X* left; *Y* cephalad, *Z* anterior. ISO 2631, SAE J-211, and the Scoliosis Research Society [13] have *X* anterior, *Y* left and *Z* cephalad.

JCS and Motion for the Spine (Fig. B4)

e_1: The axis fixed to the proximal vertebra and coincident with the Z-axis of the proximal vertebra coordinate system.
Rotation (α): flexion or extension.
Displacement (q_1): mediolateral translation.

e_3: The axis fixed to the distal vertebra and coincident with the y-axis of the distal vertebra coordinate system.
Rotation (γ): axial rotation.

Displacement (q_3): proximo-distal translation.

e_2: The floating axis, the common axis perpendicular to e_1 and e_3.
Rotation (β): lateral bending.
Displacement (q_2): antero-posterior translation.

Acknowledgements Dr. Michell Gatton (Queensland University of Technology, Australia) and Dr. Stuart McGill (University of Waterloo, Canada) made valuable suggestions on the JCS for the spine.

References

1. Bell AL, Pedersen DR, Brand RA (1990) A comparison of the accuracy of several hip center location prediction methods. J Biomech 23:617–662
2. Cappozzo A (1984) Gait analysis methodology. Hum Mov Sci 3:27–54
3. Cappozzo A, Catani F, Della Croce U, Leardini A (1995) Position and orientation of bones during movement: anatomical frame definition and determination. Clin Biomech 10:171–178
4. Cristofolini L (1997) A critical analysis of stress shielding evaluation of hip prostheses. Crit Rev Biomed Eng 25(4&5):409–483
5. Davis RB, Ounpuu S, Tyburski D, Gage JR (1991) A gait analysis data collection and reduction technique. Hum Mov Sci 10:171–178
6. Grood ES, Suntay WJ (1983) A joint coordinate system for the clinical description of three-dimensional motions: application to the knee. J Biomech Eng 105:136–144
7. Leardini A, Cappozzo A, Catani F, Toksvig-Larsen S, Petitto A, Sforza V, Cassanelli G, Giannini S (1999) Validation of a functional method for the estimation of hip joint centre location. J Biomech 32(1):99–103
8. Panjabi MM, Brand RA, White AA (1976) Three-dimensional flexibility and stiffness properties of the human thoracic spine. J Biomech 9:185–192
9. Ruff CB, Hayes WC (1983) Cross-sectional Geometry of pecos pueblo femora and tibiae—a biomechanical investigation: i. Method and general patterns of variation. Am J Phys Anthropol 60:359–381
10. Schultz AB, Benson DR, Hirsch C (1974) Force–deformation properties of human costo-sternal and costo-vertebral articulations. J Biomech 7:311–318
11. Seidel GK, Marchinda DM, Dijkers M, Soutas-Little RW (1995) Hip joint center location from palpable bony landmarks—a cadaver study. J Biomech 28(8):995–998
12. Stagni R, Leardini A, Cappozzo A, Benedetti MG, Cappello A (2000) Effects of hip joint centre mislocation on gait analysis results. J Biomech 33(11):1479–1487
13. Stokes IAF (1994) Scoliosis research society working group on 3-D terminology of spinal deformity: three-dimensional terminology of spinal deformity. Spine 19:236–248
14. White AA, Panjabi MM (1978) Clinical biomechanics of the spine. JB Lippincott Co., Philadelphia, pp 463–464
15. Wu G, Cavanagh PR (1995) ISB recommendations for standardization in the reporting of kinematic data. J Biomech 28(10):1257–1261
16. Yoshioka Y, Siu D, Cooke TDV (1987) The anatomy and functional axes of the femur. J. Bone Joint Surg 69-A:873–880

Appendix C
ISB Recommendation on Definitions of Joint Coordinate Systems of Various Joints for the Reporting of Human Joint Motion—Part II: Shoulder, Elbow, Wrist and Hand

Ge Wu, Frans C. T. van der Helm, H. E. J. (DirkJan) Veeger, Mohsen Makhsous, Peter Van Roy, Carolyn Anglin, Jochem Nagels, Andrew R. Karduna, Kevin McQuade, Xuguang Wang, Frederick W. Werner and Bryan Buchholz

Abstract In this communication, the Standardization and Terminology Committee (STC) of the International Society of Biomechanics proposes a definition of a joint coordinate system (JCS) for the shoulder, elbow, wrist, and hand. For each joint, a standard for the local axis system in each articulating segment or bone is generated. These axes then standardize the JCS. The STC is publishing these recommendations so as to encourage their use, to stimulate feedback and discussion, and to facilitate further revisions. Adopting these standards will lead to better communication among researchers and clinicians.

G. Wu (✉)
Department of Physical Therapy, University of Vermont, 305 Rathwell Building, Burlington, VT, USA
e-mail: ge.wu@uvm.edu

F. C. T. van der Helm
Department of Mechanical Engineering, Delft University of Technology, Delft, The Netherlands

H. E. J. (DirkJan) Veeger
Department of Human Movement Sciences, Institute for Fundamental and Clinical Movement Sciences, Amsterdam, The Netherlands

Department of Mechanical Engineering, Delft University of Technology, Delft, The Netherlands

M. Makhsous
Department of Physical Therapy and Human Movement Sciences, Northwestern University, Chicago, IL, USA

P. Van Roy
Experimental Anatomy, Vrije U niversiteit Brussel, Ixelles, Belgium

C. Anglin
Department of Mechanical Engineering, University of British Columbia, Vancouver, Canada

J. Nagels
Department of Orthopaedics, Leiden University Medical Center, Leiden, The Netherlands

V. Medved (ed.), *Measurement and Analysis of Human Locomotion*,
Series in Biomedical Engineering, https://doi.org/10.1007/978-3-030-79685-3

Keywords Joint coordinate system · Shoulder · Elbow · Wrist · Hand

Article Note

Corresponding author: G. Wu

Chairperson of the Standardization and Terminology Committee. The International Society of Biomechanics: G. Wu

Authored shoulder and elbow: F. C. T. van der Helm, H. E. J. (DirkJan) Veeger, M. Makhsous, P. Van Roy, C. Anglin, J. Nagels, A. R. Karduna, K. McQuade, X. Wang

Authored wrist and hand: F. W. Werner, B. Buchholz

Subcommittee chair: F. W. Werner

Introduction

In the past several years, the Standardization and Terminology Committee (STC) of the International Society of Biomechanics has been working to propose a set of standards for defining joint coordinate systems (JCS) of various joints based on Grood and Suntay's JCS of the knee joint [6]. The primary purpose of this work is to facilitate and encourage communication among researchers, clinicians, and all other interested parties.

The STC has established a total of nine subcommittees, involving nearly 30 people who have extensive experience (either research or clinical) in joint biomechanics, and had developed proposals for nine major joints in the body. These joints include: foot, ankle, hip, spine, shoulder, elbow, hand and wrist, TMJ, and whole body. The proposals are based on the ISB standard for reporting kinematic data published by Wu and Cavanagh [16]. The first set of these standards for the ankle joint, hip joint, and spine was published in Journal of Biomechanics in April 2002 [17]. A response to comments to this set of standards was later published in 2003 [1].

A. R. Karduna
Exercise and Movement Science, University of Oregon, Eugene, OR, USA

K. McQuade
Department of Physical Therapy and Rehabilitation Science, University of Maryland, Baltimore, MD, USA

X. Wang
Biomechanics and Human Modeling Laboratory, National Institute for Transport and Safety Research, Bron, France

F. W. Werner
Department of Orthopedic Surgery, SUNY Upstate Medical University, Syracuse, NY, USA

B. Buchholz
Department of Work Environment, University of Massachusetts, Lowell, MA, USA

In this publication, the proposed standards for the shoulder joint, elbow joint, and wrist and hand are included. For each joint, the standard is divided into the following sections: (1) Introduction, (2) Terminology, (3) Body segment coordinate systems, and (4) JCS and motion for the constituent joints. It is then up to the individual researcher to relate the marker or other (e.g. electromagnetic) coordinate systems to the defined anatomic system through digitization, calibration movements, or population-based anatomical relationships.

The two major values in using Grood and Suntay's JCS are: (1) conceptual, since it appears easier to communicate the rotations to clinicians when using individual axes embedded in the proximal and distal segments and (2) the inclusion of calculations for clinically relevant joint translations. Some confusion, however, has arisen over their statement that the JCS is sequence independent, whereas Euler or Cardan angle representations are not. It should be noted that the Grood and Suntay's convention, without the translations, is simply a linkage representation of a particular Cardan angle sequence; the floating axis is the second, i.e. rotated, axis in the Cardan sequence [3, 8, 10]. The angles are independent because the sequence is defined by the mechanism; a Cardan or Euler sequence is equally "independent" once the sequence is defined.

JCS for the Shoulder

Introduction

Standardization of joint motions is very important for the enhancement of the study of motion biomechanics. The International Shoulder Group (ISG) supports the efforts of the ISB on this initiative, and recommends that authors use the same set of bony landmarks; use identical local coordinate systems (LCS); and report motions according to this recommended standard.

The starting point for the shoulder standardization proposal was a paper by Van der Helm [12]. More information can be obtained at: http://www.internationalshouldergroup.org.

The standardization of motions is only described for right shoulder joints. Whenever left shoulders are measured, it is recommended to mirror the raw position data with respect to the sagittal plane ($z = -z$). Then, all definitions for right shoulders are applicable.

Rotations are described using Euler angles. For a clearer interpretation of these angles it is suggested that the coordinate systems of the proximal and distal body segments are initially aligned to each other by the introduction of 'anatomical' orientations of these coordinate systems. The rotations of the distal coordinate system should then be described with respect to the proximal coordinate system. If both coordinate systems are aligned, the first rotation will be around one of the common axes, the second rotation around the (rotated) axis of the moving coordinate systems, and the third rotation again around one of the rotated axes of the moving coordinate system. This last axis is preferably aligned with the longitudinal axis of the moving segment. This method is equivalent to the method of [6] using floating axes. They also describe the first rotation around an axis of the proximal coordinate system and the last rotation around the longitudinal axis of the moving segment. The second axis is by definition perpendicular to both the first and third rotation axis.

For joint displacements, a common point in both the proximal and distal coordinate systems should be taken, preferably the initial rotation center (or a point on the fixed rotation axis in the case of a hinge joint). For most shoulder motions the rotation center would be only a rough estimate, since only the glenohumeral joint resembles a ball-and-socket joint. The definition of the common rotation centers of the sternoclavicular joint and acromioclavicular joint are left to the discretion of the researcher. Displacements should be described with respect to the axes of the coordinate system of the segment directly proximal to the moving segment to represent true joint displacements.

Terminology

Anatomical Landmarks Used in This Proposal (Fig. C1)

Thorax: C7: Processus Spinosus (spinous process) of the 7th cervical vertebra
T8: Processus Spinosus (Spinal Process) of the 8th thoracic vertebra
IJ: Deepest point of Incisura Jugularis (suprasternal notch).
PX: Processus Xiphoideus (xiphoid process), most caudal point on the sternum.

Clavicle: SC: Most ventral point on the sternoclavicular joint.

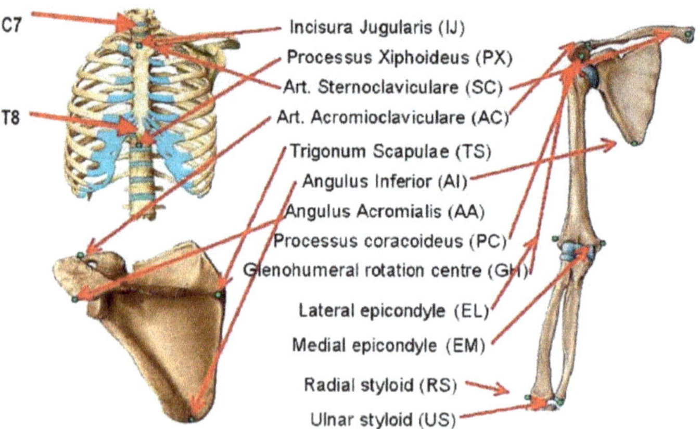

Fig. C1 Bony landmarks and local coordinate systems of the thorax, clavicle, scapula, and humerus

AC: Most dorsal point on the acromioclavicular joint (shared with the scapula).

Scapula: TS: Trigonum Spinae Scapulae (root of the spine), the midpoint of the triangular surface on the medial border of the scapula in line with the scapular spine.
AI: Angulus Inferior (inferior angle), most caudal point of the scapula.
AA: Angulus Acromialis (acromial angle), most laterodorsal point of the scapula.
PC: Most ventral point of processus coracoideus.

Humerus: GH: Glenohumeral rotation center, estimated by regression or motion recordings.
EL: Most caudal point on lateral epicondyle.
EM: Most caudal point on medial epicondyle.

Forearm: RS: Most caudal–lateral point on the radial styloid.
US: Most caudal–medial point on the ulnar styloid.

For the clavicle only two bony landmarks can be discerned: SC and AC. Hence, the axial rotation of the clavicle cannot be determined through non-invasive palpation measurements, but can be estimated on the basis of optimization techniques [13]. In contrast to Van der Helm [12], the use of the landmark AA is now proposed instead of the acromioclavicular joint (AC joint). This choice will reduce the occurrence of complications due to gimbal lock [5]. The GH is strictly speaking not a bony landmark, but is needed to define the longitudinal axis of the humerus. The GH can be estimated by regression analysis [9] or by calculating the pivot point of instantaneous helical axes (IHA) of GH motions [11, 14]. The IHA method is preferred since it is more accurate, and is also valid for patients in whom the GH has changed due to degeneration of the articular surfaces, or due to an implant. In some pathological cases it is likely that the GH cannot be accurately estimated with the IHA method due to translations in the joint. It is then, however, a question whether the regression method will be an acceptable alternative or whether different methods (such as CT or MRI) should be used.

Body Segment Coordinate Systems

Thorax Coordinate System—$X_tY_tZ_t$ (See Figs. C1 and C2)

O_t: The origin coincident with IJ.

Y_t: The line connecting the midpoint between PX and T8 and the midpoint between IJ and C7, pointing upward.

Z_t: The line perpendicular to the plane formed by IJ, C7, and the midpoint between PX and T8, pointing to the right.

X_t: The common line perpendicular to the Z_t- and Y_t-axis, pointing forwards.

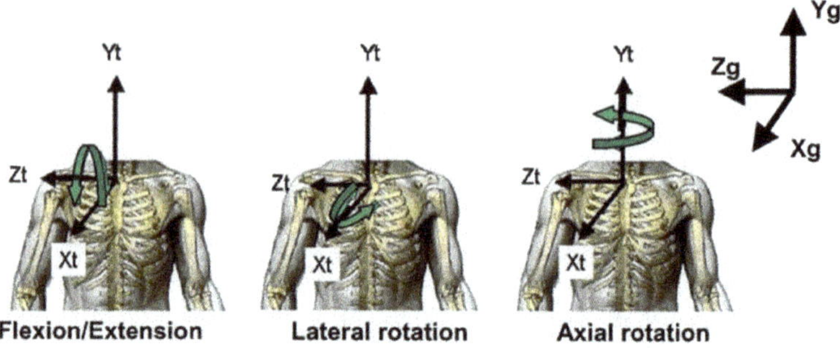

Fig. C2 Thorax coordinate system and definition of motions

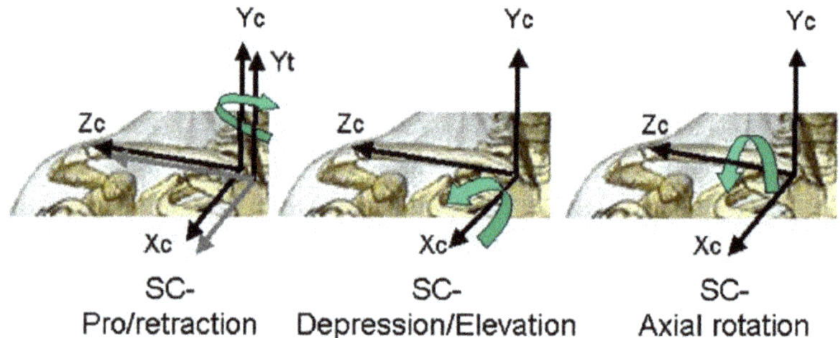

Fig. C3 Clavicule coordinate system and definition of SC motions. Y_t is the local axis for the thorax coordinate system, which is initially aligned with Y_c of the clavicle

Clavicle Coordinate System—$X_cY_cZ_c$ (See Figs. C1 and C3)

O_c: The origin coincident with SC.

Z_c: The line connecting SC and AC, pointing to AC.

X_c: The line perpendicular to Z_c and Y_t, pointing forward. Note that the X_c-axis is defined with respect to the vertical axis of the thorax (Y_t-axis) because only two bony landmarks can be discerned at the clavicle.

Y_c: The common line perpendicular to the X_c- and Z_c-axis, pointing upward.

Scapula Coordinate System—$X_sY_sZ_s$ (See Figs. C1 and C4)

O_s: The origin coincident with AA.

Z_s: The line connecting TS and AA, pointing to AA.

X_s: The line perpendicular to the plane formed by AI, AA, and TS, pointing forward. Note that because of the use of AA instead of AC, this plane is not the same as the visual plane of the scapula bone.

Y_s: The common line perpendicular to the X_s- and Z_s-axis, pointing upward.

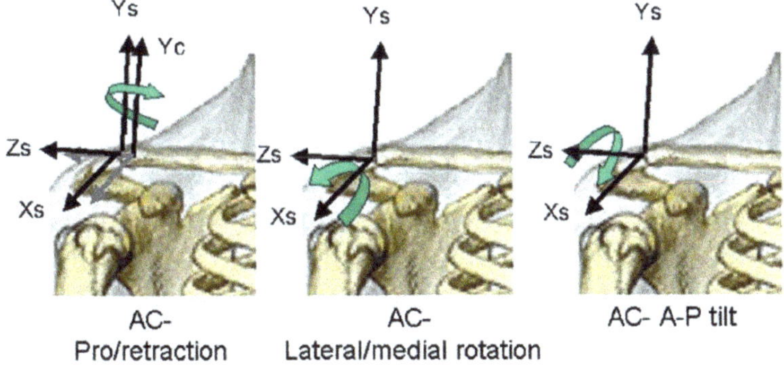

Fig. C4 Scapula coordinate system and definition of AC motions. Y_c is the local axis for the clavicle coordinate system (Please note, the origin, shown here at AC, should be placed at AA)

Humerus (1st Option) Coordinate System— $X_{h1}Y_{h1}Z_{h1}$ (See 1 and 5; See also Notes 1 and 2)

O_{h1}: The origin coincident with GH.
Y_{h1}: The line connecting GH and the midpoint of EL and EM, pointing to GH.
X_{h1}: The line perpendicular to the plane formed by EL, EM, and GH, pointing forward.
Z_{h1}: The common line perpendicular to the Y_{h1}- and Z_{h1}-axis, pointing to the right.

Humerus (2nd Option) Coordinate System— $X_{h2}Y_{h2}Z_{h2}$

O_{h2}: The origin coincident with GH.
Y_{h2}: The line connecting GH and the midpoint of EL and EM, pointing to GH.
Z_{h2}: The line perpendicular to the plane formed by Y_{h2} and Y_f (see Sect. "Forearm Coordinate System—$X_fY_fZ_f$ (see Figs. C1 and C6)"), pointing to the right.
X_{h2}: The common line perpendicular to the Z_{h2}- and Y_{h2}-axis, pointing forward.

Note 1: The second definition of humerus coordinate system is motivated by the high error sensitivity of the direction connecting EL and EM due to the short distance between them. Since it cannot be assured that the Z_{h2}-axis is equal to the joint rotation axis, its orientation depends on the position of the upper arm and forearm as well as the forearm orientation [15]. Therefore, by definition, the Z_{h2}-axis is taken with the elbow flexed 90° in the sagittal plane and the forearm fully pronated.

Note 2: We are faced with two difficulties in defining Z_h: (1) the anatomical definition of neutral humeral internal/external rotation is unclear; and (2) the numerical and practical inaccuracies in defining EL and EM may swamp the accuracy of our definition. The 1st and 2nd definitions will not agree if the true EM–EL line is rotated with respect to the forearm axis (in pronation). For the humerus, the difference will only affect the value for internal/external rotation; for the forearm it will affect all three angles to some degree, most significantly pro/supination. Our recommendation

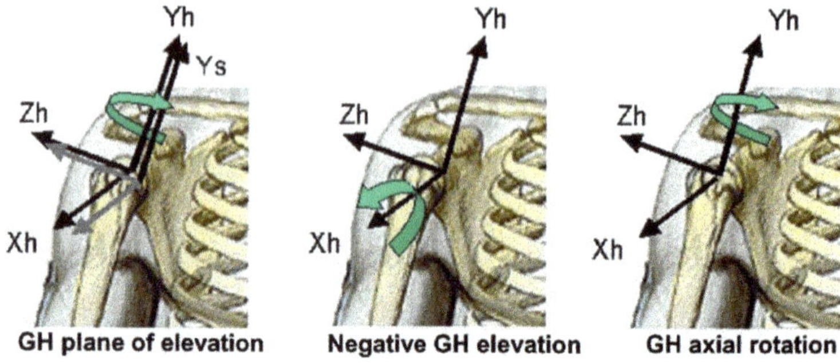

Fig. C5 Humerus coordinate system and definition of GH motions. Y_s is the local axis for the scapula coordinate system

is to use option 2 when the forearm is available for recording and otherwise to use option 1 (Fig. C5).

Forearm Coordinate System—$X_f Y_f Z_f$ (See Figs. C1 and C6)

O_f: The origin coincident with US.

Y_f: The line connecting US and the midpoint between EL and EM, pointing proximally.

X_f: The line perpendicular to the plane through US, RS, and the midpoint between EL and EM, pointing forward.

Z_f: The common line perpendicular to the X_f and Y_f-axis, pointing to the right.

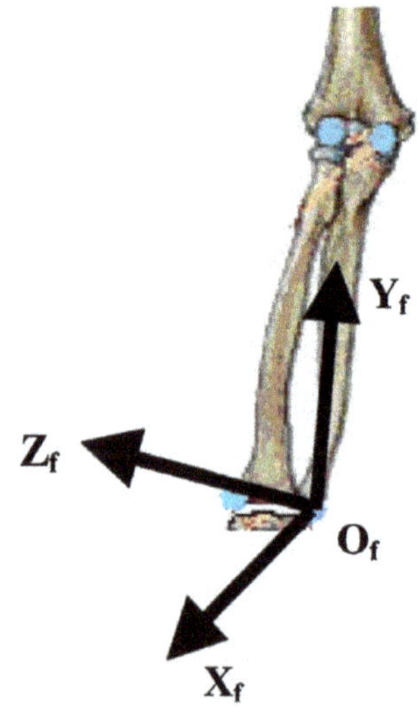

Fig. C6 Definition of forearm coordinate system

JCS and Motion for the Shoulder Complex

In the shoulder, it can be useful to report two types of rotations. One is joint rotation, i.e., rotation of a segment with respect to the proximal segment including the clavicle relative to the thorax (SC joint), the scapula relative to the clavicle (AC joint), and the humerus relative to the scapula (GH joint). The other is segment rotation, i.e., rotation of the clavicle, scapula, or humerus relative to the thorax (the non-existent thoracohumeral joint, often loosely defined as the shoulder joint). The definition of joint displacements is only useful if it is defined with respect to the proximal segment.

Many rotation orders are possible (such as X–Y–Z in Cardan angles or Y–Z–Y in Euler angles). We have chosen rotation orders so that the angles remain as close as possible to the clinical definitions of joint and segment motions. Differences are unavoidable since these clinical definitions are not consistent in 3-D. For example, although flexion and abduction each is clearly defined in 2-D, flexion followed by abduction gives a different result than abduction followed by flexion (see [2], Sect. 8.1).

In the following definitions, α is around the Z-axis, β around the X-axis, and γ around the Y-axis, irrespective of the order of rotation.

JCS and Motions of the Thorax Relative to the Global Coordinate System (Z–X–Y Order, Fig. C2)

Displacement (q): corresponds to motions of IJ with respect to the global coordinate system (X_g–Y_g–Z_g defined by Wu and Cavanagh [16]).

- $e1$: The axis coincident with the Z_g-axis of the global coordinate system. Rotation (α_{GT}): flexion (negative) or extension (positive).
- $e3$: The axis fixed to the thorax and coincident with the Y_t-axis of the thorax coordinate system. Rotation (γ_{GT}): axial rotation to the left (positive) or to the right (negative).
- $e2$: The common axis perpendicular to $e1$ and $e3$, i.e., the rotated X_t-axis of the thorax. Rotation (β_{GT}): lateral flexion rotation of the thorax, to the right is positive, to the left is negative.

JCS and Motion for the SC Joint (Clavicle Relative to the Thorax, Y–X–Z Order, Fig. C3)

Displacement (q): corresponds to translations of the common rotation center of the SC joint with respect to the thorax coordinate system.

- $e1$: The axis fixed to the thorax and coincident with the Y_t-axis of the thorax coordinate system. Rotation (γ_{SC}): retraction (negative) or protraction (positive).

*e*3: The axis fixed to the clavicle and coincident with the Z_c-axis of the clavicle coordinate system.
Rotation (α_{SC}): axial rotation of the clavicle; rotation of the top backwards is positive, forwards is negative.

*e*2: The common axis perpendicular to $e1$ and $e3$, the rotated X_c-axis.
Rotation (β_{SC}): elevation (negative) or depression (positive).

JCS and Motion for the AC Joint (Scapula Relative to the Clavicle, *Y*–*X*–*Z* Order, Fig. C4)

Displacement (q): corresponds to translations of the common rotation center of the AC joint with respect to the clavicle coordinate system.

Note: The following sequence is supported by Karduna et al. [7], who studied the six possible Euler sequences for scapular motion. They found that the proposed sequence is "consistent with both research and clinical-based 2-D representations of scapular motion". They also found that changing the sequence resulted in "significant alterations in the description of motion, with differences up to 50° noted for some angles". Since the scapular coordinate system is initially aligned with the clavicular coordinate system even though this position is never assumed anatomically, typical angle values are offset from zero (either positive or negative).

*e*1: The axis fixed to the clavicle and coincident with the Y_c-axis of the clavicle coordinate system.
Rotation (γ_{AC}): AC retraction (negative) or AC protraction (negative); the scapula is usually retracted.

*e*3: The axis fixed to the scapula and coincident with the Z_s-axis of the scapular coordinate system (scapular spine).
Rotation (α_{AC}): AC-anterior (negative) or AC-posterior (Positive) tilt; the scapula is usually tilted posteriorly.

*e*2: The common axis perpendicular to $e1$ and $e3$, the rotated X_s-axis of the scapula coordinate system.
Rotation (β_{AC}): AC-lateral (negative) or AC-medial (positive) rotation; the scapula is usually laterally rotated.

JCS and Motion for the GH Joint (Humerus Relative to the Scapula, *Y*–*X*–*Y* Order, Fig. C5)

Note: This is the one joint that is based on an Euler rotation sequence. Since $e1$ and $e3$ start in the same direction, the standard Grood and Suntay (floating-axis) equations cannot be used. Instead, an Euler decomposition is used to find the corresponding angles. As stated before, we have avoided the clinical terms flexion and abduction because flexion followed by abduction would give radically different results than abduction followed by flexion. Furthermore, these terms are only defined relative to the thorax, not the scapula (see Sect. "JCS and Motion for the Humerus Relative to

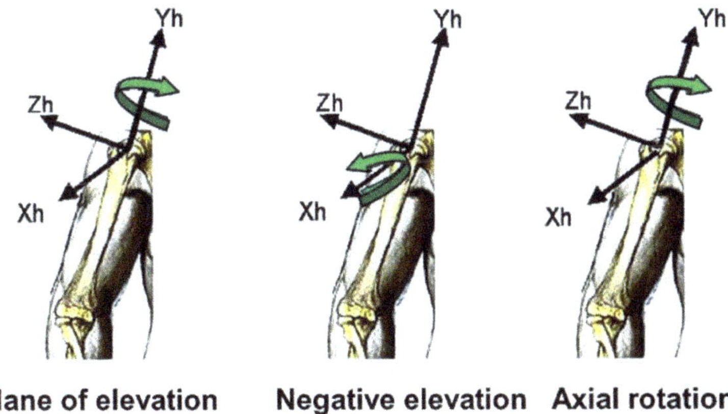

Fig. C7 Definition of thoracohumeral rotations

the Thorax (Y–X–Y Order) (Fig. C7)"). For comparison, flexion is elevation parallel to the sagittal plane and abduction is elevation in the coronal (frontal) plane.

Displacement (q): Corresponds to translations of the common rotation center of the GH joint with respect to the scapular coordinate system. In particular, we define qx = anterior/posterior translation; qy = inferior/superior translation; and qz = joint distraction.

$e1$: The axis fixed to the scapula and coincident with the Y_s-axis of the scapular coordinate system.
Rotation (γ_{GH1}): GH plane of elevation.

$e3$: Axial rotation around the Y_h-axis.
Rotation (γ_{GH2}): GH-axial rotation, endo- or internal-rotation (positive) and exo- or external-rotation (negative).

$e2$: The axis fixed to the humerus and coincident with the X_h-axis of the humerus coordinate system.
Rotation (β_{GH}): GH elevation (negative[1]).

JCS and Motion for the Clavicle Relative to the Thorax

For the motions of the clavicle no distinction between segment and joint rotations needs to be made, since the proximal coordinate system of the clavicle is the thorax. Definitions are equal to the definitions in Sect. "JCS and Motion for the SC Joint (Clavicle Relative to the Thorax, Y–X–Z Order, Fig. C3": $\alpha_c = \alpha_{SC}$; $\beta_c = \beta_{SC}$; and $\gamma_c = \gamma_{SC}$.

[1] As a consequence of the chosen direction of axes (ISB choice, but not preferred by the ISG), the second rotation elevation is by definition in the negative direction. The clinical term "elevation" corresponds to negative rotations around the $e2$-axis.

JCS and Motion for the Scapula Relative to the Thorax (*Y–X–Z* Order)

*e*1: The axis fixed to the thorax and coincident with the Y_t-axis of the thorax coordinate system. Rotation (γ_s): retraction (negative) or protraction (positive).

*e*3: The axis fixed to the scapula and coincident with the Z_s-axis of the scapular coordinate system.
Rotation (α_s): anterior (negative) or posterior (positive) tilt.

*e*2: The common axis perpendicular to *e*1 and *e*3. Rotation (β_s): lateral (negative) or medial (positive) rotation.

JCS and Motion for the Humerus Relative to the Thorax (*Y–X–Y* Order) (Fig. C7)

*e*1: The axis fixed to the thorax and coincident with the Y_t-axis of the thorax coordinate system. Rotation (γ_h): Plane of elevation, 0° is abduction, 90° is forward flexion.

*e*3: Axial rotation around the Y_h-axis.
Rotation $(\gamma_h)_2$: axial rotation, endo- or internal-rotation (positive) and exo- or external-rotation (negative).

*e*2: The axis fixed to the humerus and coincident with the X_h-axis of the humerus coordinate system.
Rotation (β_h): elevation (negative).

JCS for the Elbow

Introduction

To make a kinematic description of the elbow joint useful and practical, we use the following anatomical approximations (see Fig. C1):

1. The GH joint is a ball joint.
2. The humeroulnar joint is a hinge joint.
3. The radioulnar joint (contacting proximally and distally) is a hinge joint. The center of the capitulum on the humerus and the axes of the two radioulnar joints (proximal and distal) are on the joint axis.

A special problem is posed to the definitions of the segment coordinate systems of the ulna and radius, in that there are only a few palpable bony landmarks. Therefore, bony landmarks of other bones are needed for definitions, which result in position-dependent definitions of the segment coordinate systems.

Terminology

See Fig. C1(1) and Sect. “Terminology”.

Body Segment Coordinate Systems

Humerus Coordinate system—$X_{h1}Y_{h1}Z_{h1}$ (1st Option) or $X_{h2}Y_{h2}Z_{h2}$ (2nd Option)

See Sects. “Humerus (1st Option) Coordinate System—$X_{h1}Y_{h1}Z_{h1}$ (See 1 and 5; See also Notes 1 and 2)” and “Humerus (2nd Option) Coordinate System—$X_{h2}Y_{h2}Z_{h2}$” for a description of the two options for humerus coordinate systems. Since the forearm is obviously needed when studying the elbow, we recommend using the second definition.

Forearm Coordinate System—$X_fY_fZ_f$

See Sect. “Forearm Coordinate System—$X_fY_fZ_f$ (See Figs. C1 and C6)”.

Ulnar Coordinate System—$X_uY_uZ_u$ (Defined at Elbow Flexed 90° in the Sagittal Plane)

O_u: The origin is at US.
Y_u: The line pointing proximally from US to the midpoint between EM and EL.
X_u: The line perpendicular to the plane formed by US, EM, and EL, pointing forward.
Z_u: The common line perpendicular to the X_u- and Y_u-axis, pointing to the right.

Radius Coordinate System—$X_rY_rZ_r$ (Defined with Forearm in the Neutral Position and Elbow Flexed 90° in the Sagittal Plane)

O_r: The origin is at RS.
Y_r: The line pointing proximally from RS towards EL.
X_r: The line perpendicular to the plane formed by RS, US, and EL, pointing forward.
Z_r: The common line perpendicular to the X_r- and Y_r-axis, pointing to the right.

JCS and Motion for the Elbow Joints

Realistically, the elbow joint and radioulnar joint do not coincide with the axes of the segment coordinate systems. However, in situations where simplifications are allowed, the axis of rotation for each of these joints can be assumed to coincide with the local axes of the humerus (Z_{h1} or Z_{h2}) or ulna (Y_u). For a detailed study of the joint kinematics, the orientation of the hinge axis with respect to the proximal coordinate system should be determined; approximations of these are available from the literature. Only joint rotations with respect to the proximal segment coordinate system are defined here, as segment rotations with respect to the thorax would be meaningless.

JCS and Motion for the Elbow Joint (Forearm Relative to the Humerus, *Z–X–Y* Order)

$e1$: The axis fixed to the proximal segment and coincident with the Z_h-axis of the humerus coordinate system (preferably an approximation of the elbow flexion/extension axis).
Rotation (α_{HF}): flexion (positive) and hyperextension (negative).

$e3$: The axis fixed to the distal segment and coincident with the Y_f-axis of the forearm coordinate system.
Rotation (β_{HF}): axial rotation of the forearm, pronation (positive) and supination (negative).

$e2$: The floating axis, the common axis perpendicular to $e1$ and $e3$, the rotated X_f-axis of the forearm coordinate system.
Rotation (β_{HF}): carrying angle, the angle between the longitudinal axis of the forearm and the plane perpendicular to the flexion/ extension axis. The carrying angle occurs due to both a tilt in the humeral (flexion/extension) axis at the humeroulnar joint and an angulation of the ulna itself (see Anglin and Wyss [2], Sect. 5.6). It is therefore a passive response to elbow flexion/extension. Since the carrying angle is passive, it is rarely reported.

JCS and Motion of the Humeroulnar Joint (Ulna Relative to the Humerus, *Z–X–Y* Order)

$e1$: The axis fixed to the proximal segment and coincident with the Z_h-axis of the humerus coordinate system (preferably an approximation of the flexion/extension axis).
Rotation (α_{HU}): flexion (positive). Hyperextension is defined negative.

$e3$: The axis fixed to the distal segment and coincident with the Y_u-axis of the ulnar coordinate system.
Rotation (γ_{HU}): axial rotation of the ulna (negligible).

*e*2: The common axis perpendicular to *e*1 and *e*3, the rotated X_u-axis of the ulnar coordinate system.
Rotation (β_{HU}): carrying angle, the angle between the longitudinal axis of the ulna and the plane perpendicular to the flexion/extension axis (see Sect. "JCS and motion for the elbow joint (forearm relative to the humerus, Z–X–Y order)").

JCS and Motion for the Radioulnar Joint (Radius Relative to the Ulna, *X–Z–Y* Order)

*e*1: The axis fixed to the proximal segment and coincident with the X_u-axis of the ulnar coordinate system (describing the orientation of the pro/supination axis with respect to the ulna). It is implicitly assumed that the pro/ supination axis intersects the elbow flexion/ extension axis, although in reality this is not the case.
Rotation (β_{UR}): orientation of the pro/supination axis relative to the ulna (constant).

*e*3: The axis fixed to the distal segment and coincident with the Y_r-axis of the radius coordinate system.
Rotation (γ_{UR}): pro/supination of the radius with respect to the ulna.

*e*2: The common axis perpendicular to *e*1 and *e*3, the rotated Z_r-axis of the radius coordinate system.
Rotation (α_{UR}): abduction/adduction of the radius (negligible).

JCS for the Hand and Wrist

Introduction

Separate coordinate systems have been developed for each bone that is distal to the elbow, so that relative motion between any two adjacent segments may be described. These systems are then also applicable to global wrist motion as well as to motion of the individual components that cause the global motion. Global wrist motion is typically considered as the motion of the second and/or third metacarpal with respect to the radius (here, we use the third metacarpal) and is achieved by movement of the carpal bones with respect to the radius as well as the numerous articulations of the eight carpal bones with respect to each other. Some researchers, who only examine global wrist motion and have no need to examine carpal motion, can still use the definitions given for the radius and the metacarpal bones to describe wrist motion.

The ISB committee proposal [16] recommends that orthogonal triads be fixed at the segmental center of mass. In the hand and wrist, the center of mass is simply not known for most of the segments or bones. Data from cadaver studies do exist that describe the center of mass location for the forearm and hand as a proportion of the entire length of each of these segments. These center of mass definitions may be suitable for global wrist motions, but cannot be used to describe the kinematics of the component parts. The phalanges cannot be ignored as many researchers are examining individual movement of the carpal bones or movement of the radius with respect to the ulna. Therefore for this joint coordinate system application, the location of the orthogonal triad on each bone is primarily based on bony landmarks and is usually located at the axial center for the long bones or the volumetric centroid for the carpal bones. (CT scans might be used to define the volumetric centroid; however, this method may not be available or necessary for all applications).

Terminology

Anatomical Landmarks Used (See Figs. C8, C9 and C10)

Radius: Radioscaphoid fossa—articulation of the scaphoid with the radius
Radiolunate fossa—articulation of the lunate with the radius
Radial Styloid
Sigmoid Notch—depression in the distal radius where the ulna articulates with it
Radial Head (proximal).

Ulna: Dome of Ulnar Head (distal)
Coronoid Process.

Carpal Bones: Scaphoid
Lunate
Triquetrum
Pisiform
Trapezium
Trapezoid
Capitate
Hamate.

Metacarpals and Phalanges: Distal Head Center of Base.

Standard Wrist Positions

Neutral wrist position: Position of the wrist relative to the radius is defined as in neutral flexion/ extension and neutral radial/ulnar deviation when the third metacarpal long axis is parallel to the Y_r axis in the radius.

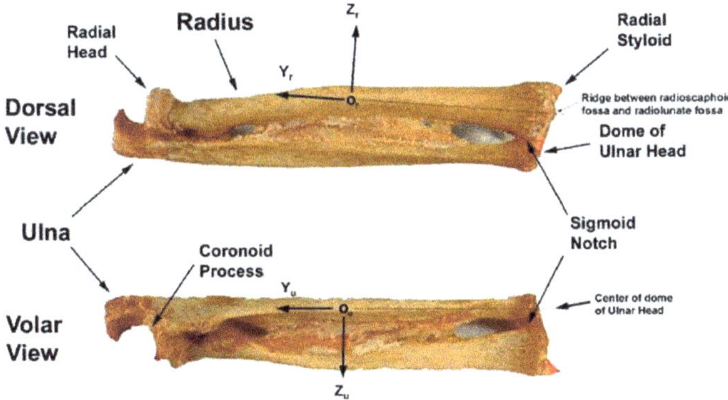

Fig. C8 View of a right forearm in neutral forearm rotation illustrating radial and ulnar coordinate systems. X-axis is pointing volarly. (For a left arm, X-axis is dorsal, Y axis is distal, Z-axis is to the right (ulnarly) in the anatomical position, so that flexion, pronation, and ulnar deviation are all positive for left and right arms.)

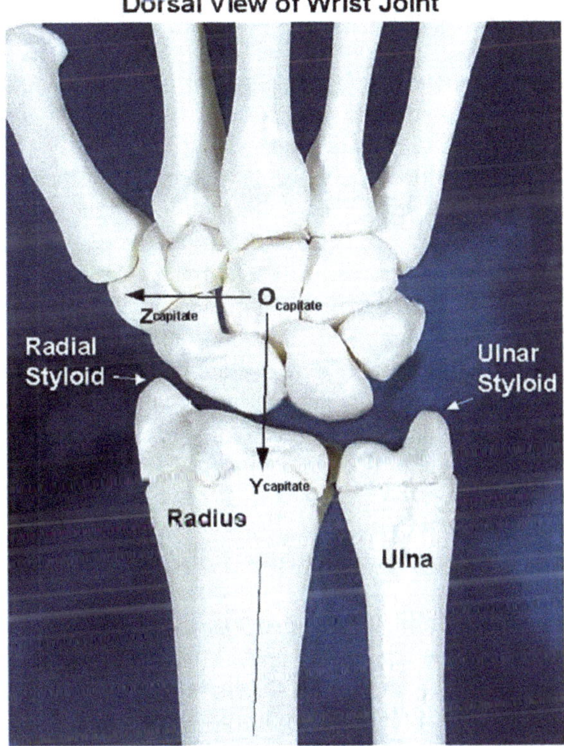

Fig. C9 Dorsal view of a right wrist joint illustrating the capitate coordinate system as an example of the carpal coordinate systems. *X*-axis is pointing volarly. (For a left arm *X*-axis is dorsal, *Y*-axis is distal, *Z*-axis is to the right (ulnarly) in the anatomical position.)

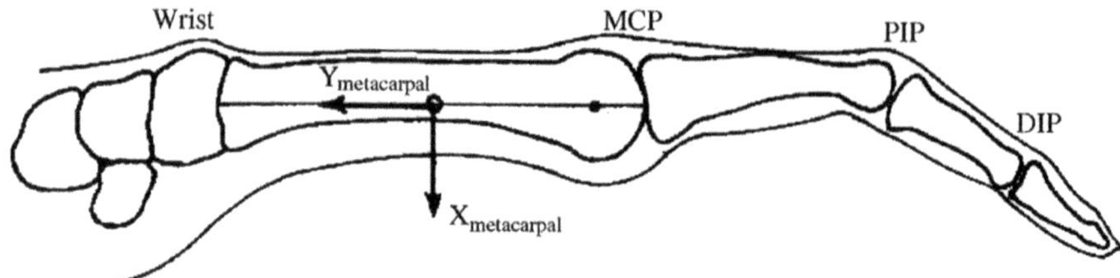

Fig. C10 Sagittal view of a right finger illustrating the metacarpal coordinate system as an example of phalangeal and metacarpal coordinate systems. *X*-axis is directed volarly and *Y*-axis is directed proximally. (For a left arm *X*-axis is dorsal, *Y*-axis is distal, and *Z*-axis is to the right in the anatomical position.)

Neutral forearm rotation: Position of the radius relative to the ulna when the elbow is flexed 90° and the thumb is pointing to the shoulder.

Body Segment Coordinate Systems

For each bone, a coordinate system is given, assuming that the forearm is initially in the standard anatomical position, with the palm forward (anterior), and the thumb lateral. The dorsum of the hand and forearm face posteriorly. In general for a right arm, the positive Y_i axis is directed proximally, the positive X_i axis is directed volarly, and the positive Z_i axis is directed to the right in the anatomical position (radially) (Figs. C8, C9 and C10). In order to have the same sign convention for clinical motion of left and right arms, for a left arm, Y_i is directed distally, X_i is directed dorsally, and Z_i is directed to the right in the anatomical position (ulnarly).

The following radius and ulna coordinate systems differ from those given in the elbow section above. Here, we are primarily concerned with studies that are based on all available bony landmarks. If a more general motion is of interest, similar to the artificial humerothoracic joint, one can use the forearm and 3rd metacarpal axes to create a simplified wrist joint.

Radius Coordinate System—$X_rY_rZ_r$

O_r: The origin is located midway between the distal radius at the level of the ridge between the radioscaphoid fossa and the radiolunate fossa, and the proximal radius at the level of the depression in the proximal radial head. If the distance to the ridge between the radioscaphoid and radiolunate fossas varies, then the location halfway between the dorsal and volar extremes of the ridge will be used to define the distal landmark on the radius. In the transverse plane it will be at the approximate center of the tubular bone (along its principal axis of inertia).

Y_r: The line parallel to the long shaft of the radius from O_r to intersect with the ridge of bone between the radioscaphoid fossa and the radiolunate fossa (midway dorsally and volarly along the ridge).
Z_r: The line perpendicular to the Y_r axis, and in a plane defined by the tip of the radial styloid, the base of the concavity of the sigmoid notch and the specified origin.
X_r: The common line perpendicular to the Y_r- and Z_r-axis.

Ulna Coordinate System—$X_u Y_u Z_u$

O_u: The origin is located midway between the distal ulna at the level of the dome of the ulnar head, and the proximal ulna at the level of the coronoid process. In the transverse plane it is at the approximate center of the tubular bone (along its principal axis of inertia).
Y_u: The line parallel to the long shaft of the ulna from O_u to intersect with the center of the dome of the ulnar head.
X_u: The line parallel to X_r when the radius is in neutral forearm rotation.
Z_u: The common line perpendicular to the X_u- and Y_u-axis.

Carpal Bones Coordinate System—$X_c Y_c Z_c$

The eight carpal bones, scaphoid, lunate, triquetrum, pisiform, trapezium, trapezoid, capitate, and hamate, will be considered simultaneously. Most researchers only report angular changes in carpal bone motion and use the neutral wrist position as a neutral reference position. The neutral wrist position is when the wrist is in neutral flexion/extension and neutral radial/ulnar deviation such that the third metacarpal long axis is parallel with the Y_r axis in the radius. These researchers define the motion relative to the radius and typically not the ulna. Therefore, the orientation of the coordinate systems for each carpal bone (Fig. C2) should be parallel with the radial coordinate system when the wrist is in the neutral wrist position. Thus, $Y_{\text{carpal bone}}$ will be parallel to Y_r and similarly for $X_{\text{carpal bone}}$ and $Z_{\text{carpal bone}}$. At present, most researchers who need to define a coordinate system origin in a carpal bone use the volumetric centroid of the bone. Therefore it is proposed that, when necessary, the origin of a coordinate system in a carpal bone be located at the volumetric centroid of the bone.

A separate coordinate system is required for the trapezium in order to describe motion at the trapeziometacarpal joint of the thumb. The coordinate system defined by Cooney et al. [4] will be adapted for this purpose: "The Y axis extends from the exact mid-point of the central ridge of the trapezial saddle to the center of the junction of the trapezium, scaphoid and trapezoid. The X axis runs in a dorsal-to-volar direction along a line perpendicular to the central ridge of the trapezium and passes through the mid-point of the dorsal surface to the proximal volar pole of the tubercle of the trapezium. The Z axis is perpendicular to the X and Y axes and nearly parallel to the central ridge of the trapezial metacarpal surface".

Metacarpals Coordinate System—$X_m Y_m Z_m$

The five coordinate systems for the five metacarpals are described in the same manner. The major differences in the metacarpals are in the shape of their bases where "contact" with the carpals is made and their relative movement capabilities. In this regard, the first metacarpal has a very large range of motion. The third metacarpal has special significance because of its use in the definition of global wrist motion. Most researchers consider either the second or third metacarpal as representative of hand motion.

O_m: The origin for each of these coordinate systems is located midway between the base and head of each metacarpal. In the transverse plane, it will be at the approximate center of the tubular bone (at its moment of inertia).

Y_m: The line parallel to a line from the center of the distal head of the metacarpal to the midpoint of the base of the metacarpal.

X_m: The X_m and Y_m-axis will form a sagittal plane that splits the metacarpal into mirror images.

Z_m: The common line perpendicular to the X_m- and Y_m-axis.

Phalanges Coordinate System—$X_p Y_p Z_p$

The 14 coordinate systems for the phalanges of the five digits can be described in a manner that is analogous to the description used for the metacarpal systems. The proximal and middle phalanges for the five digits are similar in shape as are the five distal phalanges.

JCS and Motion for the Hand and Wrist

JCS and Motion for the Interphalangeal, Metacarpophalangeal, Intercarpal, Radiocarpal, and Carpometacarpal Joints

$e1$: The axis fixed to the proximal segment and coincident with the Z-axis of the proximal segment coordinate system.
Rotation (α): flexion or extension (flexion is positive).
Displacement ($q1$): radial or ulnar translation.

$e3$: The axis fixed to the distal segment and coincident with the Y-axis of the distal segment coordinate system.
Rotation (γ): rotation (pronation–supination). Zero degrees of rotation is defined to be at the neutral forearm position. Pronation is a positive rotation. Supination is a negative rotation. Displacement ($q3$): proximal or distal translation.

$e2$: The common axis perpendicular to $e1$ and $e3$. Rotation (β): adduction or abduction, or radial or ulnar deviation (ulnar deviation is positive). Displacement ($q2$): dorsal or volar translation.

For the interphalangeal, first metacarpophalangeal, intercarpal, and radiocarpal joints, a neutral posture is defined as the position where the orientations of the proximal and distal segmental systems are aligned. For the second through fifth metacarpophalangeal joints, a neutral posture is defined as the position where the orientation of the distal segmental system is identical to that of the third metacarpal. The third carpometacarpal joint will be neutral when the third metacarpal system is aligned with the wrist system. For the first carpometacarpal (trapeziometacarpal) joint, a neutral posture will be defined as the position where the orientations of the proximal segmental system (as defined by Cooney et al. [4]) and distal segmental system are identical. The neutral posture for the second, fourth, and fifth carpometacarpal joints can be defined in an analogous manner.

JCS and Motion for the Radioulnar Joint

For the radioulnar joint, the Y-axis of the radius and ulna may not be parallel at the neutral posture. They may only diverge by a few degrees depending upon the subject. The neutral position for the radius and ulna is clinically called neutral forearm rotation. With the elbow flexed to 90°, this position can be visualized as when the thumb is pointing to the shoulder. In the standard anatomical position, the radius is supinated about the ulna.

For the radioulnar joint, we propose an intermediate coordinate system whose origin is identical with the radius coordinate system origin. The orientation of this intermediate coordinate system will be aligned with the ulnar coordinate system when the forearm is in neutral forearm rotation. The motion of the radius with respect to the ulna will then be described using the flexion/extension, radioulnar deviation, and pronation/supination definitions given above but using the intermediate coordinate system of the radius and the ulnar coordinate system. The user of this standard should define the orientation of the intermediate coordinate system relative to the anatomically based radial coordinate system.

$e1$: The axis fixed to the ulna and coincident with the Y-axis of the intermediate radial coordinate system.
Rotation (α): supination or pronation (pronation is positive).
Displacement ($q1$): proximal or distal translation.

$e3$: The axis fixed to the intermediate radial coordinate system and coincident with the Z-axis of the intermediate radial coordinate system.
Rotation (γ): flexion–extension (flexion is positive).
Displacement ($q3$): radial or ulnar translation.

$e2$: The common axis perpendicular to $e1$ and $e3$:
Rotation (β): radial–ulnar deviation (ulnar deviation is positive).
Displacement ($q2$): dorsal or volar translation.

Acknowledgements We thank Ed Chadwick, Brendan McCormack, A.C. Nicol, Bo Peterson, and Victor Waide for their past involvement in the development of the elbow joint standard.

References

1. Allard P, Cristofolini L, D'Lima D, Kirtley C, Leardini A, Rosenbaum D, Schmid O, Siegler S, Stokes I, Whittle M, Witte H, Wu G (2003) Response to Dr. Baker's letter to the editor. J Biomech 36(2):303–304
2. Anglin C, Wyss UP (2000) Review of arm motion analyses. Proc Inst Mech Eng 214(Part H):541–555
3. Baker R (2003) Letter to the editor regarding ISB recommendation on definition of joint coordinate systems for the reporting of human joint motion—part I: ankle, hip and spine. J Biomech 36(2):300–302
4. Cooney WP, Lucca MJ, Chao EYS, Linscheid RL (1981) The kinesiology of the thumb trapeziometacarpal joint. J Bone Joint Surgery 63A(9):1371–1381
5. de Groot JH, Valstar ER, Arwert HJ (1998) Velocity effects on the scapulo-humeral rhythm. Clin Biomech 13:593–602
6. Grood ES, Suntay WJ (1983) A joint coordinate system for the clinical description of three-dimensional motions: application to the knee. J Biomech Eng 105:136–144
7. Karduna AR, McClure PW, Michener LA (2000) Scapular kinematics: effects of altering the Euler angle sequence of rotations. J Biomech 33:1063–1068
8. Li J, Wyss UP, Costigan PA, Deluzio KJ (1993) An integrated procedure to assess knee-joint kinematics and kinetics during gait using an optoelectronic system and standardized X-rays. J Biomed Eng 15:392–400
9. Meskers CGM, Fraterman H, Van der Helm FCT, Vermeulen HM, Rozing PM (1998) In vivo estimation of the glenohumeral joint rotation center from scapular bony landmarks by linear regression. J Biomech 31:93–96
10. Small CF, Bryant JT, Pichora DR (1992) Rationalization of kinematic descriptors for three-dimensional hand and finger motion. J Biomed Eng 14(2):133–141
11. Stokdijk M, Nagels J, Rozing PM (2000) The glenohumeral joint rotation centre in vivo. J Biomech 33(12):1629–1636
12. Van der Helm FCT (1996) A standardized protocol for motions recordings of the shoulder. In: Veeger HEJ, VanderHelm FCT, Rozing PM (eds) Proceedings of the first conference of the international shoulder group. Shaker Publishing, Maastricht, pp 1–7
13. Van der Helm FCT, Pronk GM (1995) Three-dimensional recording and description of motions of the shoulder mechanism. J Biomech Eng 117:27–40
14. Veeger HEJ, Yu B, An KN (1996) Orientation of axes in the elbow and forearm for biomechanical modeling. In: Veeger HEJ, VanderHelm FCT, Rozing PM (eds) Proceedings of the first conference of the international shoulder group. Shaker Publishing, Maastricht, pp 83–87
15. Wang X (1996) Construction of arm kinematic linkage from external surface markers. In: Proceeding of the fourth international symposium on 3D analysis of human movement, 1–3 July, 1996, Grenoble, France
16. Wu G, Cavanagh PR (1995) Recommendations for standardization in the reporting of kinematic data. J Biomech 28(10):1257–1260
17. Wu G, Siegler S, Allard P, Kirtley C, Leardini A, Rosenbaum D, Whittle M, D'Lima D, Cristofolini L, Witte H, Schmid O, Stokes I (2002) ISB recommendation on definitions of joint coordinate system of various joints for the reporting of human joint motion—part I: ankle, hip, and spine. J Biomech 35(4):543–548

Appendix D
ISB Recommendations on the Reporting of Intersegmental Forces and Moments During Human Motion Analysis

Timothy R. Derrick, Antonie J. van den Bogert, Andrea Cereatti, Raphael Dumas, Silvia Fantozzi and Alberto Leardini

Abstract The International Society of Biomechanics (ISB) has charged this committee with development of a standard similar in scope to the kinematic standard proposed in Wu et al. [65] and [66]. Given the variety of purposes for which intersegmental forces and moments are used in biomechanical research, it is not possible to recommend a particular set of analysis standards that will be acceptable in all applications. Instead, it is the purpose of this paper to recommend a set of reporting standards that will result in an understanding of the differences between investigations and the ability to reproduce the research. The end products of this standard are (1) a critical checklist that can be used during submission of manuscripts and abstracts to insure adequate description of methods, and (2) a web based visualization tool that can be used to alter the coordinate system, normalization technique and internal/external perspective of intersegmental forces and moments during walking

---

T. R. Derrick (✉)
Iowa State University, Ames, IA, USA
e-mail: tderrick@iastate.edu

A. J. van den Bogert
Cleveland State University, Cleveland, OH, USA
e-mail: a.vandenbogert@csuohio.edu

A. Cereatti
University of Sassari, Sassari, Italy
e-mail: acereatti@uniss.it

R. Dumas
University of Lyon—IFSTTAR, Lyon, France
e-mail: raphael.dumas@ifsttar.fr

S. Fantozzi
University of Bologna, Bologna, Italy
e-mail: silvia.fantozzi@unibo.it

A. Leardini
IRCCS Istituto Ortopedico Rizzoli, Bologna, Italy
e-mail: leardini@ior.it

V. Medved (ed.), *Measurement and Analysis of Human Locomotion*,
Series in Biomedical Engineering, https://doi.org/10.1007/978-3-030-79685-3

and running so that the shape and magnitude of the curves can be compared to one's own data.

Keywords Locomotion · Gait · Methods · Torque · Biomechanics

Introduction

Progress in any field of inquiry relies on the ability of researchers to compare previously published results and replicate research. As complexity of design and analysis increases this becomes more challenging. The nature of human motion research is such that direct measurement techniques are rarely available and often inadequate to measure internal loading during activities of daily living and exercise. We often rely on layers of models to estimate these loads and apply the models in a variety of ways. Results from insilico, i.e. computer based simulations, in-vitro, i.e. anatomical specimens, and in-vivo measures are produced in specific research centers, but then reported at national and international levels, in congresses and in journals, to be shared within the scientific community. There is a need to establish a shared knowledge base, to benefit populations of interest, and ultimately to improve the life of individuals (patients, athletes, workers, etc.). In order to effectively communicate the results of these studies, calculations must be done correctly and reported clearly, with the goal that the research can be understood and replicated without ambiguity. Relevant dissemination of results must be according to standard mechanics, consistent with human body anatomy, and comprehensible by any professional involved, no matter the medical, engineering, technical or industrial background of the reader. In our field of study confusion exists on these matters, with evident errors in a number of published papers, and incomprehension and questionable interpretation of many available results. This hinders the ability of researchers to take advantage of the shared knowledge base. A number of review papers have investigated explicit protocols and techniques for human motion analysis, but only a few specific research topics such as finite element modelling [10], multi-segment foot kinematics analysis [7] and soft tissue artefact description [12] have received recommendations. In this regard, the International Society of Biomechanics (ISB) attempted standardization for the description of joint kinematics in two papers [65, 66], which have received more than 1300 and 1600 citations respectively (as of March 2019, Scopus).

Fundamental quantities of interest in human motion research are the intersegmental forces and moments acting at the joints. These forces and moments represent the net loads that act at a joint. The resultant forces should not be considered physical interactions that occur within the joint as they are often many times smaller than the actual joint contact forces, which include the contribution of muscles [56]. Both force and moment vectors are usually decomposed into three components and transformed into a relevant three-dimensional coordinate system for presentation purposes. This can be accomplished by projecting the vectors onto the corresponding axes. These axes will be referred to as the superior-inferior, anterior-posterior and medial-lateral

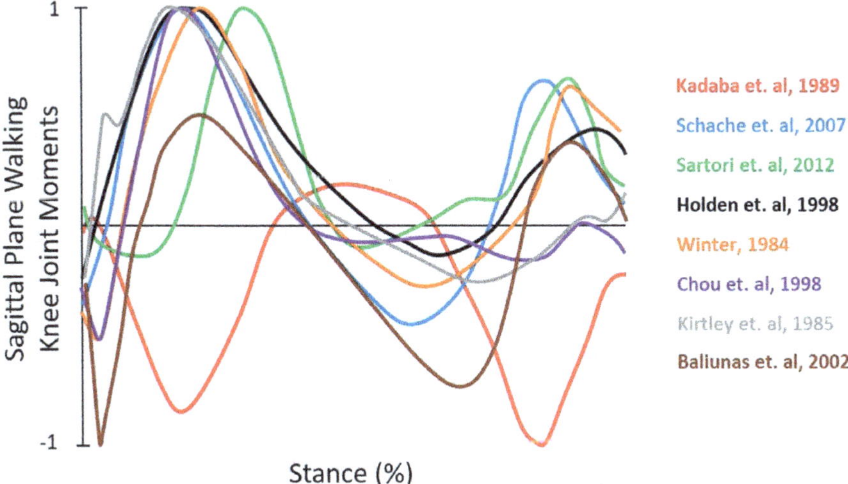

Fig. D1 A selection of sagittal plane walking knee joint moments. Each curve is normalized to its own maximum absolute value. Various coordinate systems and methods of calculation result in an assortment of curve shapes. Average citations for these research papers is 286 (Scopus, May, 2019). (See above-mentioned references for further information.)

axes. Intersegmental moments can be referred to by their action: internal-external rotation, adduction-abduction and flexion-extension; the plane in which the moment acts: transverse, frontal and sagittal; or by the axis of rotation: superior-inferior, anterior-posterior and medial-lateral. Intersegment moments can be analysed on their own or used in the further estimation of muscle forces and joint contact forces. However, intersegmental forces are not the total force acting at a joint and therefore have limited utility on their own (except for specific cases such as kinetic analysis in the prosthetic joints in amputees, Dumas et al. [23]) but are necessary for the estimation of joint moments and joint contact forces.

There are a number of decisions that need to be made in the collection and analysis stages and these must be described in any dissemination stage because they affect the calculated values and the interpretation of the results. Among these are the anthropometric modelling, joint center estimation, smoothing/filtering, method of calculation, coordinate system, evaluation perspective (internal or external), and normalization. As an example of the inherent variety of results different methodological choices can make, sagittal plane knee joint moments during walking are presented in Fig. D1 from eight research studies on healthy adults. It is presumed that these curves were all calculated correctly yet various methods and coordinate systems were utilized, thus altering the shape of the curves. Failure to adequately describe the methods will result in data that cannot be interpreted by the reader nor replicated by the research community.

The aims of the present paper are to discuss the major issues in the definition, calculation, and interpretation of intersegmental forces and moments in human motion analysis, and to make final recommendations on these matters with guidance from relevant papers in the literature. The goal is to eliminate the most frequent sources of error and confusion in the field of human motion analysis so that research can be correctly interpreted and replicated. We are not putting forth these recommendations in an attempt to standardize the methods of estimating intersegmental forces and moments, rather we hope that it is seen as an attempt to standardize the reporting of such methods, after careful consideration of procedures and calculations have been applied.

Anthropometric Model

The relationships between kinetic variables (force and moment) and kinematic variables (linear and angular velocity and acceleration) are governed by the anthropometric properties of mass and moment of inertia about the center of mass.

$$\sum \mathbf{F} = \mathbf{ma}$$

where, $\sum \mathbf{F}$ is the sum of the external forces applied to a given human body segment.
m is the segment mass,
a is the linear acceleration of the center of mass.

$$\sum \mathbf{M} = \mathbf{I}^{\mathrm{cm}}\,\dot{\boldsymbol{\omega}} + \boldsymbol{\omega} \times \mathbf{I}^{\mathrm{cm}}\boldsymbol{\omega}$$

where, $\sum \mathbf{M}$ is the sum of the external moments acting on a given human body segment,
$\mathbf{I}^{cm}$ is the inertia matrix with respect to the center of mass,
$\boldsymbol{\omega}$ is the angular velocity vector,
$\dot{\boldsymbol{\omega}}$ is the angular acceleration vector.

The summations on the left hand side of these equations include terms due to gravity, external forces, and intersegmental forces and moments. The intersegmental forces and moments are generally solved recursively [64].

Since intersegmental forces and moments are derived from these rigid body equations, their computation requires the estimation of segment mass, the position of the center of mass (CoM), and its inertia tensor (moments and products of inertia). All of these quantities must be transformed into a common coordinate system prior to estimation of the intersegmental forces and moments. Body segment inertial parameters (BSIPs) can be obtained using regression equations based on subject's segment length and body mass [18], geometrical models [32, 67], optimization of parameters via COP errors in various postures [16], or, when available, directly from subject-specific medical imaging [15, 29, 47]. These estimations of BSIPs are the basis for rigid body models. The assumption that the estimates are constant are a source of uncertainty in model output.

For most regression equations, the position of the segment center of mass is given with respect to the proximal and distal endpoints, which define the segment length [18] or with respect to a number of anatomical landmarks (Zatsiorsky et al. 1990). Consequently, the determination of the BSIPs also depend on estimation of the joint centers positions. Adjusted scaling equations have been proposed for computing 3D inverse dynamics in which BSIPs are expressed in standardized definitions of the relevant anatomical axes [24, 65, 66].

Due to specific anthropometric characteristics, BSIPs estimated in different populations can lead to different values [48]. Uncertainties in the identification of BSIPs can play a critical role in reliable joint moment estimation, especially when analyzing motor activities involving high accelerations, such as in running [40] or when the population under examination has special anthropometric features, such as amputees [54]. To the contrary, when analyzing level walking at natural speeds, a minor influence of uncertainties in BSIPs is expected, particularly at the more distal joints in the stance phase [11, 51].

Summary and Recommendations

The anthropometric model used to estimate body segment parameters must be detailed in order for results to be replicated. This includes procedures for estimating moments of inertia, mass, and center of mass locations. The sample for which regression equations were established should be consistent with the subjects being studied. This becomes especially important as linear and angular accelerations increase and for specific populations that may have substantially different BSP's (e.g. children, amputees...).

Joint Centers

To compute the intersegmental joint moments, a reduction point, that is the point with respect to which the system of forces is reduced, is required. This point is classically defined as a joint center. In most of the human movement analysis protocols proposed in the literature, adjacent bony segments are conceptually assumed to be connected by spherical pairs, and their relative motion is described by three joint angles about the three anatomical axes defining the joint coordinate system and passing through this joint center [65, 66]. Then, when the joint allows only a small rotation about one axis (resisted degrees of freedom (DoF), e.g. adduction-abduction at the knee or elbow joint), it can be assumed, in a first approximation, that the relevant moment represents the action of the main anatomical joint restraints (i.e. articular contacts and ligaments) [49]. On the other hand, in case of large rotations about one joint axis and neglecting friction (unresisted DoF, e.g., flexion-extension, adduction-abduction and internal-external rotation at the hip or shoulder joints), the resultant of articular contact forces passes through this joint center and therefore it can be assumed that the relevant moment represent the actions of the muscle-tendon units [14]. To interpret the joint moments with this rationale, it is required that the origin of the joint anatomical axes coincides with the reduction point (joint center) and that the joint moments are expressed then about these joint axes.

Joint centers are commonly defined by using regression equations from palpated external anatomical landmarks [6], using functional approaches [43], multibody kinematic optimization techniques [4] or using medical imaging techniques [19]. The latter approach, although very accurate and able to create individualized musculoskeletal models, requires access to expensive and cumbersome measurement systems, time-consuming postprocessing, highly-specific and multidisciplinary expertise, and in some cases, involves exposure to ionizing radiation. This explains why medical imaging techniques are frequently used as gold standards in biomechanics for the development and validation of models and techniques, but their use is limited in the fields of clinical movement analysis and sport applications.

It has been largely demonstrated that joint moment estimates are very sensitive to errors in the joint center and that this inaccuracy affects the calculations to a larger extent than other concurring factors such as errors on BSIPs [11]. For instance, a hip joint anterior and lateral mislocation of 30 mm, which can be expected in typical of routine gait analyses [33], can cause a mean error of about 1.43 and 1.38% bodyweight $\times$ height in the flexion–extension and abduction–adduction moment components of the corresponding range, respectively [58]. In general, the strengths of the functional method are that the calculated center of rotation is subject specific, side specific i.e. right different from left, and not influenced by the presence of bony or joint deformity. Nor is it influenced by differences in body segment proportions, as expected for gender, age, genetic traits, etc. On the other hand, its implementation requires the subject to perform, either passively or actively, a sufficiently wide joint angular excursion. Conversely, regression methods can be applied when movement restrictions are present. However, their accuracy and repeatability strongly depend on the original regression model, and are affected by uncertainties in anatomical landmark identification [52]. The hip joint center is certainly the most critical lower extremity joint due to the large distance from the palpable pelvic landmark. It has been determined that the functional approach has errors between 10 and 20 mm whereas regression methods find errors between 15 and 30 mm [13, 34, 43, 53].

Summary and Recommendations

Since joint center positions are used to define the moment arm of the forces acting on the segment under analysis, the manner in which they are identified will influence the estimation of intersegmental moments. Furthermore, because joint centers are commonly used to define segment length, they also affect body segment inertial parameters. It is therefore fundamental to use valid methods, and to clearly state these methods, for joint centers determination.

Signal Processing

Correct application and complete reporting of signal processing methods are crucial when dealing with kinematic and kinetic data. Of first concern is the sampling, which must be of an adequate rate to insure the frequencies present in the motion are completely captured. At a minimum, the sampling rate must be greater than twice the highest frequency in the signal. A sampling frequency below this threshold will not only miss higher frequencies, the higher frequencies will fold back into the data and result in a contaminated signal [25]. This minimum sampling rate insures no information is lost but if peak values need to be accurately digitized the signal must be sampled at a much higher rate (5–10 times the highest frequency in the signal) or the digitized signal must be reconstructed using resampling techniques [31]. In general, movements that contain collisions are composed of higher frequencies and therefore must be digitized at a higher rate.

Kinematic data must be differentiated to calculate velocities and again to calculate accelerations in preparation for use in the equations of inverse dynamics. This double differentiation process amplifies the time series recording by the square of the frequency [2]. Thus, high frequency noise can dominate the acceleration signal if proper processing procedures are not utilized. Smoothing techniques attempt to attenuate frequencies that comprise the noise while leaving the true signal unaffected. Various methods such as splines, time domain filters and frequency domain filters have been used to accomplish this. Selection of the frequencies that are being attenuation may be done using a set value or algorithms that objectively identify the cutoff frequencies [30, 36].

There is a concern that the frequency content of kinematic and kinetic data should be in agreement. Several researchers [8, 26, 41] have shown that a disparity in the frequency content can cause artifacts in the intersegmental moments that cannot be explained by the dynamics of the activity. This necessitates the same cutoff frequencies for kinetic and kinematic data filtering. However, this poses a problem with impact forces that can be attenuated by low-pass cutoff frequencies that are in the range necessary to reduce noise in the kinematic data. Decisions concerning the cutoff frequency need to be made based on the purpose of the experiment and the variables being measured.

Summary and Recommendations

Both kinematic and kinetic sampling frequencies must be clearly identified. The method of smoothing should be identified and the degree of smoothing (typically in the form of the frequency response) should be noted. They technique used to differentiate the data and any specialized techniques such as optimized cutoffs, resampling of data and procedures to minimize artifact should be detailed and cited.

Method of Calculation

There are two equivalent methods to describe the dynamics of a mechanical system, namely the Newton-Euler and Lagrange formulations. In terms of interpretation, the differences between the two are generally procedural rather than substantive. Note that in biomechanics, few human joints involve translational DoF greater than a few millimeters, thus, most of the time, the Lagrange equations of motion only result in moments. The Newton-Euler method is simpler and gives access to the full three-dimensional intersegmental force and moment vectors, including the moments for resisted degrees of freedom, such as knee adduction-abduction [63]. Lagrange methods are especially useful when joint models that are more complex than spherical or hinge joints are needed, but do not solve for loads associated with resisted degrees of freedom [60]. Equivalent moments result from using the Newton-Euler equations of motion projected onto the DoF axes (projection with a dot product). These moments about the joint DoF are directly related to the joint power (they just need to be multiplied by the DoF angular velocity) and can be described as "motor" or internal joint moments. In musculoskeletal modelling, these are typically the moments involved in the computation of the musculotendon forces while the other components of the intersegmental moments are assumed to represent the actions of ligaments and contact forces [20]. Although theoretically equivalent, these two methods may produce small deviations because of differences in soft tissue artifact propagation.

Both the Newton-Euler and the Lagrange equations [27] lead to inverse dynamics procedures, meaning the intersegmental forces and moments are derived from the kinematics. In forward dynamics procedures, a muscle-driven or torque-driven model is used to estimate intersegmental moments from a neural signal obtained via electromyography, optimization procedures, or a combination [9].

When inertial components are absent or negligible, a static analysis was used to roughly estimate intersegmental forces and moments [28]. This simplified method consists of multiplying the ground reaction force vector by its moment arm at each joint and has been described as the 'ground reaction technique'. This method assumes that segment accelerations and/or the body segment inertial parameters are negligible, large errors can arise if this assumption is not met [62].

Summary and Recommendations

In general, static analysis of the human body should be restricted to static or near-static situations. Newton-Euler and Lagrange formulations of intersegmental moments are mathematically equivalent but the method should be identified because their sensitivity to signal processing methods can be different. Forward or inverse dynamics procedures also need to be specified. If muscle-driven forward dynamics are used additional methods detailing the estimation of muscle forces are necessary.

Coordinate System

Intersegmental forces and moments have been presented in a variety of coordinate systems: global (also known as inertial or laboratory), proximal, distal and the joint coordinate system [55]. In general, presentation of intersegment forces and moments in the global coordinate system should be avoided. Unlike segment coordinate systems, the intersegmental forces and moments presented in the global coordinate system will be affected by changes in the direction of motion.

The joint coordinate system (JCS) is appealing if kinematics are also presented in the JCS, but caution should be used because the axes of the JCS are not orthogonal. Projection onto nonorthogonal axes is problematic when the moment norm is to be computed or when the 3D vector is to be retrieved. If a JCS is used, the projection using a dot product (as opposed to a non-orthogonal projection) is preferred [42]. Those projected moments can be multiplied by the rates of change in the JCS angles to obtain a mechanically consistent joint power analysis. Also these orthogonally projected moments obtained from the Newton-Euler method will be identical to moments obtained from the Lagrange method where the JCS mechanism is explicitly modeled.

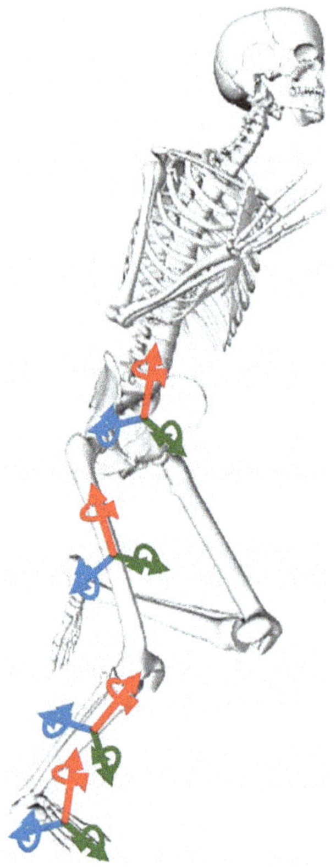

Fig. D2 Segment coordinate systems for the pelvis, thigh, leg and foot segments. For the right leg these coordinate systems are defined by positive values pointing anterior, proximal, and lateral. Positive moments are clockwise about the axis while looking in the positive direction (right hand rule)

Proximal or distal segment coordinate systems (Fig. D2) are useful in answering particular research questions. For instance, during the estimation of tibial tissue stresses, the intersegmental forces and moments would be expressed in a proximal segment coordinate system at the ankle or a distal segment coordinate system at the knee so they are in a coordinate system that is suitable for further analysis, e.g. with a finite element model [22]. Segment coordinate systems (and the joint coordinate systems derived from them) should be defined using ISB standards [65, 66]. If non-standard coordinate systems are used, they should be fully specified in terms of anatomical landmarks or other suitable definitions.

The choice of a coordinate system used to report intersegmental forces and moments can dramatically affect the interpretation of data. Note that during walking (Fig. D3) and running (Fig. D4) the sagittal plane moments and the vertical forces are similar between proximal and distal coordinate systems but there are some relatively large differences in the other planes and axes.

The choice of intersegmental coordinate systems should be consistent with the kinematics and the anthropometric model. Consistency will take two forms: mathematical and informational. Mathematical consistency is necessary to prevent inaccuracies that result from calculations with variables in more than one coordinate system.

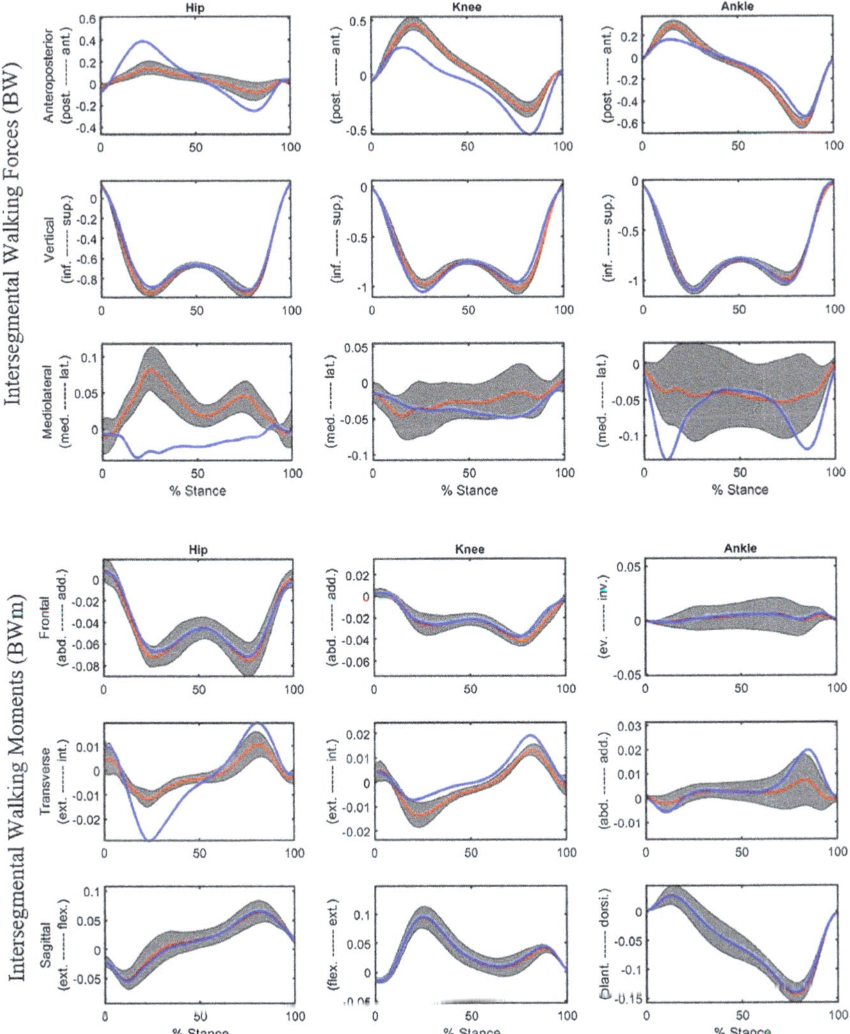

Fig. D3 Ensemble averages of intersegmental forces (top) and moments (bottom) of eight subjects and five trials of walking (1.3 m/s). Both proximal (red) and distal (blue) coordinate systems are represented. Filled area represents ± 1 standard deviation of the proximal coordinate system. Kinematic and kinetic data were low-pass filtered at 6 Hz. Segment masses estimated using Dempster [21]. Segment moments of inertia and center of mass locations estimated using Hanavan [32]. Hip joint center estimated using Bell et al. [5]. Forces and moments were estimated using inverse dynamics with the Newton–Euler equations. (For interpretation of the references to colour in this figure legend, the reader is referred to the web version of this article.)

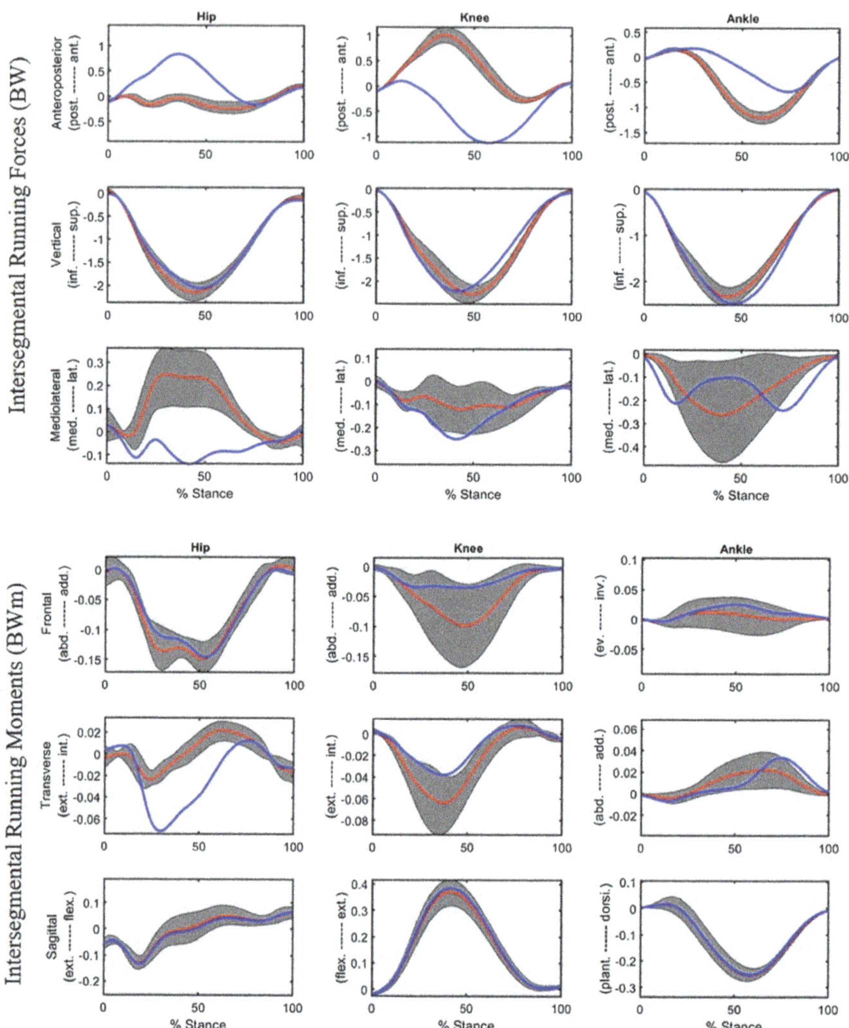

Fig. D4 Ensemble averages of intersegmental forces (top) and moments (bottom) of eight subjects and five trials of running (3.5 m/s). Both proximal (red) and distal (blue) coordinate systems are represented. Filled area represents ± 1 standard deviation of the proximal coordinate system. Running kinematic and kinetic data were low-pass filtered at 10 Hz. Segment masses estimated using Dempster [21]. Segment moments of inertia and center of mass locations estimated using Hanavan [32]. Hip joint center estimated using Bell et al. [5]. Forces and moments were estimated using inverse dynamics with the Newton–Euler equations. (For interpretation of the references to colour in this figure legend, the reader is referred to the web version of this article.)

For instance, if moments of inertia from the anthropometric model were calculated in a segment coordinate system they should not be multiple by angular accelerations in a global or joint coordinate system. Likewise, subsequent calculations using intersegment forces and moments such as joint powers and apparent joint stiffness must have a consistent coordinate system to be accurate. Care should also be taken when estimating muscle moment arms using a musculoskeletal model. The kinematics applied to the model should be in the same coordinate system as the intersegmental moments if they are to be used in a common calculation. Informational consistency suggests that all quantities presented in a paper should be in the same coordinate system unless there is a justifiable reason. This will remove ambiguity and instil confidence in the reader or reviewer that proper procedures have been followed.

Summary and Recommendations

The choice of the coordinate system (global coordinate system, proximal segment coordinate system, distal segment coordinate system, or joint coordinate system) highly influences the intersegmental forces and moments. It is therefore essential that the coordinate system used to interpret the intersegmental forces and moments be carefully considered and reported. Much thought and debate has gone into standardizing kinematic coordinate systems [65, 66] and the motivation for using JCS's for intersegmental forces and moments holds. Unless a rationale exists these previously defined kinematic coordinate systems should also be used to present intersegmental forces and moments. While describing coordinate systems, the directional signs of the forces should be defined (e.g., superior, anterior and lateral are positive) as well as for the moments (e.g., flexion, adduction, and internal rotation are positive) if these parameters are utilized. If moment signe are This is preferred over defining the x, y and z axes of the coordinate system because it gives additional information to the reader.

Internal or External Perspective

Intersegmental forces and moments can be viewed from two perspectives. From an ***external point of view*** the intersegmental forces and moments represent the result of forces acting on the body as well as centrifugal and Coriolis actions arising from motion of the body segments. From an ***internal point of view*** these variables represent forces and moments that originate from within the body and act to resist external load and to maintain posture or accelerate the segments. Anatomical structures crossing the joint such as skin, fat, fascia, muscles, ligaments, together with friction and contact between the articular surfaces produce the mechanical action. Both perspectives are equally valid and in fact the numerical value of the result is equal and opposite because the joint system is in balance. From the internal point of view an extension knee moment is present during mid-stance to prevent knee collapse but from an external point of view there exists a flexion knee moment caused by the external ground reaction force (passing posteriorly to the joint center in the static analysis perspective).

The decision to present internal or external intersegmental forces and moments is often made based on the researcher's view of the source of the moments. Internal moments are considered to be primarily caused by muscles when the joint is not near the end range of motion. This presents a problem at the knee joint because the result of all the muscles spanning the knee may produce a relatively small adduction-abduction moment, the primary source is considered to be ligaments and articular surfaces. This has led many researchers to present adduction-abduction moments at the knee using an external perspective when this is the primary variable of analysis [59]. This leads to additional confusion when the moments are referred to by their action. From an external perspective, an adduction moment at the knee is one in which the external forces are tending to cause the knee to adduct (ground reaction force passing medially to the joint center in the static analysis perspective), potentially tearing the lateral ligaments. However, from an internal perspective, an adduction moment at the knee is one in which the balance of muscles, ligaments and articular contact tend to adduct the knee. These differing perspectives can, and often do, lead to confusion for a reader trying to interpret knee function.

Summary and Recommendations

Whether intersegmental forces and moments are presented as internal or external can be determined by the research question being asked but may also be dependent on the perspective that the researcher is trying to convey. A clear statement of this perspective is essential to communicating concepts in the paper.

Normalization

In clinical movement analysis, demographic/anthropometric characteristics (i.e. age, height, body mass, gender) and the velocity ranges, with which the motor task is executed, influence the amplitude of the kinematic and kinetic variables and if not properly treated may act as confounding factors [1, 45, 46, 57]. If these variables are not equivalent between the groups that are being compared they may need to be controlled statistically (covariate analysis) or their effect removed from the analysis (normalization). When a repeated measures study design is employed such that the normalization parameters do not change over time, the normalization process will not alter the statistical results. However normalization may still be warranted so that results can be conveniently compared to other studies.

Normalization procedures reduce the variance between individuals when comparing the intersegmental forces and moments among individuals. Most commonly, forces are divided by body weight or body mass, while moments are divided by these variables or quantities that result in non-dimensional values such as body weight multiplied by height or body weight multiplied by limb or foot length. For instance, before normalization, lower extremity peak moments during gait were statistically different between males and females in ten cases, but normalizing by body mass reduced this number to 6 and normalizing by body weight times height further reduced the number to 2 [46]. Pinzone et al. [50] and Hof [35] have also shown advantages to using non-dimensional normalization. Additional reductions in residuals may be attained by considering more joint specific distance measurements or non-linear adjustments [44, 61].

Summary and Recommendations

Normalization of data is often necessary if groups are dissimilar on specific variables such as mass or height. This is especially useful in the context of gait laboratories in which individual data are frequently compared to a database. Although normalization will not change the statistical evaluation of a repeated measures study it may still be useful when comparing values to other studies. Normalization procedures need to be clearly outlined in the methods of the paper and normalization values such as average mass and height (or leg length) should be reported. Ranges of these values can also be useful so that researchers can avoid extrapolation of results. There is no simple method to normalize by the walking or running velocity therefore it is necessary to report average velocity or other variables that may influence the intersegmental forces and moments. This will assist researchers in explaining differences between studies.

Conclusions

Many options for the estimation and presentation of intersegmental forces and moments have been presented. These variations should not be considered correct or incorrect because each may be superior to the others in the context of the research paradigm and the questions being asked. However, as members of this field of research, we must insure that no ambiguity exists in the presentation of results. This is critical to an efficient evolution of a body of knowledge. We must be able to relate the curves and values presented by past researchers to the functional movement of the human body so that we can evaluate the results of the study, verify our own research data, and ultimately create new theories and form new hypotheses. Realizing that there are differences in the detail of methodological information required in differing dissemination formats and in biomechanical vs clinical journals we make two suggestions: (1) even conference abstracts should include the coordinate system used and the internal/external perspective—interpretation of the results requires this minimum amount of information and (2) take advantage of the liberal policies journals have generally adopted that allow addition information to be posted online. In partial fulfilment of the goal of improved clarity in the scientific arena, we have identified three tangible items, in addition to this article, that we hope will help to fulfill this goal:

1. Reviewer/author checklist for presentation of intersegmental forces and moments. (Appendix A).
2. Online visualization tool for comparison of typical walking and running intersegmental forces and moments with adjustable coordinate systems, normalization methods and internal/external perspectives (ISB Website).
3. Software transparency. A request to the major biomechanics software companies to make easily available the items in the checklist so that users can access this information in a single location in the software.

Declaration of Competing Interest

All authors were fully involved in the study and preparation of the manuscript. The material within has not been and will not be submitted for publication elsewhere. We declare that we have no financial or personal relationships with other people or organizations that could inappropriately influence our work.

Acknowledgements The authors would like to thank Dr. Glen Lichtwark for his assistance in the organization of this committee and the current President of the International Society of Biomechanics, Dr. Joseph Hamill, for his leadership in forming the committee.

Appendix A. Example Checklist for the Reporting of Intersegmental Forces and Moments

Anthropometric Model

Segment Masses	Dempster, 1955
Moments of Inertia	Hanavan, 1964
Joint Center Locations	Hip: Bell, 1989
COM Locations	Hanavan, 1964

Signal Processing

Kinematics	Low-pass Filter @ 6Hz
Kinetics	Low-pass Filter @ 6Hz
Differentiation	First Central Difference Method
Other	

Method of Calculation

- ☑ Newton-Euler
- ☐ Lagrange
- ☑ Inverse Dynamics
- ☐ Forward Dynamics

Internal or External Perspective

- ☑ Internal
- ☐ External

Normalization

Moments	Average Value/Range
☐ Body Mass	
☑ Body Weight	621 (449-721) N
☐ Body Mass x Body Height	
☐ Body Weight x Leg	
☐ Other	
☐ None	

Forces	Average Value/Range
☐ Body Mass	
☑ Body Weight	same as moments
☐ Other	
☐ None	

Coordinate System

- ☐ Global (Inertial)
- ☑ Proximal Segment
- ☐ Distal Segment
- ☐ Joint Coordinate System
- ☐ Other

Positive Force Directions

- ☑ Proximal ☑ Lateral ☑ Anterior
- ☐ Distal ☐ Medial ☐ Posterior

Positive Moment Directions

- ☐ Flexion ☐ Adduction ☐ Internal Rotation
- ☐ Extension ☐ Abduction ☐ External Rotation

or

- ☑ Positive moment directions are defined by right hand rule of coordinate system

Other

Speed of Locomotion	3.5 m/s

- ☑ Is the kinetic coordinate system the same as anthrometrics?
- ☑ Is the kinetic coordinate system the same as kinematics?

References

1. Andriacchi TP, Ogle JA, Galante JO (1977) Walking speed as a basis for normal and abnormal gait measurements. J Biomech 10(4):261–268
2. Antonsson EK, Mann RW (1985) The frequency content of gait. J Biomech 18(1):39–47
3. Baliunas AJ, Hurwitz DE, Ryals AB, Karrar A, Case JP, Block JA, Andriacchi, TP (2002) Increased knee joint loads during walking are present in subjects with knee osteoarthritis. Osteoarthritis Cartilage 10(7):573–579
4. Begon M, Andersen MS, Dumas R (2018) Multibody kinematics optimization for the estimation of upper and lower limb human joint kinematics: a systematized, methodological review. J Biomech Eng 140(3):030801
5. Bell AL, Brand RA, Pedersen DR (1989) Prediction of hip joint center location from external landmarks. J Biomech 20(9):913
6. Bell AL, Pedersen DR, Brand RA (1990) A comparison of the accuracy of several hip center location prediction methods. J Biomech 23(6):617–621
7. Bishop C, Paul G, Thewlis D (2012) Recommendations for the reporting of foot and ankle models. J Biomech 45(13):2185–2194
8. Bisseling RW, Hof AL (2006) Handling of impact forces in inverse dynamics. J Biomech 39(13):2438–2444
9. Buchanan TS, Lloyd DG, Manal K, Besier TF (2005) Estimation of muscle forces and joint moments using a forward-inverse dynamics model. Med Sci Sports Exerc 37(11):1911–1916
10. Burkhart TA, Andrews DM, Dunning CE (2013) Finite element modeling mesh quality, energy balance and validation methods: a review with recommendations associated with the modeling of bone tissue. J Biomech 46(9):1477–1488
11. Camomilla V, Cereatti A, Cutti AG, Fantozzi S, Stagni R, Vannozzi G (2017) Methodological factors affecting joint moments estimation in clinical gait analysis: a systematic review. Biomed Eng Online 16(1):106
12. Cereatti A, Bonci T, Akbarshahi M, Aminian K, Barré A, Begon M, Lin CC (2017) Standardization proposal of soft tissue artefact description for data sharing in human motion measurements. J Biomech 62:5–13
13. Cereatti A, Donati M, Camomilla V, Margheritini F, Cappozzo A (2009) Hip joint centre location: an ex vivo study. J Biomech 42(7):818–823
14. Challis JH, Kerwin DG (1996) Quantification of the uncertainties in resultant joint moments computed in a dynamic activity. J Sports Sci 14(3):219–231
15. Cheng CK, Chen HH, Chen CS, Lee CL, Chen CY (2000) Segment inertial properties of Chinese adults determined from magnetic resonance imaging. Clin Biomech 15(8):559–566
16. Chen SC, Hsieh HJ, Lu TW, Tseng CH (2011) A method for estimating subjectspecific body segment inertial parameters in human movement analysis. Gait Posture 33(4):695–700
17. Chou LS, Draganich LF (1998) Increasing obstacle height and decreasing toeobstacle distance affect the joint moments of the stance limb differently when stepping over an obstacle. Gait Posture 8(3):186–204
18. De Leva P (1996) Adjustments to Zatsiorsky-Seluyanov's segment inertia parameters. J Biomech 29(9):1223–1230
19. Della Croce U, Cappozzo A, Kerrigan DC (1999) Pelvis and lower limb anatomical landmark calibration precision and its propagation to bone geometry and joint angles. Med Biol Eng Compu 37(2):155–161
20. Delp SL, Anderson FC, Arnold AS, Loan P, Habib A, John CT, Thelen DG (2007) OpenSim: open-source software to create and analyze dynamic simulations of movement. IEEE Trans Biomed Eng 54(11):1940–1950
21. Dempster WT (1955) Space requirements of the seated operator, geometrical, kinematic, and mechanical aspects of the body with special reference to the limbs. Michigan State Univ East Lansing

22. Derrick TR, Edwards WB, Fellin RE, Seay JF (2016) An integrative modeling approach for the efficient estimation of cross sectional tibial stresses during locomotion. J Biomech 49(3):429–435
23. Dumas R, Brånemark R, Frossard L (2017) Gait analysis of transfemoral amputees: errors in inverse dynamics are substantial and depend on prosthetic design. IEEE Trans Neural Syst Rehabil Eng 25(6):679–685
24. Dumas R, Cheze L, Verriest JP (2007) Adjustments to McConville et al. and Young et al. body segment inertial parameters. J Biomech 40(3):543–553
25. Edwards WB, Derrick TR, Hamill J (2017) Time series analysis in biomechanics. In: Handbook of human motion. Springer International Publishing, pp 1–24
26. Edwards WB, Troy KL, Derrick TR (2011) On the filtering of intersegmental loads during running. Gait Posture 34(3):435–438
27. Eberhard P, Schiehlen W (2006) Computational dynamics of multibody systems: history, formalisms, and applications. J Comput Nonlinear Dyn 1(1):3–12
28. Fantozzi S, Garofalo P, Cutti AG, Stagni R (2012) 3D joint moments in transfemoral and transtibial amputees: when is the "Ground Reaction Vector Technique" an alternative to inverse dynamics? J Mech Med Biol 12(04):1250061
29. Ganley KJ, Powers CM (2004) Anthropometric parameters in children: a comparison of values obtained from dual energy x-ray absorptiometry and cadaver-based estimates. Gait Posture 19(2):133–140
30. Giakas G, Baltzopoulos V (1997) A comparison of automatic filtering techniques applied to biomechanical walking data. J Biomech 30(8):847–850
31. Hamill J, Caldwell GE, Derrick TR (1997) Reconstructing digital signals using Shannon's sampling theorem. J Appl Biomech 13(2):226–238
32. Hanavan Jr EP (1964) A mathematical model of the human body (No. AFIT-GAPHYS-64–3). Air Force Aerospace Medical Research Lab Wright-Patterson AFB OH
33. Hara R, McGinley J, Briggs C, Baker R, Sangeux M (2016). Predicting the location of the hip joint centres, impact of age group and sex. Sci Rep 6:37707
34. Harrington ME, Zavatsky AB, Lawson SEM, Yuan Z, Theologis TN (2007) Prediction of the hip joint centre in adults, children, and patients with cerebral palsy based on magnetic resonance imaging. J Biomech 40(3):595–602
35. Hof AL (1996) Scaling gait data to body size. Gait Posture 3(4):222–223
36. Jackson KM (1979) Fitting of mathematical functions to biomechanical data. IEEE Trans Biomed Eng 2:122–124
37. Holden JP, Stanhope SJ (1998) The effect of variation in knee center location estimates on net knee joint moments. Gait Posture 7(1):1–6
38. Kadaba MP, Ramakrishnan HK, Wootten ME, Gainey J, Gorton G, Cochran GVB (1989) Repeatability of kinematic, kinetic, and electromyographic data in normal adult gait. J Orthop Res 7(6):849–860
39. Kirtley C, Whittle MW, Jefferson RJ (1985) Influence of walking speed on gait parameters. J Biomed Eng 7(4):282–288
40. Krabbe B, Farkas R, Baumann W (1997) Influence of inertia on intersegment moments of the lower extremity joints. J Biomech 30(5):517–519
41. Kristianslund E, Krosshaug T, van den Bogert AJ (2012) Effect of low pass filtering on joint moments from inverse dynamics: implications for injury prevention. J Biomech 45(4):666–671
42. Kristianslund E, Krosshaug T, Mok KM, McLean S, van den Bogert AJ (2014) Expressing the joint moments of drop jumps and sidestep cutting in different reference frames—does it matter? J Biomech 47(1):193–199
43. Leardini A, Cappozzo A, Catani F, Toksvig-Larsen S, Petitto A, Sforza V, Cassanelli G, Giannini S (1999) Validation of a functional method for the estimation of hip joint centre location. J Biomech 32:99–103
44. Lee D, Oh SE, Lee IK, Sim T, Joo SB, Park HJ, Mun JH (2015) Comparison of three normalization methods for 3D joint moment in the asymmetric rotational human movements in golf swing analysis. J Biosyst Eng 40(3):289–295

45. Lelas JL, Merriman GJ, Riley PO, Kerrigan DC (2003) Predicting peak kinematic and kinetic parameters from gait speed. Gait Posture 17(2):106–112
46. Moisio KC, Sumner DR, Shott S, Hurwitz DE (2003) Normalization of joint moments during gait: a comparison of two techniques. J Biomech 36(4):599–603
47. Mungiole M, Martin PE (1990) Estimating segment inertial properties: comparison of magnetic resonance imaging with existing methods. J Biomech 23(10):1039–1046
48. Nguyen TC, Reynolds KJ (2014) The effect of variability in body segment parameters on joint moment using Monte Carlo simulations. Gait Posture 39(1):346–353
49. O'Connor JJ, Lu TW, Wilson DR, Feikes J, Leardini A (1998) Review: diarthrodial joints-kinematic pairs, mechanisms or flexible structures?. Comput. Methods Biomech Biomed Eng 1:123–150
50. Pinzone O, Schwartz MH, Baker R (2016) Comprehensive non-dimensional normalization of gait data. Gait Posture 44:68–73
51. Rao G, Amarantini D, Berton E, Favier D (2006) Influence of body segments' parameters estimation models on inverse dynamics solutions during gait. J Biomech 39(8):1531–1536
52. Sangeux M (2015) On the implementation of predictive methods to locate the hip joint centres. Gait Posture 42(3):402–405
53. Sangeux M, Pillet H, Skalli W (2014) Which method of hip joint centre localisation should be used in gait analysis? Gait Posture 40(1):20–25
54. Sawers AB, Hahn ME (2010) The potential for error with use of inverse dynamic calculations in gait analysis of individuals with lower limb loss: a review of model selection and assumptions. JPO: J Prosthet Orthot 22(1):56–61
55. Schache AG, Baker R (2007) On the expression of joint moments during gait. Gait Posture 25(3): 440–452
56. Scott SH, Winter DA (1990) Internal forces of chronic running injury sites. Med Sci Sports Exerc 22(3):357–369
57. Senden, R, Meijer K, Heyligers IC, Savelberg HHCM, Grimm B (2012) Importance of correcting for individual differences in the clinical diagnosis of gait disorders. Physiotherapy 98(4):320–324
58. Stagni R, Leardini A, Cappozzo A, Benedetti MG, Cappello A (2000) Effects of hip joint centre mislocation on gait analysis results. J Biomech 33(11):1479–1487
59. Telfer S, Lange MJ, Sudduth AS (2017) Factors influencing knee adduction moment measurement: a systematic review and meta-regression analysis. Gait Posture 58:333–339
60. van den Bogert AJ, Geijtenbeek T, Even-Zohar O, Steenbrink F, Hardin EC (2013) A real-time system for biomechanical analysis of human movement and muscle function. Med Biol Eng Comput 51:1069–1077
61. Wannop JW, Worobets JT, Stefanyshyn DJ (2012) Normalization of ground reaction forces, joint moments, and free moments in human locomotion. J Appl Biomech 28(6):665–676
62. Wells RP (1981) The projection of the ground reaction force as a predictor of internal joint moments. Bull Prosthetics Res 10:15–19
63. Winter DA (1984) Kinematic and kinetic patterns in human gait: variability and compensating effects. Hum Mov Sci 3(1–2):51–76
64. Winter DA (2009) Biomechanics and motor control of human movement. Wiley & Sons
65. Wu G, Siegler S, Allard P, Kirtley C, Leardini A, Rosenbaum D, Schmid O (2002) ISB recommendation on definitions of joint coordinate system of various joints for the reporting of human joint motion—part I: ankle, hip, and spine. J Biomech 35(4):543–548
66. Wu G, Van der Helm FC, Veeger HD, Makhsous M, Van Roy P, Anglin C, Werner FW (2005) ISB recommendation on definitions of joint coordinate systems of various joints for the reporting of human joint motion—Part II: shoulder, elbow, wrist and hand. J Biomech 38(5):981–992
67. Yeadon MR (1990) The simulation of aerial movement –II. A mathematical inertia model of the human body. J Biomech 23(1):67–74

Appendix E
Standards for Reporting EMG Data

Authors are advised that the following protocols must be observed and supplied in the Methods section of all submitted manuscripts. To avoid delay or return of manuscripts, the requirements below should be considered when preparing the manuscript.

Electrodes:

Reports on surface recording of EMG should include:

- electrode material (e.g., Ag/AgCl)
- electrode geometry (discs, bars, rectangular)
- size (e.g., diameter, radius, width • length)
- use of gel or paste, alcohol applied to cleanse skin, skin abrasion, shaving of hair, etc.
- interelectrode distance
- electrode location, orientation over muscle with respect to tendons, motor point and fibers direction.

Intramuscular wire electrodes should be described by:

- wire material (e.g., stainless steel)
- if single- or multi strand
- insulation material
- length of exposed tip
  method of insertion (e.g., hypodermic needle)
- depth of insertion
- if single or bipolar wire
- location of insertion in the muscle
- interelectrode distance
- type of ground electrode used, location.

Needle electrodes and their application should be described and include material, size of conductive contact points at the tip, depth of insertion and accurate location in the muscle.

V. Medved (ed.), *Measurement and Analysis of Human Locomotion*,
Series in Biomedical Engineering, https://doi.org/10.1007/978-3-030-79685-3

Amplification:

Amplifiers should be described by the following:

- if single, differential, double differential, etc.
- input impedance
- Common Mode Rejection Ratio (CMRR)
- signal-to-noise ratio
- actual gain range used.

Filtering of the raw EMG should be specified by:

- low and/or high pass filters
- filter types (e.g., Butterworth, Chebyshev, etc.) –low and/or high pass cut-off frequencies.

Since the power density spectra of the EMG contains most of its power in the frequency range of 5–500 Hz at the extremes, the journal will not accept reports in which surface EMG was filtered above 10 Hz as a low cut-off, and below 350 Hz as the high cut-off; e.g., 10–350 Hz is preferred for *surface* recording. Filtering in the band of 10–150 Hz or 50–350 Hz, for example, is not acceptable as portions of the signal's power above 150 Hz and below 50 Hz are eliminated. This should be kept in mind when designing a study's protocol. Exceptions will be made only in rare cases that carry full scientific justification.

Intramuscular recording should be made with the appropriate increase of the high frequency cut-off to a minimum 450 Hz. A bandpass filter of 10–450 Hz is therefore required.

Needle recording should have a bandwidth of 10–1,500 Hz.

Rectification: A note should be made if full or half-wave rectification was carried out.

EMG Processing: There are several methods of EMG processing. *Smoothing* the signal with a low pass filter of a given time constant (normally 50–250 ms) is best described as "smoothing with a low-pass filter of x ms". Alternatively, one can describe it as a "linear envelope" or "the Mean Absolute Value", while giving time constant type and order of the low-pass filter used. Also acceptable is determination of the "Root Mean Square" or RMS.

Authors should include the time period over which the average RMS was calculated.

Integrated EMG is sometimes reported, but the signal is actually integrated over time, rather than just smoothed. Such procedure allows observation of the accumulated EMG activity over time, and should be presented with information as to whether time or voltage was used to reset the integrator and at what threshold it was reset.

Power Density Spectra presentation of the EMG should include:

- time epoch used for each calculation segment
- type of windows used prior to taking the Fast Fourier Transform (FFT) (e.g., Hamming, Hanning, Tukey, etc.)

- taking the algorithm (e.g., FFT)
- number of zero padding applied in the epoch and the resultant resolution
- equation used to calculate the Median Frequency (MDF), Mean Frequency (MNF), etc.
- the muscle length or fixed joint angle at the time of recording.

Other processing techniques, especially novel techniques, are encouraged if accompanied by full scientific description.

Sampling EMG into the Computer:

Computer processing of the EMG is encouraged if authors observe these important factors:

1. It is advisable that the raw EMG (e.g., after differential amplification andband-pass filtering) be stored in the computer before further analysis in case modification of the protocol is required in the future. In this case, the minimal acceptable sampling rate is at least twice the highest frequency cut-off of the bandpass filter, e.g., if a bandpass filter of 10–350 Hz was used, the minimal sampling rate employed to store the signal in the computer should be 700 Hz (350×2), and *preferably higher* to improve accuracy and resolution. Sampling rates below twice the highest frequency cut-off will not be accepted.
2. If smoothing, with a low-pass filter was performed with hardware prior to sampling and storing data in the computer, the sampling rate could be drastically reduced. Rates of 50–100 Hz are sufficient to introduce smoothed EMG into the computer.
3. It is also advisable that authors consider recording the raw EMG (prior to band-pass filtering) in the computer; in such cases a sampling rate of 2500 Hz or above could be used. Yet, to avoid aliasing of high-frequency noise, bandpass filtering (written in software) in the range prescribed above should be performed prior to any further processing of the signal. This approach allows authors to perform EMG recording with minimal hardware and maximal flexibility. Yet, it may be at the expense of computer memory space and speed.
4. Number of bits, model, manufacturer of A/D card used to sample data intothe computer should be given.

Normalization: In investigations where the force/torque was correlated to the EMG, it is common to normalize the force/torque and its respective EMG, relative to the values at maximal voluntary contraction (MVC). Authors should be aware that obtaining true MVC from subjects requires some preliminary training. Without training, the MVC could be as much as 20–40% less of that obtained after appropriate training. The journal, therefore, will not accept reports in which subjects were not properly trained to elicit true MVC.

Normalizing the force/torque with respect to its MVC is commonly performed with MVC as 100% of the force/torque, and other force levels are expressed as the appropriate percentage of MVC. Similarly, the EMG associated with 100% MVC is designated as 100%. Both force/torque and EMG normalization should include other relevant information such as joint angle(s) and/or muscle length(s) in isometric contractions, and range of joint angle, muscle length, velocity of shortening/elongation, and load applied for non-isometric contractions.

Normalization of data collected from one experimental condition with respect to other contractile conditions can be performed for comparative purposes and will be accepted by the journal only if full description is given.

In sum, the following information should be provided when normalizing data:

- how subjects were trained to obtain MVC –joint angle or muscle length
  - angles of adjoining joint, e.g., for studies on elbow flexion, the position of the wrist and shoulder joints should be provided
  - rate of rise of force
  - velocity of shortening/elongation
  - changes in muscle length
  - ranges of joint angle/muscle length in non-isometric contraction –load appied in non-isometric contractions.

EMG Crosstalk:

Authors should demonstrate that significant effort was made to determine that EMG crosstalk from muscles near the muscle of interest did not contaminate the recorded signal. Selecting the appropriate electrode size, interelectrode distance and location of recordings over the muscle should be carefully planned, especially when working on area where many narrow muscles are tightly gathered (e.g., forearm), or when working with superficial/thin muscles (e.g., trapezius). The work of Winter et al. [3] and Fuglevand et al. [1] should be consulted if doubts exist. Care also should be employed when recording surface EMG from areas with subcutaneous adipose tissue as it is known that adipose tissue enhances crosstalk [2].

[1] Fuglevand AJ, Winter DA, Patala AE, Stashuk D (1992) Detection of motor units' action potentials with surface electrodes—influence of electrode size and spacing. Biol Cybern 67:143–153

[2] Fuglevand AJ, Winter DA, Patala AE, Stashuk D (1992) Detection of motor units' action potentials with surface electrodes—influence of electrode size and spacing. Biol Cybern 67:143–153

[3] Winter DA, Fuglevand AJ, Archer SE (1994) Crosstalk in surface electromyography: theoretical and practical estimates. JEMG Kinesiol 4:15–26

Index

V. Medved (ed.), *Measurement and Analysis of Human Locomotion*,
Series in Biomedical Engineering, https://doi.org/10.1007/978-3-030-79685-3

GPSR Compliance
The European Union's (EU) General Product Safety Regulation (GPSR) is a set of rules that requires consumer products to be safe and our obligations to ensure this.

If you have any concerns about our products, you can contact us on

ProductSafety@springernature.com

In case Publisher is established outside the EU, the EU authorized representative is:

Springer Nature Customer Service Center GmbH
Europaplatz 3
69115 Heidelberg, Germany

www.ingramcontent.com/pod-product-compliance
Ingram Content Group UK Ltd.
Pitfield, Milton Keynes, MK11 3LW, UK
UKHW021833270726
14058UKWH00001B/121

* 9 7 8 3 0 3 0 7 9 6 8 7 7 *